Infrastructural Development, Corruption, Xenophobia, and Colonization in Central and Southeastern Europe

Imperialism and Colonialism in Central and Eastern Europe

Investigating colonialism and imperialism within the history of Central and Eastern Europe is challenging for two reasons. Firstly, national historiographies treat imperial past as their opposition and deny any continuity with its features. Secondly, theories and methodologies developed on the example of maritime/colonial empires can only serve as starting points for research dealing with Central and Eastern Europe's landlocked empires and their successor states. From a theoretical perspective, it is often problematic to simply transfer the key concepts developed for the interpretation of modern maritime empires, to Central and Eastern Europe with its peculiar history of state formation and differentiated rule. Empires in Central and Eastern Europe had their own imperial terminology, which require careful assessment and often translation into the language of global imperial endeavors.

Inspired by the current of New Imperial History we invite new proposals for books addressing the history of Central and Eastern Europe through the lens of imperialism and colonialism making the region comparable with other areas but also revealing its specific features, that way even contributing to a rethinking of empire. The series investigates the (post)imperial and (post)colonial past and present of Central and Eastern Europe from the 16th century to the present. By 'Central and Eastern Europe' we refer to (post-) Habsburg Central Europe, (post-) Ottoman South-Eastern Europe, the European provinces of the Russian Empire or the satellite countries of the Soviet Union and the territories of historical Prussia. The series includes books written by scholars inspired by various disciplines including cultural studies, cultural anthropology, historical geography, area studies, and international law, but with a clear focus on history.

Infrastructural Development, Corruption, Xenophobia, and Colonization in Central and Southeastern Europe

Edited by
Silvia Marton and Andrei-Dan Sorescu

CEU PRESS

Central European University Press
Budapest–Vienna–New York–Amsterdam

The publication of this book is made possible by a grant from the Romanian Ministry of Research, Innovation and Digitization, CNCS - UEFISCDI, under the "Exploratory Research Projects" scheme (project number PN-III-P4-PCE-2021-0399).

Cover illustration: Source: *Călăuza Căilor Ferate Române* (1913)
Cover design: Coördesign, Leiden
Lay-out: Crius Group, Hulshout

ISBN 978 90 4857 727 9
e-ISBN 978 90 4857 730 9 (pdf)
e-ISBN 978 90 4857 731 6 (accessible ePub)
https://doi.org/10.5117/9789048577279

An electronic version of this book is freely available thanks to the libraries supporting CEU Press's Opening the Future model. More information and links to the open access ebook can be found at CEU.openingthefuture.net.

Contents

Researching Infrastructural Development, Corruption, and Colonization in Central and Southeastern Europe: An Introduction 7
Silvia Marton & Andrei-Dan Sorescu

Part I Conceptual Framework

1. On Hegemony and Connectivity: A Short History of Infrastructure-Political Struggles and their Discourses 29
Malte Fuhrmann

2. Transportation Infrastructure and Corruption Scandals 61
Silvia Marton

3. Writing the History of Critical Infrastructures: Some Conceptual Remarks 87
Jens Ivo Engels

4. The Infrastructural Power of the "Colonial" as a Concept 105
Andrei-Dan Sorescu

Part II (Transportation) Infrastructure, Corruption Debates, and Colonization

5. Accusations of Tyranny: Mobilizing the Ottoman Imperial Past in Service of West European Informal Empire in the Early Greek State 135
Alex R. Tipei

6. British Investors and the "Kustendjie Harbor Dues Issue" 159
Constantin Ardeleanu

7. Steel and Wood: Railway Construction and Rent-Seeking 185
Gábor Egry

8. Hoping and Coping: Local Anxieties and Railway Development in Romania 219
Toader Popescu

9. Regional Competition, Economic Interests, and Military Stumbling-Blocks in the Making of the Moldavian Railway Network 239
Mihai Chiper

10. Conduits of Violence: Railway Infrastructure and Student Anti-Semitism in Interwar Romania 267
Raul Cârstocea

Index 295

Researching Infrastructural Development, Corruption, and Colonization in Central and Southeastern Europe: An Introduction

Silvia Marton & Andrei-Dan Sorescu

Abstract: This introductory chapter argues for the relevance of considering the conceptual nexus of "colonization," "corruption," and xenophobia in conjunction with the anxieties generated by infrastructural development in nineteenth-century Central and Southeastern Europe. The need for attracting capital and expertise generated scandals across the region and saw historical actors deploy and renegotiate this conceptual vocabulary. By connecting empires and fostering the emergence of nation-states, infrastructure was perceived as both a necessity and a liability, and the promise of progress was doubled by the creation of asymmetries. The chapter thus provides a methodological agenda and analytical framework for the volume and outlines the many ways in which its chapters address these interconnected issues.

Keywords: Infrastructure, Xenophobia, Corruption scandals, History of empires, History of nation-states

In December 1871, polymath and politician B. P. Hașdeu (1838–1907) held a speech at a public rally, meant to tackle the "Strousberg-Bleichroeder question" and the entanglements it had created. Hașdeu, who had made a name for himself in the 1860s as a polemicist on the "Jewish question," now addressed what he felt was the crux of a scandal that was ever growing in intensity. The construction of Romania's main railway network had been

Marton, Silvia, and Andrei-Dan Sorescu, eds. *Infrastructural Development, Corruption, Xenophobia, and Colonization in Central and Southeastern Europe*. Amsterdam: Amsterdam University Press/Central European University Press, 2026.
 https://doi.org/10.5117/9789048577279-1

awarded to two concessionaires—one Prussian, the other Austrian—and public opinion, in no small part influenced by interventions such as Hașdeu's, had been inflamed. Not only was a generalized sense of Teutophobia in the air after the fall of France the year before,[1] but so was a demand for rescinding contracts with the entrepreneurs, who were falling short of what they were obliged to deliver. In a more abstract sense, however, the "truly important element behind the Strousberg-Bleichroeder question," as per Hașdeu, "was nationality." For him, alluding to the uproar that Romania's anti-Semitic measures had created throughout Europe just a few years earlier, "nationality" was the bottom line, regardless of whether "foreigners and those who feel in foreign ways ["*străinește*"] will call us 'barbarians' and 'savages' as they always have whenever Romanians vigorously attempted to defend their nationality."[2] What Hașdeu was railing against here was the similarity he perceived between earlier admonishments received in connection with Romania's "uncivilized" conduct toward its Jewish minority and the European backlash that rescinding the railway concessions would trigger.

But why was the Strousberg-Bleichroeder contract so controversial? For one thing, coming to terms with the fact that Romania lacked the capital, expertise, and manpower to build a rail network without foreign involvement was symbolically transgressive from the nationalist point of view: Britain, Hașdeu proclaimed, was exclusivist and exclusionary when it came to awarding such positions to foreigners, and, "far from being a sign of barbarism and savagery, this exclusivism is the most shining proof of a nation's high civilization."[3] This was meant not only as a jab against what Hașdeu saw as a Western double standard but also as a defense of a salutary barrier against the "pacific means of conquest that is the economy." If "de-nationalizing

1 Andrei Sorescu, "The Political Language of Moral Panics: Teutophobia and the Image of a Besieged Romania in 1871," in *Penser le XIXe siècle: nouveaux chantiers de recherche*, ed. Silvia Marton and Constanţa Vintilă-Ghiţulescu (Iași: Editura Universității "Alexandru Ioan Cuza" din Iași, 2013), 153–173; Andrei Sorescu, "The Perils (and Promise) of German Colonization: Civilizational Hierarchies and Anxieties in Nineteenth-Century Romania," *East European Politics and Societies* 39, no. 3 (2025): 717–740.

2 B. P. Hașdeu, "Discursul d-lui B. P. Hașdeu ținut la întrunirea publică din 5 decembrie 1871, în casa d-lui Gr. Băleanu asupra cestiunii Strussberg-Bleichroeder," *Columna lui Traian* 2, no. 46 (108), 8 December 1871: 179–180, *apud* B. P. Hașdeu, *Opere. IV. Publicistica politică, 1858–1904*, ed. Stancu Ilin and I. Oprișan (Bucharest: Editura Fundației Naționale pentru Știință și Artă; Univers Enciclopedic, 2007), 1497–1507, here 1498. For the public and parliamentary debates on the "Jewish question," see Silvia Marton, *La construction politique de la nation: La nation dans les débats du Parlement de la Roumanie (1866–1871)* (Iași: Institutul European, 2009), 111–123, 292–320.

3 Hașdeu, "Discursul d-lui B. P. Hașdeu," 1499.

[a country] through war is altogether more difficult," Hașdeu saw in the presence of German and Austrian capital and infrastructure a first step toward this more achievable means of takeover.[4] Just as British merchants in India had managed to dupe the natives into subjection, so did, as per Hașdeu, Jewish merchants in Moldavia corrupt the boyars of Iași: Romania was still in a phase of "national childhood," especially when faced with "the mature German nation."[5] This comparison between Romanians and Indians as hapless victims of empire was grounded in older fears of German expansionism redirecting emigration and colonization toward the Lower Danube, with a presumed "Jewish invasion" as "the most natural vanguard of Germanism." Taken together, all of this made Hașdeu rhetorically wonder how the 1868 Strousberg concession had ever been awarded in the first place, while nevertheless saluting the Chamber for transforming it into a joint-stock company under closer control.[6] Fears of "German" colonization and anti-Semitic xenophobia, however far-flung and fanciful, were thus grounded in the ultimately material question of infrastructure, and brought into play by the latter's role as an attractor and amplifier of public controversy.

This opening vignette seeks to provide a point of entry for the present volume. Our book originates in a collective research project that has examined the historical relevance of transportation infrastructure for the self-imaginings of nation-building and the narratives of modernization in the Romanian Principalities/Romania, from the 1840s to the 1920s.[7] Focusing on a series of major debates surrounding transportation infrastructure construction in nineteenth-century Romania has proven fruitful for recovering the contested meanings and layers of key concepts such as "colonization," "corruption," "anti-Semitism," and "xenophobia," as well as for opening up wider, complementary inquiries.[8]

However, while this book's historiographical and conceptual concerns originate *in*, and still include, a significant focus *on* Romania, the scope of the analysis offered here aims to be wider, allowing us to horizontally

4 Hașdeu, "Discursul d-lui B. P. Hașdeu," 1500.

5 Hașdeu, "Discursul d-lui B. P. Hașdeu," 1500.

6 Hașdeu, "Discursul d-lui B. P. Hașdeu," 1501.

7 "Colonial Anxieties, Corruption Scandals and Xenophobia in Nineteenth-Century Infrastructure Development in Romania (CanCor)," grant of the Ministry of Research, Innovation, and Digitization, CNCS-UEFISCDI, project number PN-III-P4-PCE-2021-0399, within PNCDI III), hosted by the New Europe College—Institute for Advanced Study in Bucharest (2022–2024 https://cancor.nec.ro/).

8 Silvia Marton and Andrei Sorescu, "Introduction: Colonial Anxieties, Corruption Scandals, and Xenophobia in Nineteenth-Century Infrastructure Development in Romania," *East European Politics and Societies* 39, no. 3 (2025): 680–690.

expand conclusions far beyond this initial testing ground and to give more depth to its conceptualizations. Our intervention is thus at once conceptual, inasmuch as it focuses on recovering a historical lexicon that reflected the *outillage mental* of our broad array of actors—and theoretical, inasmuch as the very nexus of concepts we thus recover suggests an integrated analytical framework. In sum, this volume seeks to broaden the inquiry into the historical layers of the conceptual nexus that brought together these key notions in other settings in Central and Southeastern Europe,[9] with a focus on transportation infrastructure projects as prime sources for polemics and anxieties. Indeed, all of our chapters share a common methodological and thematic concern with the core issues of conceptual vocabulary, both as employed by historical actors themselves *and*, critically, by researchers in the present.

Infrastructural Development under Pressure

Indeed, this assemblage of themes is something that insistently recurred in our historical sources. Not only has juxtaposing them been fruitful from our own analytical hindsight, but, crucially, the convergence of colonization, corruption, and xenophobia was something that, in varying proportions, was apparent to historical actors themselves. The different chapters in this volume are thus empirically grounded in examining how anxieties over "colonizing" or "corrupting" Others became part and parcel of the historical, material-technical, institutional, and discursive production of transportation infrastructure—in short, of its phenomenological and performative nature.

Colonial anxieties manifested themselves not only in an inter-imperial context but also, from the standpoint of emerging nation-states, as an overlooked facet of a broader fear of peripheralization. That infrastructural expansion *qua* symbol and vector of modernity was a double-edged sword seemed clear enough: the price that historical actors were willing to pay for being drawn into the transcontinental networks of capitalist expansion was, thus, open to negotiation, though their often-precarious position and relative

9 In using this terminology, we are aware of the politics of region-making and of how such ways of drawing boundaries—though historicizable in and of themselves—do not necessarily reflect our historical actors' sense of geographical space or belonging. As the case studies in the present volume cover both of these (sub)regions as typically understood at present, we have opted for this double-barreled denomination for the area in question. For an in-depth discussion, see Balázs Trencsényi and Diana Mishkova, eds., *European Regions and Boundaries: A Conceptual History* (New York: Berghahn, 2017).

lack of bargaining power needs to be unpacked and reassessed. Foreign capital and expertise could be seen as desirable, yet also potentially dangerous—and this played into, indeed heightened, xenophobic attitudes. These could include a more general nativist reaction and ideological drive toward reaching economic autarky, but also more specific forms of xenophobic discourse, targeting either individual capitalists and entrepreneurs, foreign laborers, or any (ethnic) group that could be imagined acting as a "fifth column" in league with foreign interests. The image of the "foreigner" in the abstract as a "corrupting" force was thus often the source of anxieties and of accusations against those already present on the territory of a nascent nation-state or those who were perceived to threaten to "colonize" it. This could also hold true, at the same time, in an imperial context—bearing in mind, however, that added layers of complexity from competing nation-building ambitions clashed with discourses of dynastic loyalty *and* anti-"foreigner" sentiment.

All of the contributions to our volume share a number of core thematic foci, and first among them, as a common denominator, is infrastructural development across the broader region. Transportation infrastructure had (and still has) both a creative and a disruptive nature.[10] In Central and Southeastern Europe in the nineteenth century, public discourse depicted it as the material and symbolic embodiment of progress, development, and civilization, and even of emancipation and (state) independence; in a word, a metonym for modernity.[11] From a historical perspective, the construction of transportation infrastructure cut across the divides between international and national politics, financial and economic policies, administrative capacity and state- and empire-building, military strategies, societal

10 Nikhil Anand, Akhil Gupta, and Hannah Appel, eds. *The Promise of Infrastructure* (Durham: Duke University Press, 2018).

11 For an examination of European history from the perspective of infrastructure's web of connections and ruptures, see Per Högselius, Arne Kaijser, and Erik Vleuten, *Europe's Infrastructure Transition: Economy, War, Nature* (London: Palgrave Macmillan, 2018), 1–17, 24–46. For the contribution of electrical infrastructure to nationalist modernism in Spain in the context of the reconfiguration of its post-imperial national identity, see Daniel Pérez-Zapico, "Electrical Futures for a Regenerated Spain: Electricity, Engineering and National Reconstruction after the 1898 'Disaster,'" *History and Technology* 39, no. 1 (2023): 91–125. See also the review article: Andreas Greiner, "On the Road: New Approaches to Infrastructure and Society," *Contemporary European History* 35 (2026): 1–10. For a recent overview of the concept "infrastructure," see Aaron Pinnix, Axel Volmar, Fernando Esposito, and Nora Binder, introduction to *Rethinking Infrastructure across the Humanities*, ed. Aaron Pinnix, Axel Volmar, Fernando Esposito, and Nora Binder (Bielefeld: transcript Verlag, 2023), 11–24.

transformation, and nation-building.[12] Through their material, usage, and discursive visibility, railways significantly transformed people's lives, urban, rural, and natural landscape, and even the representation of time and space.[13]

The historical exploration of transport infrastructure in the context of technological and administrative modernization in Central and Southeastern Europe across the nineteenth century has received intensive scholarly attention. Recent studies have focused on state- and empire-consolidation after the Crimean War, which gave the decisive impetus for strategic and military railway construction in Southeastern Europe.[14] Researchers have examined economic and urban (under)development[15] or transport and rail interconnectivity.[16] Importantly, they have also highlighted the entanglements between infrastructure construction and the Great Powers' ambitions of economic or political expansion—sometimes through railway

12 Jan Musekamp, *Shifting Lines, Entangled Borderlands: Mobilities and Migration along the Prussian Eastern Railroad* (Bloomington: Indiana University Press, 2024); Felix Jeschke, "Dracula on Rails: The Pressburgerbahn between Imperial Space and National Body, 1867–1935," *Central Europe* 10, no. 1 (2012): 1–17; Dietmar Neutatz, *Die Moskauer Metro: Von den ersten Plänen bis zur Grossbaustelle des Stalinismus (1897–1935)* (Cologne: Böhlau Verlag, 2001).

13 Wolfgang Schivelbusch, *The Railway Journey: The Industrialization of Time and Space in the Nineteenth Century* (Oakland: University of California Press, 1986); Dirk van Laak, *Lifelines of Our Society: A Global History of Infrastructures*, trans. Erik Butler (Cambridge, MA: MIT Press, 2023 [2018]).

14 Candan Badem, introduction to *The Routledge Handbook of the Crimean War*, ed. Candan Badem (Abingdon: Routledge, 2023), 1–4; Peter H. Christensen, *Germany and the Ottoman Railways: Art, Empire, and Infrastructure* (New Haven: Yale University Press, 2017); Sean McMeekin, *The Berlin–Baghdad Express: The Ottoman Empire and Germany's Bid for World Power* (Cambridge: Belknap Press of Harvard University Press, 2010); Frithjof Benjamin Schenk, *Russlands Fahrt in die Moderne: Mobilität und sozialer Raum im Eisenbahnzeitalter* (Stuttgart: Franz Steiner Verlag, 2014); Thomas Grob, Boris Previsic, and Andrea Zink, eds., *Erzählte Mobilität im östlichen Europa: (Post)Imperiale Räume zwischen Erfahrung und Imagination* (Tübingen: Francke, 2013).

15 Ralf Roth and Paul van Heesvelde, eds., *The City and the Railway in the World from the Nineteenth Century to the Present* (Abingdon: Routledge, 2022); Eduard J. Alvarez-Palau, Alfonso Díez-Minguela, and Jordi Martí-Henneberg, "Railroad Integration and Uneven Development on the European Periphery, 1870–1910," *Social Science History* 45, no. 2 (2021): 261–89; Basil C. Gounaris, *Steam over Macedonia, 1870–1912: Socio-Economic Change and the Railway Factor* (Boulder: East European Monographs, 1993).

16 Ralf Roth and Henry Jacolin, eds., *Eastern European Railways in Transition: Nineteenth to Twenty-first Centuries* (Abingdon: Routledge, 2013); Radu Mârza, ed., *Shaping Modernity: The Railway Journey across Two Centuries. Proceedings of the International Conference: The Railway Journey from the 19th to the 21st Centuries: A Cultural History of Railway Transport and Mobility* (Cluj-Napoca: Editura Mega, 2022); Florian Riedler and Nenad Stefanov, eds., *The Balkan Route: Historical Transformations from Via Militaris to Autoput* (Berlin: De Gruyter, 2021). For the Danube as a connecting infrastructure, see Constantin Ardeleanu, *The European Commission of the Danube, 1856–1948: An Experiment in International Administration* (Leiden: Brill, 2020).

imperialism. A notable emphasis has, in turn, been placed on technological and industrial know-how transfer from the West to the East.[17] All of the above tie into broader discussions of capitalist expansion into the region as a semi-periphery[18] and were reflected in nineteenth-century debates over free trade versus protectionism, with transportation infrastructure as the symbolic backbone of both desired development and potentially dangerous intrusion.[19]

That infrastructural development came with its own pressures and difficulties is also, correspondingly, a recurring feature of our analyses. Infrastructural development in the nineteenth century had outsize visibility in its technological materiality. It involved huge financial efforts and investments, complex planning, and sophisticated technological and industrial means of production. It triggered ever-widening public reactions, well beyond a handful of decision-makers and elites, and attracted public scrutiny as decision-makers were pressured to deliver and to find solutions to technical and political quandaries. They were even held accountable, in a dawning age of "transparency," by a growing literate public that, even in the absence of universal suffrage, made ultimately political claims on infrastructural development as a public good. And, not least, transportation infrastructure contributed to territorial reconfiguration, in that it both created and transcended state borders or regional disparities and acted as a conduit for the hopes and anxieties of progress.

17 Rang-Ri Park, *La Société de construction des Batignolles: des origines à la Première guerre mondiale, 1846–1914* (Paris: Presses de l'Université Paris-Sorbonne, 2005); Rang-Ri Park-Barjot, "La Société de construction des Batignolles en Russie (1851–1914)," in *La France et les Français en Russie: Nouvelles sources et approches (1815–1917)*, ed. Annie Charon, Bruno Delmas, and Armelle Le Goff (Paris: Publications de l'École nationale des chartes, 2011), 479–494; Alexandre Kostov, "Entre l'influence occidentale et les efforts nationaux: le choix des systèmes et du financement des chemins de fer dans les Balkans (1860–1912)," in *La France, l'Europe et les Balkans. Crises historiques et témoignages littéraires*, ed. Roumania L. Stantchéva, Alain Vuillemin, Oguniana Hrissimova, and Jean-Pierre Arrignon (Arras: Artois Presses Université, 2002), 11–20; Jonathan S. McMurray, *Distant Ties: Germany, the Ottoman Empire, and the Construction of the Baghdad Railway* (Westport: Praeger. 2001).

18 Christopher Clay, *Gold for the Sultan: Western Bankers and Ottoman Finance, 1856–1881: A Contribution to Ottoman and to International Financial History* (London: I. B. Tauris, 2000); Darina Martykánová and Juan Pan-Montojo, eds., *Misioneros del capitalismo. Aventureros, hombres de negocios y expertos transnacionales en el siglo XIX* (Granada: Comares, 2023); Constantin Ardeleanu, *Un drum de fier prin "Estul Sălbatic": Dobrogea otomană în zorii capitalismului* (Bucharest: Humanitas, 2025).

19 See a brief discussion of this in Andrei-Dan Sorescu, "The Infrastructural Power of the 'Colonial' as a Concept" in this volume.

Infrastructural Development and Its Narratives

One assumption shared by all of our contributors is that infrastructure is not only a material construct but also socially and politically constructed; historical actors ascribed meaning and values and projected their expectations onto it.[20] Chapters focus on specific flashpoints of controversy by relating to major infrastructure projects across a period from the 1840s to the 1920s in Central and Southeastern Europe. This was during what Dirk calls the "classical era of infrastructure,"[21] in a region where such projects were, for the most part, built from scratch and when institutional actors, legal frameworks, policies, knowledge, or practices were equally undergoing rapid and crucial transformations. Common to all chapters, therefore, is that they explore the historical, material-technical, institutional, *and* discursive production of transportation infrastructure. Discourses, narratives, and language also give form to infrastructure, as much as iron, concrete, or intricate technological processes. The study of historical actors' discourse, representations, and decision-making is of particular concern for us, as is their agency and capacity to negotiate and to adapt to their environment. These, then, are the ways in which our contributors keep track of how actors substantiated and accounted for the pressures of infrastructural development and how language itself was molded by the sense of urgency and acceleration typical of the period, *qua* "classic age of infrastructure." Infrastructure building—to borrow one of our reviewers' felicitous metaphors—provides us with a magnifying glass for how "colonization," xenophobia, and "corruption" were conceptualized in the past.

Both from a spatial and a chronological standpoint, our volume examines a coherent, if not always fully cohesive, set of case studies. They represent an infrastructural *Sattelzeit* on the edges of Europe—one slightly out of synch with the initial explosion and expansion in the West, yet never wholly so—as was equally the case with the conceptual apparatus employed by these actors, who were reflexively and acutely aware of their geopolitical, civilizational, and technological positions. Moreover, these narratives of how infrastructure could serve to connect the national and local to the imperial and continental bring issues of scale to the forefront: not, to be sure, in a linear or "nesting" way, but with a critical and reflexive edge,

20 Hannah Appel, Nikhil Anand, and Akhil Gupta, "Introduction: Temporality, Politics, and the Promise of Infrastructure," in Anand, Gupta, and Appel, *The Promise of Infrastructure*, 1–38.
21 Van Laak, *Lifelines of Our Society*, Part I, "The Classical Era of Infrastructure."

questioning the logics of how these units and dimensions interacted and were imagined to interact.

All of this was particularly salient in Central and Southeastern Europe, where asymmetrical power relations—conceptualized by the historical actors themselves as local underdevelopment and technological lag—were part of their mindset. The pivotal role that transnational capital and expertise had in the nineteenth century for the financing and construction of infrastructure works—notably in regions that were relative late-comers to capitalist industrialization and state capitalism—not only opened up financial and technological opportunities but also sharpened tensions over (potential or real) dependency, informal empire, loss of sovereignty, and even colonization, as evidenced by chapters in this volume. Our collected contributions address controversies over financial and human capital and technology transfers, as well as the question of who was to take the initiative for infrastructural expansion, be it private finance and private business or an increasingly capitalist state. Moreover, while the infrastructural ambitions of nation-states and empires were also on the agenda, they sometimes clashed with (transnational) capitalist or private interests or local interests and actors and implicitly led to the reevaluation of the peripheral nature of some regions.

Infrastructural Development, Colonization, and Corruption

To return to our initial observation, then, the nexus between the "colonial" and the infrastructural and between "Othering" and "corruption" was articulated as the sum of the tensions we have briefly outlined above. While no single contributor to our volume addresses all of their possible permutations, the volume cumulatively charts a significant number, both for each individual case study and for the overarching analytic endeavor of historicizing and cataloguing these complex interactions. Thus far, no existing collection has considered the interaction between our key concepts of "colonization" and "corruption," as evidenced in the corpus generated by controversies surrounding transportation infrastructure construction.

The volume discusses the "progressive," "disruptive," "dangerous," and sometimes overtly "colonial" characteristics of transportation infrastructure. We make the case for a legitimate conceptual stretching of "colonization"[22]

22 David Collier and James E. Mahon, Jr., "Conceptual 'Stretching' Revisited: Adapting Categories in Comparative Analysis," *The American Political Science Review* 87, no. 4 (1993): 845–855. We

that allows us to explain various power dynamics of individual or group interests, of economic exploitation, but also of national and international politics and (imagined and/or real) colonial encroachment in infrastructural development in Central and Southeastern Europe. Paying attention to both normative understandings of the term and the definitions employed by historical actors, the authors identify the targets of (anti-)colonial discourse, including the Great Powers, capitalism (as a human network and as a vision of the world[23]), private entrepreneurs, or multinational banking networks.

This book contributes to the growing literature on colonialism in Europe as a paradigm transcending binaries between empires and nation-states and complements the existing focus on overseas colonialism that has long meant the overshadowing of continental Europe in general as a potential colonial space. The present volume engages with broader debates on economic imperialism after the end of formal empire[24] and with research on "soft power" and colonialism in Central and Southeastern Europe.[25] Of particular interest for us is the way in which narratives of national emancipation from imperial domination were subverted by fears of continued subordination or the rise of new, potentially even more serious forms of (economic) dependence, as created by the construction of infrastructure. This, in turn, may help broaden the geographical scope of our collective inquiry: the spatial proximity or distance of perceived threats may well go beyond the confines of the region, narrowly defined.

Aside from the role of infrastructure as a conduit for modernization, development, and state building, the volume approaches the history and role of infrastructure through the lens of debates on and accusations of

prefer to speak of "colonization" rather than "colonialism" not only due to the normative weight of the latter category as a tool for analysis but also due to its relative absence in historical actors' discourse until the early twentieth century; "colony" and "colonization," on the other hand, were far from absent as terms; see also Sorescu, "The Promise."

23 Ralf Roth and Günter Dinhobl, eds., *Across the Borders: Financing the World's Railways in the Nineteenth and Twentieth Centuries* (Abingdon: Routledge, 2008); Youssef Cassis, Giuseppe De Luca, and Massimo Florio, eds., *Infrastructure Finance in Europe: Insights into the History of Water, Transport, and Telecommunications* (Oxford: Oxford University Press, 2016).

24 Zak Cope and Immanuel Ness, eds., *The Oxford Handbook of Economic Imperialism* (Oxford: Oxford University Press, 2022).

25 Stephen G. Gross, *Export Empire: German Soft Power in Southeastern Europe, 1890–1945* (Cambridge: Cambridge University Press, 2018); Murat Özyüksel, *The Berlin–Baghdad Railway and the Ottoman Empire: Industrialization, Imperial Germany and the Middle East* (London: I. B. Tauris, 2016); Timothy Snyder, introduction to *The Balkans as Europe, 1821–1914*, ed. Timothy Snyder and Katherine Younger (Rochester: Boydell & Brewer, 2018), 1–10; Alex R. Tipei, *Unintended Nations: France's Empire of Civilization, Southeast Europe, and the Post-Napoleonic World* (Montreal: McGill University Press, 2025).

corruption. Most major infrastructure projects engendered controversies and even scandals; they questioned regulatory standards, political and administrative practices and (political) institutions, and the collusion of business and politics and of public priorities and private gains. They also spotlight the blurred and shifting lines between public and private interests drawn by the elitist and often centralized nature of infrastructure decision-making in the nineteenth and early twentieth centuries. Though Central and Southeastern Europe are still conspicuously absent from studies on the conceptual history of political corruption[26] or from research on contemporary political corruption,[27] the volume contributes to an already well-defined body of literature on the history of corruption[28] by including Central and Southeastern Europe as a subject of analysis—analyzed from a constructivist standpoint, rather than associated *a priori* with "endemic" and/or transhistorical "corruption." The constructivist perspective on corruption also provides methodological and conceptual tools to further explore the many ways in which "corruption" as idea, discourse, and practice played a role in the history of transportation infrastructure and, with it, in the processes of state- and nation-building or of empire consolidation or erosion through the nineteenth and early twentieth centuries in Central and Southeastern Europe.

The first four chapters by Malte Furhmann, Silvia Marton, Jens Ivo Engels, and Andrei-Dan Sorescu lay the conceptual framework for the historical exploration of the complex connections transportation infrastructure, corruption debates, xenophobia, and colonization had in the nineteenth century.

Examining several present-day and mid-nineteenth-century controversies over transportation infrastructure in Ottoman and Bulgarian lands, Malte Fuhrmann's conceptual chapter discusses four legitimizing and

26 Bruce Buchan and Lisa Hill, *An Intellectual History of Political Corruption* (London: Palgrave Macmillan, 2014).

27 Paul M. Heywood, ed., *Routledge Handbook of Political Corruption*, 1st ed. (Abingdon: Routledge, 2015); Arnold Heidenheimer and Michael Johnston, eds., *Political Corruption: Concepts and Contexts*, 3rd ed. (Abingdon: Routledge, 2002); Simona Piattoni, ed., *Clientelism, Interests, and Democratic Representation: The European Experience in Historical and Comparative Perspective* (Cambridge: Cambridge University Press, 2001).

28 Jens Ivo Engels, "La nueva historia de la corrupción: Algunas reflexiones sobre la historiografía de la corrupción política en los siglos XIX y XX," *Ayer: Revista de Historia Contemporanea* 115, no. 3 (2019): 23–49; Ronald Kroeze, Pol Dalmau, and Frédéric Monier, eds., *Corruption, Empire and Colonialism in the Modern Era: A Global Perspective* (London: Palgrave Macmillan, 2021); Ricard Torra-Prat, Joan Pubill-Brugués, and Arndt Brendecke, eds., *Corruption, Anti-Corruption, Vigilance, and State Building from Early to Late Modern Times* (Abingdon: Routledge, 2024).

delegitimizing rhetorical tropes employed to discuss infrastructure development and politics: boasting, shaming, smokescreening, and scapegoating. Fuhrmann underscores that all these tropes are contingent upon the "promise of infrastructure"[29]—that is, on everyday life experiences and expectations of the future—and in part on alleged or actual corruption. Fuhrmann explains that infrastructural boasting is done by officeholders who claim to have performed well and who derive their legitimacy from this achievement—often within a framework of authoritarian rule lacking participatory support. Infrastructural shaming, by contrast, intends to delegitimize officeholders on the grounds that they have failed to make good on their promises of infrastructure-driven prosperity. Finally, infrastructural smokescreening is performed to obfuscate and deflect charges of inefficiency or corruption, while scapegoating functions as a strategic Othering along the same lines. The case studies examined by Fuhrmann illustrate that these rhetorical ideal types of PR management transcended historical discontinuities and functioned across the democratic/non-democratic divide.

Silvia Marton's chapter examines how infrastructure projects acted as catalysts for the historical construction of corruption-related categories, such as immoral public official behavior, abuse, favoritism, embezzlement, or rent-seeking. A pamphlet published in 1875 during ongoing controversies surrounding railway concessions in Romania serves as a starting point for Marton's chapter, highlighting the powerful performative capacity that infrastructure polemics had in unravelling conceptions of "corruption" and questioning public norms and ideals. This also came, Marton demonstrates, with ascribing blame and Othering corruption in the era of liberal nationalism, when railway construction was intertwined with nation-building, the politics of identity, and ideas of Othering in the region.

Jens Ivo Engels outlines the relevance of historical approaches to the recent and highly interdisciplinary debates around "critical infrastructures." He explains the versatile usages of criticality in historical research. He maintains that, since criticality is a relational property measured by comparison with and in relation to the surrounding society, politics, or state, it is constructed not only technologically but also socially and culturally through historical actors assigning to it meanings, values, or expectations. Moreover, Engels emphasizes that, besides the various functions infrastructure performs, its criticality also stems from social attributions (failure, happiness, etc.) that evolve over time.

29 Anand, Gupta, and Appel, *The Promise of Infrastructure*.

Andrei-Dan Sorescu considers not only how anxieties over "colonization" functioned in tandem with infrastructure scandals but also how the "colonial" was, as a conceptual category, infrastructural in its importance for nineteenth-century Romanian nationalism. Aiming to go beyond the Romanian case study, however, his chapter asks the broader question of how this stacks up against more common understandings of "anti-colonial nationalism." To this end, it unpacks the logic whereby some Others were perceived as "colonizers," the temporality and timing of anti-colonial anxieties versus the attainment of state independence, and the fraught question of which metropole, if any, was to be emulated by a country that had never been a colony, properly speaking.

The themes and the historical cases treated vary in the book's chapters. The focus on the historical actors' discourse and praxis of infrastructural development allows all contributors to examine, to various degrees, the conceptual nexus of our key concepts—namely, "colonization," "corruption," and xenophobia.

Alex R. Tipei focuses on the early Greek state in the aftermath of the 1831 assassination of Ioannis Kapodistrias (1776–1831), the first governor of an independent Greece. She explores how rival Greek politicians used the concept of tyranny as a shorthand to denote all that was corrupt and to assimilate their rivals with the Ottoman past. Tipei explains that the aim of their accusations and of their specific rhetorical repertoire was to win the support of West European allies *and* to discredit their rivals. This came against the backdrop of infrastructure modernization controversies and during a period best understood as one of transition from formal (Ottoman) to informal (West European) empire. Greek leaders often characterized their local political and economic competitors as corrupt and tyrannical in the letters, articles, and pamphlets that they wrote for Francophone correspondents and readers. Over time, these terms grew increasingly coded as un-modern (or un-European) and Ottoman (or Muslim).

Constantin Ardeleanu analyzes the disputes between the Ottoman authorities and the Danube and Black Sea Railway (DBSR) Company for building a canal between the Danube and the Black Sea in the 1860s, showing the difficulties of business ventures in a peripheral region. The railway and the port works in the Ottoman town of Kustendjie aimed to reshape transport patterns in Southeastern Europe. The chapter examines the capitalists' project to build a canal between the Danube and the Black Sea to bypass the dangerous bottleneck of Sulina in the Danube Delta. The port was also critical for the Romanian state after it acquired the province of Dobrudja in 1878. Based on diplomatic and company archives, Ardeleanu deciphers how Ottoman, British, and Romanian diplomats, central and regional authorities, investors, and representatives of the DBSR Company mobilized both financial figures and rhetorical tropes in the intricate and multi-level negotiations. Building a canal between the Danube and the Black Sea thus meant oscillating between mutual recriminations, blame-shifting, procrastination tactics, and profit-oriented or geopolitical priorities.

Gábor Egry illustrates how the discursive consensus over the strategic importance of railways for society, commerce, and (state) modernization allowed ample room for complex and asymmetrical negotiations between concession-seeking groups, local notables, politicians, and investors. His chapter does so by placing railway construction by Austrian and Hungarian capitalists in the late Ottoman Empire side by side with railway projects in Transylvania—a distant and somewhat backward region of dualist Hungary—between 1890 and 1914. His case studies include not only the Transylvanian Saxon railway ventures and concessions from the Ottoman lands, which were critical in that they had value as marketable rights, but also the role played by local and transnational networks in successfully exploiting the rail lines. Surprisingly, Egry demonstrates that such cases show that rent-seeking played a more important role than colonial extraction, and he shows how peripheral actors' strategies led to the transfer of capital *from* the core, rather than merely extraction.

Focusing on late nineteenth-century Romania, Toader Popescu discusses the opportunities and restrictions that railway infrastructure was perceived to bring for municipalities once connected to the main national network. Railway and municipal administrations had to deal with this complex interaction without the benefit of a specific doctrine or the support of previous similar practical experiences. This generated accusations of corruption, great expectations, and significant anxieties at the local level. The latter sprang from the unpredictable effects that this new system might have had on established urban structures, as well as from the relatively small influence that local communities were allowed in the national decision-making process.

Mihai Chiper investigates the path dependencies created by the under-allocation of resources for infrastructural development in the Moldavia province of Romania across the long nineteenth century. By showing how decision-making in Bucharest prioritized railway building in Wallachia—the country's southern province—Chiper demonstrates how projections of "backwardness" served to justify centralist policies and how, beginning with the very layout of the rail network, Moldavia was systematically disadvantaged. The perspective of the center, in turn, is complemented by a careful reading of regional actors' initiatives and agency and their attempts to speak against the economic, political, or even military motivations invoked against the expansion of Moldavian railways. Moreover, the international entanglements created by the need for connecting the national rail network to those in neighboring empires are also given their due importance. The chapter integrates these levels of analysis by constantly taking stock of the language and logics of accusations fielded against Bucharest, the capital—allegations in which corruption and intranational xenophobia played no small part in defining the nature of contestation.

Finally, Raul Cârstocea highlights the continuities of the debates that began with nineteenth-century railway construction in the very different context of a significantly expanded Romanian state after the First World War. He shows how a "native" version of modern anti-Semitism in Romania—developed notably during the Strousberg scandal of the 1870s and imbricated with allegations of "corruption" and colonial anxieties—continued and mutated into the 1920s. He explores how anti-Semitism transformed from parliamentary rhetoric to street violence against Jews. At the forefront of the latter development were student organizations that rose to prominence against the background of yet another major state project: that of expanding higher education for the purposes of nationalizing a suddenly very diverse country. Railway infrastructure now assumed a vital role in connecting the different provinces of a still loosely integrated Greater Romania, while also revealing a surprising one as a prominent conduit of violence. As students primarily traveled by train, trains and railway stations increasingly became sites of anti-Semitic violence, especially when student delegations would attend congresses at other universities. Numerous police reports from the 1920s attest both to individual attacks against Jews on trains and to organized violence in and around railway stations.

The source base covered by the present volume showcases both the linguistic and geographical complexity of the case studies under investigation, as well as the complexity of the analytical angles required for an integrated analysis along the lines outlined above. In more concrete terms, archival sources (be they from central or local authorities, private enterprises, or professional bodies) are complemented by parliamentary debates, the press, and pamphlets. Moreover, the sources reflect not only the perspectives of local actors but also those of actors external to the region, thereby offering a complementary angle and an even wider web of connections and representations.

*

Our appreciation goes to the institutions whose financial and logistical support made this publication possible. The book originates in the research grant "Colonial Anxieties, Corruption Scandals and Xenophobia in Nineteenth-Century Infrastructure Development in Romania (CanCor)" (principal investigator Silvia Marton) of the Executive Unit for Financing Higher Education, Research, Development and Innovation of Romania (as part of the agency's basic and frontier research scheme, PN-III-P4-PCE–2021–0399). Our collective research was generously hosted by the New Europe College—Institute for Advanced Study in Bucharest (from June 2022 to December 2024). We would like to thank our colleagues who contributed to two events focused on method, sources, and conceptualization: "Colonial Anxieties, Corruption Scandals and Modernization in Nineteenth-Century Infrastructure Development" (October 2023) and "Transportation Infrastructure, (Anti) Corruption, and Power Relations in the Nineteenth and Twentieth Centuries" (May 2024). We are grateful to the chapter authors and all participants for their valuable contributions to these events.

We are particularly grateful to our colleagues in the research project—Constantin Ardeleanu, Raul Cârstocea, and Toader Popescu—for their intellectual insights and for their friendship. From among our friends and colleagues, we would especially like to thank Frédéric Monier for his deep understanding of this volume's core aims and for his firm belief in our research agenda. Toader Popescu's contribution to designing the maps that serve to illustrate some of the volume's chapters has been invaluable. Finally, we would like to extend our thanks to the editors of the Imperialism and Colonialism in Central and Eastern Europe book series and to Jen McCall at the CEU Press for their enthusiastic support for our endeavor and to the anonymous reviewers for their valuable suggestions.

Bibliography

Alvarez-Palau, Eduard J., Alfonso Díez-Minguela, and Jordi Martí-Henneberg. "Railroad Integration and Uneven Development on the European Periphery, 1870–1910." *Social Science History* 45, no. 2 (2021): 261–289, https://doi.org/10.1017/ssh.2021.1.

Anand, Nikhil, Akhil Gupta, and Hannah Appel, eds. *The Promise of Infrastructure*. Durham: Duke University Press, 2018.

Ardeleanu, Constantin. *The European Commission of the Danube, 1856–1948: An Experiment in International Administration*. Leiden: Brill, 2020.

Ardeleanu, Constantin. *Un drum de fier prin "Estul Sălbatic": Dobrogea otomană în zorii capitalismului.* Bucharest: Humanitas, 2025.

Badem, Candan. Introduction to *The Routledge Handbook of the Crimean War,* edited by Candan Badem. Abingdon: Routledge, 2023, 1–4.

Buchan, Bruce, and Lisa Hill. *An Intellectual History of Political Corruption.* London: Palgrave Macmillan, 2014.

Cassis, Youssef, Giuseppe De Luca, and Massimo Florio, eds. *Infrastructure Finance in Europe: Insights into the History of Water, Transport, and Telecommunications.* Oxford: Oxford University Press, 2016.

Christensen, Peter H. *Germany and the Ottoman Railways: Art, Empire, and Infrastructure.* New Haven: Yale University Press, 2017.

Clay, Christopher. *Gold for the Sultan: Western Bankers and Ottoman Finance, 1856–1881: A Contribution to Ottoman and to International Financial History.* London: I. B. Tauris, 2000.

Collier, David, and James E. Mahon, Jr. "Conceptual 'Stretching' Revisited: Adapting Categories in Comparative Analysis." *The American Political Science Review* 87, no. 4 (1993): 845–855. https://doi.org/10.2307/2938818.

Cope, Zak, and Immanuel Ness, eds. *The Oxford Handbook of Economic Imperialism.* Oxford: Oxford University Press, 2022.

Engels, Jens Ivo. "La nueva historia de la corrupción: Algunas reflexiones sobre la historiografía de la corrupción política en los siglos XIX y XX." *Ayer: Revista de Historia Contemporanea* 115, no. 3 (2019): 23–49. https://doi.org/10.55509/ayer/115-2019-02.

Gounaris, Basil C. *Steam over Macedonia, 1870–1912: Socio-Economic Change and the Railway Factor.* Boulder: East European Monographs, 1993.

Greiner, Andreas. "On the Road: New Approaches to Infrastructure and Society." *Contemporary European History* 35 (2026): 1–10. https://doi.org/10.1017/S0960777325000177.

Grob, Thomas, Boris Previsic, and Andrea Zink, eds. *Erzählte Mobilität im östlichen Europa: (Post-)Imperiale Räume zwischen Erfahrung und Imagination.* Tübingen: Francke, 2013.

Gross, Stephen G. *Export Empire: German Soft Power in Southeastern Europe, 1890–1945.* Cambridge: Cambridge University Press, 2018.

Hașdeu, B. P. "Discursul d-lui B. P. Hașdeu ținut la întrunirea publică din 5 decembrie 1871, în casa d-lui Gr. Băleanu asupra cestiunii Strussberg-Bleichroeder." *Columna lui Traian* 2, no. 46 (108), December 8, 1871: 179–180. *Apud* B. P. Hașdeu, *Opere. IV. Publicistica politică, 1858–1904*, edited by Stancu Ilin and I. Oprișan. Bucharest: Editura Fundației Naționale pentru Știință și Artă; Univers Enciclopedic, 2007, 1497–1507.

Heidenheimer, Arnold J., and Michael Johnston, eds. *Political Corruption: Concepts and Contexts*. 3rd ed. Abingdon: Routledge, 2002.

Heywood, Paul M., ed. *Routledge Handbook of Political Corruption*. 1st ed. Abingdon: Routledge, 2015.

Högselius, Per, Arne Kaijser, and Erik van der Vleuten. *Europe's Infrastructure Transition: Economy, War, Nature*. London: Palgrave Macmillan, 2018.

Jeschke, Felix. "Dracula on Rails: The Pressburgerbahn between Imperial Space and National Body, 1867–1935." *Central Europe* 10, no. 1 (2012): 1–17, https://doi.org/10.1179/1479096312Z.0000000001.

Kostov, Alexandre. "Entre l'influence occidentale et les efforts nationaux: le choix des systèmes et du financement des chemins de fer dans les Balkans (1860–1912)." In *La France, l'Europe et les Balkans : Crises historiques et témoignages littéraires*, edited by Roumania L. Stantchéva, Alain Vuillemin, Oguniana Hrissimova, and Jean-Pierre Arrignon, 11–20. Arras: Artois Presses Université, 2002.

Kroeze, Ronald, Pol Dalmau, and Frédéric Monier, eds. *Corruption, Empire and Colonialism in the Modern Era: A Global Perspective*. London: Palgrave Macmillan, 2021.

Marton, Silvia, and Andrei Sorescu. "Introduction: Colonial Anxieties, Corruption Scandals, and Xenophobia in Nineteenth-Century Infrastructure Development in Romania." *East European Politics and Societies* 39, no. 3 (2025): 680–690. https://doi.org/10.1177/08883254251352114.

Martykánová, Darina, and Juan Pan-Montojo, eds., *Misioneros del capitalismo: Aventureros, hombres de negocios y expertos transnacionales en el siglo XIX*. Granada: Comares, 2023.

Mârza, Radu, ed. *Shaping Modernity: The Railway Journey across Two Centuries. Proceedings of the International Conference: The Railway Journey from the 19th to the 21st Centuries: A Cultural History of Railway Transport and Mobility*. Cluj-Napoca: Editura Mega, 2022.

McMeekin, Sean. *The Berlin–Baghdad Express: The Ottoman Empire and Germany's Bid for World Power*. Cambridge: Belknap Press of Harvard University Press, 2010.

McMurray, Jonathan S. *Distant Ties: Germany, the Ottoman Empire, and the Construction of the Baghdad Railway*. Westport: Praeger. 2001.

Musekamp, Jan. *Shifting Lines, Entangled Borderlands: Mobilities and Migration along the Prussian Eastern Railroad*. Bloomington: Indiana University Press, 2024.

Neutatz, Dietmar. *Die Moskauer Metro: Von den ersten Plänen bis zur Grossbaustelle des Stalinismus (1897–1935)*. Cologne: Böhlau Verlag, 2001.

Özyüksel, Murat. *The Berlin–Baghdad Railway and the Ottoman Empire: Industrialization, Imperial Germany and the Middle East*. London: I.B. Tauris, 2016.

Park, Rang-Ri. *La Société de construction des Batignolles: des origines à la Première guerre mondiale, 1846–1914*. Paris: Presses de l'Université Paris-Sorbonne, 2005.

Park-Barjot, Rang-Ri. "La Société de construction des Batignolles en Russie (1851–1914)." In *La France et les Français en Russie: Nouvelles sources et approches (1815–1917)*, edited by Annie Charon, Bruno Delmas, and Armelle Le Goff, 479–494. Paris: Publications de l'École nationale des chartes, 2011.

Pérez-Zapico, Daniel. "Electrical Futures for a Regenerated Spain: Electricity, Engineering and National Reconstruction after the 1898 'Disaster.'" *History and Technology* 39, no. 1 (2023): 91–125. https://doi.org/10.1080/07341512.2023.2183804.

Piattoni, Simona, ed. *Clientelism, Interests, and Democratic Representation: The European Experience in Historical and Comparative Perspective*. Cambridge: Cambridge University Press, 2001.

Pinnix, Aaron, Axel Volmar, Fernando Esposito, and Nora Binder. Introduction to *Rethinking Infrastructure across the Humanities*, edited by Aaron Pinnix, Axel Volmar, Fernando Esposito, and Nora Binder, 11–24. Bielefeld: Transcript Verlag, 2023.

Riedler, Florian, and Nenad Stefanov, eds. *The Balkan Route: Historical Transformations from Via Militaris to Autoput*. Berlin: De Gruyter, 2021.

Roth, Ralf, and Günter Dinhobl, eds. *Across the Borders: Financing the World's Railways in the Nineteenth and Twentieth Centuries*. Abingdon: Routledge, 2008.

Roth, Ralf, and Henry Jacolin, eds. *Eastern European Railways in Transition: Nineteenth to Twenty-first Centuries*. Abingdon: Routledge, 2013.

Roth, Ralf, and Paul van Heesvelde, eds. *The City and the Railway in the World from the Nineteenth Century to the Present*. Abingdon: Routledge, 2022.

Schenk, Frithjof Benjamin. *Russlands Fahrt in die Moderne: Mobilität und sozialer Raum im Eisenbahnzeitalter*. Stuttgart: Franz Steiner Verlag, 2014.

Schivelbusch, Wolfgang. *The Railway Journey: The Industrialization of Time and Space in the Nineteenth Century*. Oakland: University of California Press, 1986.

Snyder, Timothy. Introduction to *The Balkans as Europe, 1821–1914*, edited by Timothy Snyder and Katherine Younger, 1–10. Rochester: Boydell & Brewer, 2018.

Sorescu, Andrei. "The Perils (and Promise) of German Colonization: Civilizational Hierarchies and Anxieties in Nineteenth Century Romania." *East European Politics and Societies:* 39, no. 3 (2025): 717–740. https://doi.org/10.1177/08883254251355736.

Sorescu, Andrei-Dan. "The Political Language of Moral Panics: Teutophobia and the Image of a Besieged Romania in 1871." In *Penser le XIXe siècle: nouveaux chantiers de recherche*, edited by Silvia Marton and Constanţa Vintilă-Ghiţulescu, 153–173. Iași: Editura Universității "Alexandru Ioan Cuza" din Iași, 2013.

Tipei, Alex R. *Unintended Nations: France's Empire of Civilization, Southeast Europe, and the Post-Napoleonic World*. Montreal: McGill University Press, 2025.

Torra-Prat, Ricard, Joan Pubill-Brugués, and Arndt Brendecke, eds. *Corruption, Anti-Corruption, Vigilance, and State Building from Early to Late Modern Times*. Abingdon: Routledge, 2024.

Trencsényi, Balázs, and Diana Mishkova, eds. *European Regions and Boundaries: A Conceptual History*. New York: Berghahn, 2017.

Van Laak, Dirk. *Lifelines of Our Society: A Global History of Infrastructures*, translated by Erik Butler. Cambridge: MIT Press, 2023 [2018].

About the Authors

Silvia Marton is associate professor at the Faculty of Political Science, University of Bucharest, and a senior researcher at the New Europe College—Institute for Advanced Study, Bucharest. Her scientific interests include the history and sociology of political corruption and nation-state building in Central and Southeastern Europe. She is principal investigator of "Transnational histories of 'corruption' in Central-South-East Europe (1750–1850)" (ERC–2022-AdG no. 101098095), hosted by the New Europe College—Institute for Advanced Study, Bucharest.

Andrei-Dan Sorescu is a cultural and intellectual historian of nineteenth-century Europe, with a focus on Romania in a transnational context. His current research interests include the role of self-comparison in nation-building, the historical semantics of the "colonial," the transnational and global dimensions of anti-Semitism, and the critical study of temporalities.

Part I

Conceptual Framework

1. On Hegemony and Connectivity: A Short History of Infrastructure-Political Struggles and their Discourses

Malte Fuhrmann

Abstract: This chapter focuses on the roots of discourses employed to discuss controversial infrastructure. I argue that they are based on the tacit acceptance of *developmentality*, a variant of *gouvernementalité* entailing a close nexus between hegemony, development, traffic infrastructure, and happiness, which evolved during enlightened absolutism and has remained relatively stable throughout modernity. After introducing the popular rhetorical forms of infrastructure-related political debate using contemporary examples, I will trace them through history, demonstrating how these tropes entered the political scene during the late eighteenth and nineteenth centuries and how they contributed to actively legitimizing and delegitimizing officeholders, focusing on the period 1839–1914.

Keywords: Roads, Railways, Legitimacy, Ottoman Empire, Bulgaria

To the memory of the victims of Pamukova, Tempi, and Novi Sad
In solidarity with those attempting to set these and related injustices right

At the time of writing, in spring 2025, Serbia, Greece, and Turkey were witnessing the largest protests in their history. In various ways, the claims the protesters made were all tied to transport infrastructure and their expectations regarding their governments' building, operating, or supervising of such means of communication. The protests began following November 1,

Marton, Silvia, and Andrei-Dan Sorescu, eds. *Infrastructural Development, Corruption, Xenophobia, and Colonization in Central and Southeastern Europe*. Amsterdam: Amsterdam University Press/Central European University Press, 2026.
 https://doi.org/10.5117/9789048577279-2

2024, when the concrete canopy over the main entrance to Novi Sad Station in Serbia collapsed, killing 16 people. A broad societal alliance laid direct blame on the government, its cronies, and its secret deals with Chinese infrastructure constructors and financiers. The movement became the most serious challenge to President Aleksandar Vučić's (Serbian Progressive Party, SNS) rule since he came to office in 2017. It came at a particularly unfortunate moment for him, as Novi Sad Station had just been renovated and reopened as part of a prestigious modernization of the Belgrade–Budapest railway line to 200 kmph speeds by Chinese companies and Serbian subcontractors. The inauguration of the Novi Sad–Subotica section had been scheduled for just three weeks after the disaster, with the Hungarian president and Chinese officials planning to attend. Unable to placate the protesters by prosecuting former ministers and officials for criminal negligence, Vučić postponed the railway inauguration indefinitely.[1] The president was caught in a dilemma: If he had proceeded with the inauguration, the accompanying mass protests would have proven that a considerable part of the electorate was far from happy with him. Yet failing to open the new infrastructure would have meant that Vučić could not deliver on the promises he made in previous PR campaigns, which had portrayed the faster Novi Sad–Subotica railway line as the ultimate key to progress and happiness. This dilemma was not lost on the opposition movement, which criticized him on both accounts.[2]

The persistent Serbian protests rekindled dissatisfaction in Greece on the second anniversary of the country's worst railway accident. On February 28, 2023, an intercity train collided head-on with a freight train near Tempi, killing 57 people. A report revealed clear signs that regulations had been violated and that gross negligence was afoot: the introduction of a modern signaling system on the line was long overdue, but trains continued to operate at high speeds nonetheless; the freight train had likely been transporting a nondisclosed highly explosive chemical, decisively worsening the magnitude of the tragedy; and the haste with which the site was covered up forestalled the possibility of fully reconstructing the causes or even recovering all the bodies. Although the underlying reasons for the crash remained partially unclear, this did not prevent protesters from taking action. As one participant of the largest demonstration in Athens on February 28, 2025, framed it:

1 Katerina Baletic, "Serbian Prosecution Charges 13 over Railway Station Disaster," *Balkan Insight*, December 30, 2024, https://balkaninsight.com/2024/12/30/serbian-prosecution-charges–13-over-railway-station-disaster/bi/.

2 PVP, "Kada će krenuti brzi voz od Novog Sada do Subotice?" *N1*, December 9, 2024, https://n1info.rs/vesti/kada-ce-krenuti-brzi-voz-od-novog-sada-do-subotice/.

"Either the state is so inadequate it cannot run a modern railway system, or we should believe the conspiracy theories and everything they say about a cover-up. One or the other is true."[3]

Boasting and Shaming

While Turkey had also experienced several large, deadly accidents in the railway sector—such as Pamukova (2004, 38 dead), Çorlu (2018, 25 dead), and Ankara (2018, nine dead)[4]—none of these sparked a nationwide solidarity movement. But when, just a few days after the record-setting demonstrations in Belgrade and Athens, the mayor of Istanbul, Ekrem İmamoğlu, was arrested on implausible charges of corruption and aiding and abetting terrorists, Istanbul also witnessed the largest gatherings in its history. While this confrontation stemmed from political rivalry—İmamoğlu's ambition to succeed Recep Tayyip Erdoğan as Turkish president in the next election and the latter's alleged use of the judiciary to block him—transport infrastructure nonetheless has played an important role here too.

In March 2024, İmamoğlu enjoyed a sweeping victory in his reelection as Istanbul mayor. He had initially won by a slim majority in 2019, becoming the first Kemalist mayor after 25 years of various Islamist mayors. His clear majority five years later surprised everyone. This success was the result not only of voters' satisfaction with his achievements but also of a highly successful PR campaign to promote them. Throughout the city, huge billboards financed by the municipality recounted İmamoğlu's good deeds. Prominent among them were infrastructural accomplishments: the amount invested into infrastructure per district, the length of new freshwater pipes, and, above all, transport infrastructure. The mayor took great pride in the number of metro lines inaugurated during his tenure. The overall PR campaign followed the slogan, "We ended wastefulness and began (public) service." The words "wastefulness" and "service" in Turkish (*israf* and *hizmet*) are ambiguous. They can be used as equivalents to widespread political and economic concepts, but they also have Islamic undertones—the former being a sin and the latter an obligation. In this way,

3 Helena Smith, "Rage in Greece as Second Anniversary of Train Disaster Prompts Mass Protests," *The Guardian*, February 28, 2025, https://www.theguardian.com/world/2025/feb/28/rage-in-greece-as-second-anniversary-of-train-disaster-prompts-mass-protests.

4 Turkey. Ministry of Transport and Infrastructure. Transport Safety Investigation Center, "Demiryolu Kaza İnceleme Raporları," accessed April 3, 2026, https://ulasimemniyeti.uab.gov.tr/demiryolu.

the mayor could also appeal to religious voters appalled by the nepotism of the nationally ruling AKP (Justice and Development Party) without offending his own CHP (Republican People's Party) laicist core voters. While the moral condemnation of "wastefulness" falls short of the legal accusation of "embezzlement," the latter is certainly implied.

The two catchwords—"wastefulness" and "service"—also triggered memories of earlier municipal PR campaigns. In 2019, a few weeks after taking office, İmamoğlu had the municipality's countless superfluous company cars assembled in a large square and aerial photos were taken. On this occasion, he declared that the CHP municipality would sell these cars to end the old AKP administration's wastefulness.[5] In the following years, the administration constantly stressed the achievement (or public service) of having secured new foreign credits to restart the city's dormant metro construction sites in order to alleviate commuters' woes.[6] The political messages and images conveyed by focusing on issues of transport infrastructure were potent, with the rhetorical tropes of "boasting" of one's own accomplishments while "shaming" one's opponent over their failures at their core. If we follow İmamoğlu's PR, we see the former AKP municipal cronies being driven around by chauffeurs in cars financed by taxpayers, while ordinary commuters were condemned to suffer in overcrowded buses. By contrast, under the new administration, public employees are no longer chauffeured around town for their leisurely pursuits but work relentlessly for the public good until they have achieved new and better transport means for all. Just a few years after the new administration has taken office, both public employees and ordinary Istanbulites will commute together stress-free on fast and efficient metro lines between home and work. This image of a happier future within easy reach convinced many voters.

İmamoğlu's use of the terms "wastefulness" and "service" was also a provocation, as *hizmet* (service) is one of the key terms in President Erdoğan's rhetorical repertoire. In his initial years, Erdoğan contrasted his government with the ineffective and unstable governments of the 1990s. By contrast, the government now supposedly provided public services of use to all citizens, regardless of ideological stance. For example, he often referred to the transport sector—especially the new high-speed train lines, bridges,

5 Meral Danyıldız, "İmamoğlu AKP'nin saltanatını sergiledi: Hangi hakla bu araçlara bindiniz!" *BirGün*, September 5, 2019. https://www.birgun.net/haber/imamoglu-akp-nin-saltanatini-sergiledi-hangi-hakla-bu-araclara-bindiniz-267465.

6 Doç. Dr. Pelin Alpkökin (@alpkokin_pelin). "Proje eksiklikleri ve uygunsuzluklara ragmen...," X, March 22, 2024. https://x.com/alpkokin_pelin/status/1771046411951366425.

and divided roads. By appropriating the term *hizmet*, İmamoğlu implied that the president had now become redundant and there was someone who could get the job done better. As if to prove that the propagandistic battle for the Turkish electorate would be decided in part over infrastructure issues, President Erdoğan issued a presidential decree on March 21, 2025—two days after İmamoğlu's arrest—that allowed the central government to plan a new metro extension without even consulting with the Greater Istanbul municipality.[7]

In this way, the governments of Serbia, Greece, and Turkey faced their most serious challenge from popular movements in recent memory. While these conflicts were not limited to the construction and maintenance of effective and safe mobility infrastructure, such concerns constituted a vital element of the debate. For mobility infrastructure to become such a central site for mobilization, it must have a prehistory. Indeed, while the scale of contentions over infrastructural claims is new, they draw on a rhetorical repertoire that has evolved over a long period.

In this chapter, I wish to draw attention to the rhetorical roots of these discussions. As Silvia Marton and others in this volume discuss in more detail, the constitution of the "public," the media it uses to communicate and organize, and the technical expertise it can rally for its cause are subject to considerable historical change.[8] By contrast, I believe that if we focus on the claims that political stakeholders make when discussing transport infrastructure, we will notice their continuity. I argue that these basic discursive constructs for framing transport infrastructure issues have a long history. İmamoğlu boasting of his achievements and Serbian protesters shaming Vučić for the criminal negligence at the Novi Sad construction site are in line with how people have discussed controversial infrastructure matters for almost 200 years. While "boasting" and "shaming" are two of the main discursive forms used to discuss controversial infrastructure politics—and potentially mark them as scandalous—other important tropes are "smokescreening" and "scapegoating." I will introduce these all with contemporary examples. I will then go back further in history to demonstrate when and why transport infrastructure matters became a major site of politics or, more abstractly, of legitimizing and delegitimizing public officeholders. I will then return to the four discursive tropes mentioned—boasting, shaming, smokescreening, and scapegoating—and

7 Turkey. Ministry of Transport and Infrastructure, "İstanbul'a 4,5 kilometrelik yeni metro hattı," March 21, 2025, https://www.uab.gov.tr/haberler/istanbul-a-4-5-kilometrelik-yeni-metro-hatti.
8 Silvia Marton, "Transportation Infrastructure and Corruption Scandals" in this volume.

demonstrate how they entered the political scene during the nineteenth century, focusing on examples from the late Ottoman Empire and Bulgaria between roughly 1839 and 1914. But first, let us finish the introduction of the four main tropes of infrastructural discussion.

Smokescreening

On May 8, 2012, the Berlin airport authority declared that the newly built Berlin Brandenburg Airport, due to be inaugurated only three weeks later, faced problems that meant its inauguration would be postponed indefinitely. As is common when infrastructural measures fail, a veil of silence fell upon the affair. In the following months and years, no clear explanation of the problems was given; the board members made only occasional, rather tight-lipped announcements about technical problems and issued several new inauguration deadlines that continued to be missed. The wider public was caught by surprise. In previous years, they had often seen the Berlin mayor, Klaus Wowereit (in office 2001–2014, Social Democratic Party of Germany, SPD), smiling for the cameras at the construction site, shaming critics of the project, and promising timely completion. After the cancellation of the inauguration, Wowereit reacted with annoyance whenever he was asked about the airport, stating that no one could expect him, a trained lawyer, to understand the airport's technical details—despite his being chairman of the airport's supervisory board.[9] In hindsight, his behavior, both before and after the failed inauguration, can be considered variants of "smokescreening." By this term, I mean deliberately covering up the fact that infrastructure development is not yielding the desired results. Two variants of smokescreening can be discerned here: an active or loud variant and a passive or silent version. Whereas the public first experienced a clear, optimistic, and boastful message about the airport, the mayor later publicly admitted his ignorance and went silent on the situation. Wowereit resigned in 2014, and the airport finally began operations in 2020.

It is important to remember that although public discourse treats it otherwise, infrastructural construction going wrong is the rule rather than the exception (although the Berlin Airport is an extreme example). As Timothy Mitchell points out, "It was an important aspect of the politics of

9 Florian Opitz, dir. *Capital B: Wem gehört Berlin?* Episode 5, "Die Stadt als Beute," Arte video, November 6, 2023, https://www.arte.tv/de/videos/087960-005-A/capital-b-wem-gehoert-berlin-5-5/. Here 18:00–21:00 and 30:00–32:35.

technical expertise that these failures and adjustments were overlooked, in fact actively covered up."[10] Developing modern transport infrastructure is a complex matter. Besides engineering knowledge, countless topographical, hydrographical, metrological, logistical, legal, financial, political, social, and economic factors must be taken into account. A more dynamic understanding of infrastructure, such as the one suggested by Jens Ivo Engels in this volume, takes the malleability of infrastructure—its inherent need to adapt and be adapted through trial and error—into account.[11] This is, however, not the understanding these popular political tropes evoke; here, infrastructure appears as a turnkey object that ought to fulfill its assumed purpose from the scheduled day of completion. Anything else appears as failure. Therefore, if outside observers realize that something is going wrong, they are often at a loss as to how to explain the problems. And this original perplexity is augmented if the process is hidden behind a smokescreen of either implausible counterclaims or sheer silence. As the breadth of issues and material that would need to be screened is demanding, if not overwhelming, even for experts, outsiders—who are neither experts nor privy to the relevant information—will most often either resort to incidental knowledge or, more likely, their general worldview to find an answer to what went wrong.

This leads outside observers to resort to hypothesizing. I consciously avoid the derogatory term "speculating" here, as I believe that an interested outsider has no other way (except perhaps blind faith in experts and officeholders) to form an opinion on infrastructural matters that may affect their daily commute, personal safety, health, prosperity, or several other crucial issues. If even a seasoned politician such as Wowereit, who as chairman of the supervisory board was obliged to oversee the process, admitted to not knowing what went wrong, it is understandable that the wider public was even more in the dark. How could an airport that was originally budgeted at 2 billion euros finally cost the taxpayers 7 billion instead and take 14 years to complete when an inauguration had seemed imminent after only six? As in the case of the 2023 Tempi disaster, the public was left to ask: Was this a case of corruption, incompetence, or sheer negligence? Or was there a more sinister scheme behind it all?

10 Timothy Mitchell, *Rule of Experts: Egypt, Techno-Politics, Modernity* (Oakland: University of California Press, 2002).

11 Jens Ivo Engels, "Writing the History of Critical Infrastructures: Some Conceptual Remarks" in this volume.

Scapegoating

When a debate ensues over the failure to meet the promise of infrastructural happiness,[12] the smokescreen and the multilayered nature of infrastructural matters described above allow the interested parties to play a blame game with considerable leverage. All kinds of agents and factors can be blamed for the failure—from political opponents, past greedy speculators, and domestic and international conspiratorial networks to foreign agents, racially defined enemies, inclement weather, and even the divine. In this way, accusations of infrastructural failure, sabotage, or corruption can be "Othered" or, more precisely, projected outward. By doing so, the myth of a political leader, administrator, or company doing its best to provide effective, safe, and healthy infrastructure—and of the subjects or customers welcoming this activity unreservedly—can be salvaged. This makes infrastructure issues particularly prone to conspiracy theories, which brings us to our fourth element: scapegoating.

In 2013, the Gezi Uprising in Turkey, which officially counted over 3.6 million participants, targeted the AKP government's rule. It especially criticized its police brutality and intolerance of divergent lifestyles, as well as its urban and infrastructural planning, including Istanbul's third ring road and new airport, both of which necessitated the destruction of wide swaths of forest. The uprising was ended through massive police violence. When Prime Minister Recep Tayyip Erdoğan (in office 2003–2014) attended the groundbreaking ceremony for the new Istanbul Airport the following year, he offered an explanation for the uprising. He claimed that with the new airport, Istanbul was destined to become a transcontinental hub for both freight and passengers, threatening both Hamburg Airport's role as a leading freight hub and Frankfurt Airport's profits, then the third largest in Europe. The new Berlin Airport had supposedly been a failed attempt to prevent the rise of Istanbul Airport. In an effort to sabotage its potential rival, the Frankfurt Airport operating company, Fraport, had submitted a tender to operate the new Istanbul Airport. When Fraport failed to win the tender, Angela Merkel's government (CDU, in office 2005–2021) allegedly initiated the Gezi Uprising as a last-ditch effort to save German aviation profits, hoping that the ecologist protest would prevent the airport's construction.[13]

12 On the nexus between infrastructure and happiness, see below.

13 Aziz Üstel, "İstanbul 'İstemezük' Havalimanı," *Star*, June 8, 2014, https://m.star.com.tr/yazar/istanbul-istemezuk-havalimani-yazi-893293/. At the time of the tender, Fraport had already been peacefully operating Turkey's third-largest airport, Antalya, for 15 years without

Ridiculous as such claims are, presupposing the manipulability of millions to serve the profits of foreign capital and reducing large investments elsewhere to the sole purpose of damaging Turkey, these conspiracy theories convinced a sufficiently large part of the electorate to rally around Erdoğan and condemn the protests.

As mentioned, these rhetorical forms—boasting, shaming, smokescreening, and scapegoating—are popular elements of infrastructure-related political debate. As with all widespread tropes, their success is based on a set of common assumptions that are tacitly accepted by the public. To demonstrate, first, the ideological roots of modern infrastructure thinking and, second, the ubiquity of these tropes, I now turn to their historical dimension.

The Origins of Developmentality

Since the eighteenth century, a particular variety of *gouvernementalité* has emerged—one which entails a close nexus between hegemony, development, traffic infrastructure, and happiness, as well as practical and epistemic resistance toward it. I have dubbed this variety of infrastructural thinking *developmentality*.[14] It can be traced throughout official and other publications, literature, and the wider public; infrastructure policy; the history of construction and operation of infrastructure; and the history of corruption, sabotage, desertion, strikes, and abuse of transport infrastructure. While in general, *developmentality* can be revealed to underlie infrastructure thought and practice most easily when employing a praxeological approach,[15] I will, for the purpose of this chapter, limit myself to discourse in the narrow definition of the word—that is, as a speech act.

The basic assumption upon which *developmentality* and all discourses deriving from it are based is that transport infrastructure—roads, railways, seaports, airports, and metros—makes people happy. This can be explained

any apparent sabotage. While the Port of Hamburg's turnover is the third largest in Europe, the airport is rather insignificant.

14 The term is first used in Debal Deb, *Beyond Developmentality: Constructing Inclusive Freedom and Sustainability* (London: Earthscan, 2009), and soon after by Jon Harald Sande Lie, *Developmentality: An Ethnography of the World Bank-Uganda Partnership* (New York: Berghahn, 2015). My usage of the term differs from Deb's. While closer to Lie's, I developed it independently to signify the link between a belief in transport infrastructure as a harbinger of progress and happiness and the legitimation and delegitimation of power.

15 I am referring to the methodology of Arlette Farge and Alf Lüdtke.

by the time and context of *developmentality*'s origins. It arose during the eighteenth century, reflecting two contemporaneous processes: an unprecedented expansion in the construction of country roads (*chaussées*) and canals in northern Europe,[16] together with new "enlightened" ideas regarding the purpose of the state.[17] Courts, bureaucracies, and armies were no longer to serve some esoteric purpose, their own interests, or their own reproduction. They were now declared instruments of the general good, also labeled "happiness," *béatitude*, or *Glückseligkeit*.

In the eighteenth-century era of absolute monarchies in most European states and oligarchic systems in the others, the equation of the public good with good mobility infrastructure proved a godsend for rulers' legitimacy. Monarchs and mercantile elites could legitimize themselves through the happiness they provided via good roads, rather than through power sharing. We find, among many others, James Anderson claiming in 1790: "Diminish the expense of carriage but one farthing, and … you form as it were a new creation, not only of stones and earth, and trees and plants, but of men also, and what is more, of industry and happiness."[18]

Another common term associated with transport infrastructure is "civilization," understood as an ongoing process of material progress. For example, one nineteenth-century author noted: "The art of road-making, and postal regulation, and the establishment of mail coaches, were the most satisfactory means of advancing civilization."[19]

Roads, it was argued, did not simply serve the state's purposes, such as exercising administrative and military control or extracting resources, taxes, and recruits from distant areas. Improved connectivity was now assumed to offer many benefits to the subjects: easing the burden of traveling, offering the opportunity for aesthetic contemplation, augmenting the prospect of prosperity, and providing access to goods and services previously beyond reach. For example, as early as May 10, 1724, the Austrian emperor Charles VI explained in detail that his road-building plans were to the benefit of the commoners:

16 Andreas Helmedach, "Infrastrukturpolitische Grundsatzentscheidungen des 18. Jahrhunderts am Beispiel des Landverkehrswesens: Großbritannien, Frankreich, Habsburgermonarchie," *Comparativ* 6, no. 2 (1996): 41–62.

17 Pankaj Mishra, *Age of Anger: A History of the Present* (New York: Farrar, Straus and Giroux, 2017).

18 Quoted in Jo Guldi, *Roads to Power: Britain Invents the Infrastructure State* (Cambridge, MA: Harvard University Press, 2012), 192–193.

19 *Blackwood's* ca. 1850, Guldi, *Roads to Power*, 194.

> The roads should be put into a "permanently durable state" so that 'through increased introduction of commerce the reputation and growth of Our hereditary kingdoms and lands as well as their native loyal obedient subjects' will be increased and also [for] the development and increased introduction of trade and change as well as to the benefit and relief of Our subjects and so that everyone in this Our Archduchy of Austria below the Enns may travel and roam with greater ease and without troublesome danger.[20]

This origin in enlightened absolutism laid the basis for the prominent role that infrastructure has played in an officeholder's balance sheet in propaganda and public opinion until today, especially in political systems with autocratic leaders—hence the latter's infrastructural "boasting." The equation of mobility infrastructure with happiness, as well as the elevation of its provision to a public service, does not amount to an ideology in itself. Instead, these elements inscribe themselves as legitimizing tropes into all kinds of political systems, especially those focusing on the persona of a singular leader. While the ideologies and fashions of the day influence the exact wording, the equating of mobility infrastructure with happiness and the association of infrastructure with progress and civilization resurface in countless different places and eras. For example, on November 1, 1937, the Turkish president Mustafa Kemal Atatürk (in office 1923–1938) pronounced: "Railways are a holy torch that illuminates a country with the light of civilization and prosperity. To achieve the aims of our railroad construction policy, which we have pursued with diligence and consistency since the first years of the Republic, it is being implemented incessantly and successfully."[21] Around roughly the same time, the General Secretary of the Soviet Communist Party, Joseph Stalin (in office 1922–1952), bragged: "And when we advance the USSR with the automobile and the farmer with the tractor, let the respectable capitalists, who adorn themselves with their civilization, ask themselves, how they will catch up with us."[22]

However, if roads and railways legitimize those in power, the opposite is also true: should leaders prove incapable of providing infrastructure that

20 Andreas Helmedach, *Das Verkehrssystem als Modernisierungsfaktor: Straßen, Post, Fuhrwesen und Reisen nach Triest und Fiume vom Beginn des 18. Jahrhunderts bis zum Eisenbahnzeitalter* (Munich: Oldenbourg, 2002), 68.

21 Quoted in Filiz Çolak, "Atatürk Döneminde Türkiye Cumhuriyeti'nin Ulaşım Politikasına Genel Bir Bakış," *Turkish Studies* 8, no. 2 (2013): 345–364.

22 Quoted in L. A Bronštejn and B. N. Budrin, *Planirane i otčitane na avtomobilnija transport*, trans. B. Janovski (Sofia: Nauka i Izkustvo, 1950), 3.

makes their citizens happy, they lose their legitimacy. This fact is not lost on political opponents, which is why they will resort to infrastructural "shaming"—that is, criticizing an officeholder's ineptitude in providing infrastructure, the officeholder's megalomania and wastefulness, or the officeholder's corruption and embezzlement (including a failure to stop such practices among subalterns).

"Infrastructural shaming" even played a minor part in the fall of Eastern European communism. The banned movie *Little Vera*, set in 1988 in Zhdanov, USSR (at the time of writing, Mariupol in occupied Ukraine), showed long shots of outdated, decrepit port infrastructure, amplifying the overall feeling of anguish.[23] More straightforwardly, in the summer of 1989, the East German rock band *Herbst in Peking* sang about Nicolae Ceaușescu's most controversial road project, the Transfăgărășan, while burning the Romanian president in effigy on stage—a hitherto unheard-of act of defiance.[24]

The Pioneers of Boasting in Southeast Europe

While the eighteenth-century emergence of *developmentality*—due to the overlap of road and canal building and new theories of state—occurred predominantly in northern Europe, it had spread to the Mediterranean and Eastern European spheres by the early nineteenth century. In the Ottoman Empire and its successor states, the importance of transport infrastructure for the relationship between the ruler and the ruled became apparent by the late 1830s, even though the Ottoman administration initially envisaged public participation as nothing more than limited consultations with provincial elites. Some historians have stressed French influence, while others have seen the Austrians—most notably Imperial Chancellor Klemens von Metternich (in office 1821–1848)—at work in this, but that is beside the point here.[25] The Gülhane Edict of 1839, often referred

23 Vasili Pichul, "Little Vera (part 1/9) eng subs." GiaCancer, March 31, 2013. YouTube video, 14:59, https://www.youtube.com/watch?v=Ab9XnGaA46M.

24 "Nicolae Ceauşescu is my name/ The world of power is my game/ And all this power I'm willing to keep/ Although my people got nothing to eat/ Go to the mountains take a glance/ There is a highway for you to see/ You can use it only in summer/ But it is for my immortality." "Herbst in Peking—Immortality | 'The Autobiography of Nicolae Ceaușescu' a film by Andrei Ujica," DieAnderenBands, November 27, 2010. YouTube video, 3:28, https://www.youtube.com/watch?v=BS9fURbBTn4.

25 İlhan Tekeli and Selim İlkin, *Cumhuriyetin harcı*, vol. 3, *Modernitenin altyapısı oluşurken* (Istanbul: İstanbul Bilgi Üniversitesi Y., 2004), 2–7; Şerif Mardin, *The Genesis of Young Ottoman Thought: A Study in the Modernization of Turkish Political Ideas* (Princeton: Princeton University

to as a pseudo-constitution of the Ottoman Reform Era (a local variant of Enlightened Absolutism), mentions infrastructure as a state obligation implicitly, while the 1856 Reform Edict mentions it explicitly.[26] New roads were to be part of a "New Deal" between Istanbul and the provinces. By the 1830s, steamship transport had revolutionized Mediterranean connectivity practically overnight.[27] In order for landlocked Ottoman provinces to profit from this, new roads (and, in the long run, railways) were to provide the possibility of transporting cash crops more speedily, reliably, safely, and in greater volume to the ports, thus making inland landowners prosperous and happy. However, while the Ottoman reformist administration was successful in creating high expectations among the provincial elites, it was slow to actually implement any terrestrial transport improvements. Lack of funding, absence of administrative and engineering expertise, embezzlement, and local resistance all played a role in this.

Nonetheless, there were some instances of successful infrastructure implementation in the provinces. This is where we find the oldest evidence of infrastructural "boasting." Only recently, the Veliko Tărnovo municipality in Bulgaria inaugurated a statue of the local master builder Nikola Ivanov Fiçoğlu, or Fičev, better known as Kolju Fičeto, in front of the *konak* (an Ottoman governor's office) he had built. Accompanying the statue is a quote attributed to Fičeto that needs some explanation. Under Governor General Midhat Pasha (in office 1864–1868), this region—then named the Danube Province of the Ottoman Empire—became a laboratory for successful expenditure cutting and local outsourcing of roadbuilding.[28] One of Midhat's most opulent projects—among his many improvements to transport infrastructure—was a bridge over the Jantra River at Bjala. At

Press, 1962), 180; Florian Riedler, "Crossroads Edirne: Building Modern Infrastructures on Ancient Routes," in *The Heritage of Edirne in Ottoman and Turkish Times: Continuities, Disruptions and Reconnections*, ed. Florian Riedler and Birgit Krawietz (Berlin: De Gruyter, 2020), 438–442.

26 Emrah Çetin, *Tanzimat'tan II: Meşrutiyet'e Anadolu'da Karayolu Ulaşımı* (Ankara: Türk Tarih Kurumu, 2017), 33.

27 Malte Fuhrmann, *Port Cities of the Eastern Mediterranean: Urban Culture in the Late Ottoman Empire* (Cambridge: Cambridge University Press 2020), 49–62; Constantin Ardeleanu, *Steamboat Modernity: Travel, Transport, and Social Transformation on the Lower Danube, 1830–1860* (Budapest: Central European University Press, 2024).

28 Florian Riedler, "'Ottoman Infrastructure Governors' as Agents of Technological and Economic Modernization," in *Transforming Southeast Europe during the Long 19th Century: Persons and Personalities as Agents of Modernization in the Ottoman and Post-Ottoman Space*, ed. Boriana Antonova-Goleva and Ivelina Masheva (Paderborn: Brill Schöningh, 2024), 323–352; Milen Petrov, "Tanzimat for the Countryside: Midhat Paşa and the Vilayet of Danube, 1864–1868" (PhD diss., Princeton University, 2006), 198–207.

276 meters long and more than 9 meters wide, the limestone-clad bridge was amply referential to both Central European and Ottoman traditions of architecture and ornaments. The Polish provincial chief engineer, Ljudmil Rola, estimated the construction cost at 3 million *kuruş*. Kolju Fičeto is said to have publicly challenged this. Addressing Midhat, the master builder supposedly said (as has been retold countless times in Bulgaria and is now inscribed in stone): "Pasha Efendi, if I do not build the bridge at Bjala my way for less than 700,000 *kuruş*, you may have my head." Fičeto then carried out the project with the aid of local artisans and corvée workers and, as the anecdote goes, managed to stick to his budget. The tale became popular because it seemed to prove that the Bulgarian people had the potential to "pull themselves up by their own bootstraps": the local master builder, who had never received formal education as an architect,[29] proved more successful than a foreign expert. Neither the Ottoman Empire nor the Danube Province ever emancipated itself from foreign capital or expertise—especially for more complex projects, such as railway or port construction, as described in Constantin Ardeleanu's chapter.[30] Total autochthony was never Midhat Pasha's intent, and even the later Bulgarian Principality remained dependent on foreign credits for its infrastructure policy. Nonetheless, Kolju Usta (and, implicitly, Midhat) became the figurehead for the trope of Bulgaria's autochthonous path to infrastructural modernity—one without foreign experts and construction companies—which is reiterated to this day despite ample historical evidence to the contrary.

The second early occurrence of boasting remembered until today likewise lauds autochthony but frames it within a far more authoritarian ideology. The 1880s proved to be the turning point in Ottoman road construction. Between 1879 and 1888, the network of macadamized roads in the Ottoman lands rose from 900 to 10,400 km. This pace of construction would never again be achieved before the demise of the empire. Of the 5,000 km of new roads in the early 1880s, 1400 were supposedly built in the province of Sivas alone.[31] An infrastructural boast from this time has survived in public memory to this day in the motto of the Turkish General Directorate of Highways: "If you can't reach it, you don't own it." The phrase is attributed to the Hamidian bureaucrat Halil Rifat Pasha during his time as governor

29 Felix Kanitz, *Donau-Bulgarien und der Balkan: Historisch-geographisch-ethnographische Reisstudien aus den Jahren 1860–1875*, vol. 2 (Leipzig: Hermann Fries, 1877), 31–33.

30 Constantin Ardeleanu, "British Investors and the 'Kustendjie Harbor Dues Issue'" in this volume.

31 Nurettin Birol, *Halil Rıfat Paşa dönemi ve icraatı (1827–1901)* (Ankara: Cedit Neşriyat, 2009), 108.

general of Sivas (1882–1885). Halil Rifat apparently took his motto as an incentive to pursue a relentless road construction program in the landlocked province, employing corvée, fiscal exploitation, and corporal (possibly even capital) punishment to a degree that his mentor Midhat never had. Just as in the Kolju Usta anecdote, this was a variant of the "pulling up by the bootstraps" story and entailed the shaming of foreign engineers. When later serving as Aydın's governor general, Halil Rifat believed the provincial head engineer to be lazy and corrupt and sarcastically suggested sending the man to Sivas: "There is no road left to build there, he can continue his easy-going life over there..."[32]

The Origins of Shaming

While Kolju Fičeto, Midhat, and Halil Rifat are early examples of successful men in Ottoman infrastructure building who knew well how to market their achievements, their successes came after a long period of inertia in this sector. Following the 1839 elevation of transport infrastructure to a public duty, no notable successes would follow for over twenty years. This gap between the rhetorically high value of infrastructure and its practical neglect gave rise to the first manifestations of infrastructural shaming—even before boasting. Andreas D. Mordtmann—an Istanbul resident of Hamburg origin, chargé d'affaires for the Hanseatic states (in office 1846–1859), and ardent critic of the Ottoman reformist bureaucracy[33]—was among the first to discover the possibilities of offensive developmental criticism. He noticed that calling out the disliked bureaucrats for their failure to deliver infrastructure was an effective way of shaming them. His offending articles appeared anonymously in 1857 in a German newspaper, beyond the reach of potential Ottoman legal action. While Mordtmann's intention might also have been to discredit the Ottoman bureaucrats' international reputation, he would have known that the Ottoman court and wider public closely followed what the international press wrote about their part of the world, and vice versa, as we know from other occasions.[34]

32 Birol, *Halil Rıfat Paşa dönemi*, 191.

33 Christoph K. Neumann, "'Ein Osmane' und die Osmanen: Andreas David Mordtmann d.Ä. als Beobachter des geistigen Lebens im Osmanischen Reich seiner Zeit," in *Daheim in Konstantinopel: Deutsche Spuren am Bosporus ab 1850 / Memleketimiz Dersaadet: 1850'den itibaren Boğaziçi'ndeki Alman izleri*, ed. Erald Pauw (Istanbul: Pagma, 2014), 93–107.

34 Fuhrmann, *Port Cities of the Eastern Mediterranean*, 234–243.

Mordtmann cited several reasons why the Ottoman government had failed in its road-building agenda. Most notably, it had not managed to finish a mere 30-kilometer road from the Marmara Sea to Bursa after six years, even though the pro-government *Journal de Constantinople* constantly lauded the work's progress and Mordtmann himself had seen three Wallachian engineers and over 1200 men at work there. The reasons he cited were:

1. Concerns over national security: Mordtmann claimed that the government did not seriously intend to improve roads, as it believed these could be of advantage to invading armies—a mentality he found antiquated.
2. Reservations toward commerce among the wider population: As trade was dominated by non-Muslims, the predominantly Muslim population did not believe they would benefit from its intensification.
3. Abuse of the corvée: Following the French eighteenth-century model, country roads were to be built by the local population. Although the corvée had been officially abolished, the regulations still specified exactly how many days an adult man was obliged to work on the roads, how far the worksite could be from his home, and that he would be paid six piasters per day. The organizers of the Bursa Road, however, recruited workers from a far wider radius than was legal and were paying as little as two piasters. Accordingly, the motivation to abscond or at least underperform was high.
4. Time conflict with agricultural tasks: Corvée road construction took place predominantly in the summer months, when the rural population was also busiest with agricultural labor. In Bursa, many men did not wish to leave silk production unattended for several days on end and thus absconded.
5. Inclement weather
6. Embezzlement and mismanagement: Mordtmann claimed that the head of the Bursa Municipality had first opposed the project but was then made its treasurer. As an example of embezzlement, Mordtmann cited an official report in which it was stated that over 80 oxen had frozen to death in October—a usually mild month in the Marmara region.[35]

Mordtmann's combination of extensive local knowledge as a resident and reader of Ottoman Turkish, his condescending attitude—which (despite his knowledge and language skills) makes his text appear as prejudiced as those of Orientalist and ill-informed writers from Western Europe—and

35 Andreas D. Mordtmann sen., *Anatolien: Skizzen und Reisebriefe aus Klein-Asien, 1850–1859*, ed. Franz Babinger (Hanover: Lafaire, 1925), 287–292.

his eagerness to embarrass reformist officeholders make it difficult to ascertain the veracity of his claims. As with the contemporary examples, veracity is not the initial concern, only whether the narrative manages to catch on or not. His detailed knowledge of the project, the many examples of bad planning, the failure to account for local sensibilities, and the presence of greedy and corrupt subaltern officials make for a "thick" description sufficient to shame the leading government bureaucrats, who had promised the population new connectivity and happiness but were not even capable of building a 30-kilometer road.

At the time, Mordtmann's intervention must have seemed something like a cry in the wilderness. But over time, dissatisfaction with the Sublime Porte's failure to provide infrastructure and the continued existence of the corvée grew more vociferous, as will be addressed later.

The Origins of Smokescreening

This brings us to the next point: the counter-information campaigns that stakeholders undertook to tackle their poor public image, best illustrated by the Smyrna–Aydın Railway Company. Despite problems in the road-building sector, the 1850s Ottoman administration also began its first railway projects. Unlike state-built roads, these relied on partnerships with British investors. The first project to be launched, the Smyrna–Aydın Railway, was the object of especially extravagant propaganda. At the 1858 laying of the cornerstone for Smyrna's (Izmir's) first railway station at La Punta (now Alsancak), British Ambassador Stratford Canning (in office 1841–1858) promised that the Ottoman Empire would soon be crisscrossed by "a network of iron communication" and "the consolidation of European interests with those of Turkey ... a progressive diminution of abuses, prejudices and national animosities."[36] When the railway was finished halfway to the town of Ayasoluk (now Selçuk) near Ephesos (Efes) in 1862, Hyde Clarke, chairman of the Smyrna–Aydın Railway, claimed that the country's best prospects for development lay "in direct relations with the Capitalists of London." He argued that resident foreigners should do "[their] duty as Europeans for the advancement of the people ... [and] contribute to the prosperity of this

36 Yakup Bektaş, "The British Technological Crusade to Post-Crimean Turkey: Electric Telegraphy, Railways, Naval Shipbuilding and Armament Technologies." (PhD diss., University of Kent at Canterbury, 1995), 129; Philip Mansel, *Levant: Splendour and Catastrophe on the Mediterranean* (London: John Murray, 2010), 161.

flourishing country… and before ten years the population and commerce would double." The governor general of the province of Aydın, Mehmed Reşid, spoke in Ayasoluk of "the prosperity and civilization of the empire."[37] Furthermore, the ceremonies were orchestrated to suggest the endeavor was but one element of a larger friendship and alliance between the Ottoman and the British Empires: locomotives were decorated in both the Union Jack and the Star and Crescent, and the imperial march, the *Mecidiye*, was played.[38]

These images, however, resemble the thick smokescreen of confidence that Klaus Wowereit established around the Berlin Airport construction site. In fact, the pompous ceremonies for the Smyrna–Aydın Railway primarily served to downplay the immense difficulties that had emerged. Construction had stalled both because of financial problems and because the surveyors had failed to account sufficiently for the difficult terrain. The builders faced a geological landscape that was either too hard (a major tunnel east of Ephesos, where today's highway runs, proved impossible to excavate, so the line was rerouted to a pass west of the ancient city), too soft (the terminus at Izmir's La Punta was built on marshland lacking solid ground for foundations), or too wet (five bridges on the original stretch through Izmir were destroyed during the first heavy rain, with many tracks and embankments inundated).[39] In and around Izmir, especially, the company faced numerous disputes over compensation for expropriated land amid claims that the previous landowners were trying to "milk" the company beyond their due, countered by charges that employees of the Smyrna–Aydın Railway had engaged in speculative land acquisition near the sites of proposed stations.[40] Finally, the original construction company went bankrupt in 1860 and had to be replaced. The PR backlash was considerable. A company representative in Istanbul at the time summarized Ottoman public sentiment as follows: "The Smyrna and Aydın Railway was nothing but waste, robbery, and speculation administered by rogues and swindlers, with the complicity and connivance of fools and knaves in England."[41] After ten years, the last leg was finally inaugurated, although the project remained incomplete.[42]

37 *Smyrna Mail* (Izmir), September 23, 1862, 2–3.

38 Bektaş, "The British Technological Crusade," 128.

39 Elvan Cobb, "Railway Crossings: Encounters in Ottoman Lands" (PhD diss., Cornell University, 2018), 86–105.

40 Cobb, "Railway Crossings," 67–79.

41 Quoted in Bektaş, "The British Technological Crusade," 132; see also Cobb, "Railway Crossings," 14.

42 Cobb, "Railway Crossings," 105–125.

The Rise of Resistance against Infrastructural Development

One can only imagine all the animated discussions, frustration, and animosity these early, badly coordinated infrastructure endeavors gave rise to. At least in the eyes of part of the population, they failed to deliver the happiness that connectivity was meant to bring. This was because they either exploited the local population through excessive levies or corvée duties, demanded inflated prices for access, delivered sub-standard infrastructure, or routed roads or railways too far from a sizable portion of the population. While Mordtmann's criticism of infrastructural inadequacy in the 1850s had remained isolated, toward the 1870s, a new generation of subjects and residents in the Ottoman lands felt more self-confident in rejecting the impositions the grand infrastructural schemes brought upon them.

Around 1865, resistance against the Trabzon–Erzurum Road had already reached a climax. Local merchants in Rize, having no interest in the promotion of a rival port, organized the villagers of their district to refuse the corvée and demonstrate against it. Presumably in the same context, the provincial French chief engineer and his son were murdered, prompting the central government to send additional troops into the area.[43] While such outbreaks of militancy against mid-nineteenth-century Ottoman infrastructure policy seem to have been rare according to present research, dissent and public criticism of transport development became increasingly widespread. Indeed, infrastructure matters were among the key issues around which a nascent Ottoman civil society and free press coalesced. For example, when the Sublime Porte issued a concession to the Brussels-based investor Baron Maurice de Hirsch to establish a railway network for Ottoman Europe,[44] the Edirne *meclis* (council of local notables) protested in 1871. The projected route between Istanbul and Edirne was badly chosen, avoiding several major centers of population and commerce, while also proving excessively long: at 320 kilometers, it was more than 50 percent longer than the road.[45]

Not only roads and railways demanded new, grand-scale infrastructure that changed the Ottoman landscape. The steamship revolution in global commerce had turned Izmir into an enormous entrepôt for the world economy's trade with West Asia. As of 1868, a new 3.5-kilometer-long

43 Fulya Özkan, "Winding Road to Modernization: Trabzon–Erzurum–Bayezid Road in the Late Ottoman World," *Comparative Studies of South Asia, Africa and the Middle East* 34, no. 1 (2014): 199–201; Tekeli and İlkin, *Cumhuriyetin harcı*, vol. 3, 110; Owen, *The Middle East*, 120.

44 Matthias B. Lehmann, *The Baron: Maurice de Hirsch and the Jewish Nineteenth Century* (Stanford: Stanford University Press, 2022).

45 Riedler, "Crossroads Edirne," 459–460.

waterfront was under construction, effectively changing the face of the city. The new quays were to allow up to 300 ships to dock simultaneously and added two additional streets in front of the existing ones. Naturally, the project also met with opposition in the city, as the operation was massive and invasive. Some previous owners of valuable seaside property now found themselves two blocks inland. Besides local property interests and aesthetic concerns, diplomatic interests also clashed over the quays; after the company contracted to develop the project passed from British to French ownership, the British consul fomented opposition against it. Once arguments about property rights and public space had been exhausted, the partisans claimed the project was detrimental to hygiene in an effort to gather broader support. When construction temporarily resulted in stagnant pools between the new quay walls and the old shoreline, the critics felt vindicated. A German deaconess and resident of the city wrote: "Smyrna is growing unhealthier by the day. In every house there are ill people. The reason for this change lies in the Turkish attitude. Dams and streets are built and drains interrupted, which spill over into the streets, form mud, and pollute the air. Before these civilizing attempts, we had a heavenly, healthy climate, and now Smyrna in winter is barely inhabitable."[46]

The author of the above-mentioned words, Minna Grosse, was a firsthand witness of the construction process but was obviously not following local politics closely. She serves as an example of how—even when there is a fairly open debate over infrastructural matters that could be followed in accessible media, such as the city's French-language newspapers—hypothesizing can be quite detached from the actual background of planning. The construction company was not Turkish but French—Frères Dussaud from Marseille. Delays were mostly due to the bankruptcy of the original Quays Company and the Dussauds' subsequent takeover. Uninformed of the British-French diplomatic tension over the construction process and the economic woes suffered by the original Quays Company, however, Grosse resorted to Orientalist prejudice to explain why the quay construction had so far provided more misery than happiness. The new quays were eventually finished—but it took eight years.[47]

The 1870s were in many respects a period of infrastructural uproar in the Ottoman Empire. Not only in Izmir but in Istanbul as well, locals were no

46 BA, Berlin, Dossier AA R 901/39576 (Ev. Gem. Smy, vol 3), 62: 10. "Bericht über die Diakonissen-Stationen im Morgenlande" (July 1870–June 1872), 24.

47 Sibel Zandi-Sayek, *Ottoman Izmir: The Rise of a Cosmopolitan Port, 1840–1880* (Minneapolis: University of Minnesota Press, 2012), 115–149.

longer willing to accept that every modern construction was a blessing. As Grosse had before, Istanbulites voiced their misgivings through a framework of health. The *Dersaâdet Su Şirketi* (the Constantinople Water Company), headed by Kâmil Bey, the Master of Ceremonies at the Palace, and the foreign subject Ternau, was issued a concession to supply a large section of the city with running water from distant Lake Terkos. The *Progrès d'Orient* reported extensively on the downsides of the endeavor. Ternau had ruthlessly exploited his connections to the court to obtain the concession. Not only would the water have to be pumped 170 meters uphill, which would be impossible without unfeasible investments, but most importantly, the *Union Médicale d'Orient* had issued a report declaring it unsanitary. Ternau took the author of the report to court for libel. As both of them were German subjects, they met in a German consular court presided over by Mordtmann.[48] The hypothesis that Terkos water was unhygienic seems to have had a long-lasting impact. When the company finally managed to pump water to downtown Istanbul in 1882—though it had not yet begun supplying households—it offered free tastings on its premises and distributed postcards stating, "Just like all the rest of the world, Istanbul drinks company water."[49]

The Origins of Scapegoating

As the examples of the Berlin Airport and the Izmir Railway have shown, using continuous, proud boasting about a soon-to-come bright future as a smokescreen for internal problems can be a successful discursive strategy only to a point. Once deadlines are missed or other visible failures become evident—the last-minute cancellation of the Berlin Airport inauguration or the destruction of the Izmir railway bridges in a flood—a new discursive strategy becomes necessary. This is where we return to explanations such as the one offered earlier for why some Istanbulites did not embrace the prospect of a new mega-airport in place of their forests: they were allegedly "trolls" set in motion by Angela Merkel. Scapegoating has been a fixture of infrastructure politics since the onset of modern transport. As mentioned, the fact that infrastructure implementation is a complex process leads to particularly large gray zones in interpretation. This, in turn, offers those politically responsible for such processes—most notably

48 *Progrès d'Orient* (Constantinople), August 15, 22, 27, 1874.

49 Vesile Necla Geyikdağı, *Foreign Investment in the Ottoman Empire: International Trade and Relations 1854–1914* (London: I. B. Tauris, 2011), 111.

politicians—considerable leeway in redirecting criticism of an infrastructure's failure to bring happiness; that is, they can point the finger elsewhere.

We can first notice this tactic clearly in Bulgaria, which had become a semi-independent vassal state of the Ottoman Empire in 1878. The Bulgarian prime minister Petko Karavelov (first term of office 1884–1886) introduced a bill into parliament in 1884 to create a state monopoly on railways. This in itself was a popular and uncontroversial move. Bulgarians had not been satisfied with the early private railway lines initiated during the Ottoman era. What was more controversial was how to counter them. For example, compensating the private British-owned Ruse–Varna Railway for the nationalization of its assets could prove expensive for the new state, especially as the railway was of little value but had British diplomacy to back up its claims (see fig. 1). Some Bulgarian politicians even argued that the new state should abandon railway development altogether. Nonetheless, Karavelov successfully bypassed such qualms. He argued: "There is an authority, Čuprov, who states that with all railways built via concessions, 54% of the equipment are stolen and [only] 46% employed for their actual purpose.... We know the Russians, the Romanian [railways]."[50]

While such reasoning might have been particularly persuasive in the wake of Baron Hirsch's extremely high profits accrued from building the Orient Railways,[51] there is, of course, no rule that embezzlement is peculiar to private enterprises. Mordtmann had earlier demonstrated how the Ottoman system of road-building corvée—which the Bulgarian Principality continued—had been highly corrupt due to the officials involved. In fact, in liberal ideology, public enterprises are considered particularly ineffective and prone to squandering, as the property belongs to no one in particular. However, the prime minister successfully deflected possible criticism of large-scale state investment by turning corruption into a peculiarity of private enterprise, using a rather wanton statement by a so-called expert as alleged hard evidence. This is one of the first examples of scapegoating in Southeast European infrastructural debates.

While Karavelov's derision toward private developers in favor of a state monopoly seems relatively harmless, both hypothesizing and scapegoating

50 Quoted in Hristo Stanišev, *Istorija na stroežite i săobštenijata v Bălgarija ot osvobošdenieto do kraja na 1939g* (Kompas, 1948), 92. By "the Russians," he is referring to the financial group Ginsburg, Poljakov, Utin, and Černi, which had been active in Romanian railway construction and had tried but failed to use Russian diplomatic pressure to gain a concession in Bulgaria. By "the Romanian [railways]," he is invoking the Strousberg concession and its aftermath, as discussed in Marton, "Transportation Infrastructure and Corruption Scandals."

51 Lehmann, *The Baron*, 89–150.

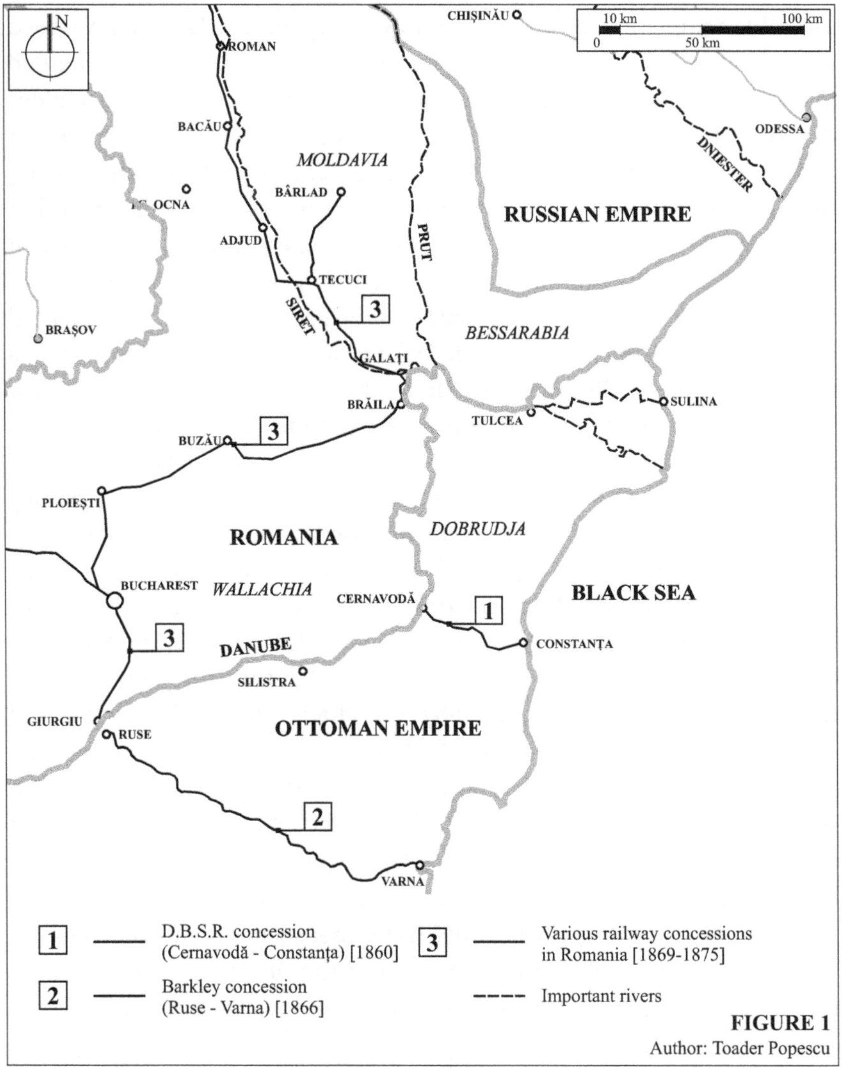

FIGURE 1
Author: Toader Popescu

Figure 1. Railway lines in the Danube – Black Sea Region (1875)

would soon evolve into an art—an art that made the most of the considerable room for interpretation offered by the various interrelated factors in infrastructure development. The Orient Railways—the operating company originally founded by Hirsch and later dominated by the Deutsche Bank—successfully resisted the Bulgarian railway state monopoly, citing international treaties and a Sultanic concession from before Bulgaria's 1885 annexation of Eastern Rumelia—where the Orient Railways operated—as a legal basis for defying Bulgarian law. Negotiations for the Bulgarian state's amicable acquisition of the Orient Railways' operating rights at an agreed price were completed in the late 1890s. However, the takeover scheduled for January 1899 was canceled by Bulgaria at the last minute, as violent demonstrations against the supposedly high price toppled the government.[52]

The Bulgarian state's inability to implement its laws and enforce nationalization seemed strange to some observers, who hypothesized that some sinister, arcane plan must be at work. When negotiations between the Orient Railways and the state resumed in 1907, the *Balkanska Tribuna* wrote: "It is time, we believe, to ask: Does the Prince really have a personal interest in the expropriation of the Orient Railways? Does he have shares in the company?"[53] Revisiting the failed takeover of 1899, the *Balkanska Tribuna* voiced the opinion that the united opposition against the state takeover had been odd, as the price had not been unreasonable. The author also wondered why the matter had not been debated afterward: "Some thick fog settled upon it and … withdrew the issue from public attention."[54] He reminded readers that the signing of the contract had caused the company's shares to drop and added that the prince's mother, Clémentine of Orléans, had supposedly been one of the largest shareholders of the Orient Railways. Clémentine had died in early 1907. The journalist therefore hypothesized that Prince Ferdinand (reigned 1887–1918) had inherited her shares and was now actively sabotaging the deal between the government and the company to protect his material interests. Nonetheless, the writer had to admit that there was no proof that Clémentine had held Orient Railways shares (or instigated the riots against the government in 1898–1899); this had simply been a recurrent rumor at the time of the treaty negotiations.[55] The example shows nicely how justified unease about smokescreening and the impermeable nature of infrastructure decision-making can be followed up by claims of a purely speculative and unproven nature. In fact, a non-transparent decision-making process is often enough for the public to accept invented or arcane background "facts" as a plausible explanation.

At approximately the same time, wild speculative hypothesizing and scapegoating reached new heights in the Ottoman Empire, as the end of the Hamidian era and the arrival of popular government brought with

52 Malte Fuhrmann, "The 'Veins and Arteries of the Country': Imagined and Actual Exclusion from Railway Connectivity in Bulgaria, 1878–1908." *Journal of Transport History* 45, no. 3 (2024): 646–670.

53 HADB OR 603: *Balkanska Tribuna*, ca. November 26, 1907 (translated).

54 HADB OR 603: *Balkanska Tribuna*, ca. November 26, 1907 (translated).

55 HADB OR 603: *Balkanska Tribuna*, ca. November 26, 1907 (translated).

them a new culture of public debate alongside labor unrest. In the summer of 1908, strikes broke out across almost all sectors, including railways and ports. One of the most controversial was a short strike on the Anatolian Railways, another company dominated by the Deutsche Bank. The mostly foreign-owned infrastructure companies in the Ottoman Empire joined forces to pressure the minister of public works, Gabriel Noradunkyan Efendi, into drafting a law banning strikes in infrastructural companies.[56] It was passed by Parliament on October 10, 1908.[57] To promote as a patriotic act what must have been seen by many workers as kowtowing to foreign imperialist interests, Noradunkyan publicly claimed that the instigators of the strike were foreigners who should be dismissed and replaced with Ottoman subjects. This was a conscious act of scapegoating. Among the Anatolian Railway labor activists, the head of the syndicate and his deputy were both Ottoman subjects, while the secretaries were Ottoman and Austrian, and the treasurer was German. Among the divisional heads, there were five Ottomans, three Hellenes, three Italians, two Austrians, and one French citizen.[58] The Anatolian Railway Company publicly reproduced the myth of foreign agents provocateurs[59] while actually firing both foreign and local labor activists.[60]

The Anatolian Railways were not popular among passengers and freight owners alike. Many claimed the company consciously neglected rolling stock and public demand. During a 1911 parliamentary debate, deputies voiced their grievances against the company at length. The harshest words were expressed by the Ankara deputy and minister of the interior, Mehmed Talât Bey (Committee of Union and Progress [CUP] in office 1909–1911 and 1913–1918), who would soon become infamous as the instigator of the genocide perpetrated against the Ottoman Armenians.[61] Echoing a widespread sentiment, he described the railway company—dominated by the Berlin-based Deutsche Bank—as a relic of Sultan Abdülhamid's tyrannic era:

56 HADB OR 254, 75: Helfferich to Gwinner, Berlin, September 16, 1908; 79, Biedermann to Gwinner, Paris, September 21, 1908; after 79, Helfferich to Anatolian Railways, September (?), 1908; 88, Hueguenot to Gwinner, Constantinople, September 30, 1908; 90, Neeff to Gwinner, Pera, October 1, 1908.

57 HADB OR 254, 93: Hueguenin to Deutsche Bank, Constantinople, October 10, 1908 (telegram).

58 HADB OR 254, 33: Procès-Verbal de l'Assemblée..., August 13, 1908.

59 HADB OR 00254, 92: Kautz to Anatolian Railway Administrative Council, Constantinople? October 1908.

60 HADB OR 0022, 181.

61 On his early parliamentarian career and first term as minister see Hans-Lukas Kieser, *Talaat Pasha: Father of Modern Turkey, Architect of Genocide* (Princeton: Princeton University Press, 2018), especially 61–106.

> Whatever accommodation was made under the *ancien régime*, it was done to the detriment of the people. Nobody could object. Not only could one not consider the general interest of the country; it was not even possible to think of one's own life and property and to defend it. In those times, [the officials] were only taking care of their own pockets. Of course they did, because the nation could not open its mouth. The government ruled despotically. They never thought about the future of the country, they wanted to fill their pockets. Companies in countries like ours are like this: they drink the juice of the orange, they throw the waste, the pulp, to us. But let us be fair. This Ottoman nation that has overthrown the Hamidian administration will not stay hapless against the railway tariffs.[62]

No resolution followed the vociferous debate, even though the speaker of the Chamber of Deputies urged the delegates to agree on some common course of action.[63] While this appears on the surface to be the unprofessional maneuvering of an upstart cast of politicians, in this case, the historian can actually reveal that malintent and arcane forces were indeed at play. In reality, Talât's venomous words served only to scapegoat the *ancien régime* officials for their connivance with the Anatolian Railways—even though the archival record reveals that several bureaucrats had warned Abdülhamid II against collaborating with the Deutsche Bank consortium.[64] In fact, in 1909, Talât and Cavid—another leading CUP member and finance minister—had enjoyed an exclusive tour of the provinces organized by the Anatolian Railways as part of the company's promotional activities. The two politicians were personally accompanied by the director of the company, Edouard Huguenin.[65] Even more influential was Talât's and Cavid's relationship with the company's second director and Deutsche Bank board member Karl Helfferich. Helfferich, in turn, had supported both the CUP and the Ottoman budget through donations, bribes, and loans that were not expected to be repaid. These were intended to secure

62 General Assembly of the Ottoman Empire, Chamber of Deputies, *Meclisi Mebusan Zabıt Ceridesi*, Protocol of 29th session, 17 Mart 1327 (March 30, 1911); HADB OR 0022, 198 ff.: Protocol of Meclis-i Mebusan debate on railways, March 31, 1911, based on *Stamboul*, March 30, 1911, in French.

63 General Asembly of the Ottoman Empire, Chamber of Deputies, *Meclisi Mebusan Zabıt Ceridesi*, Protocol of 29th session, 17 Mart 1327 (March 30, 1911); HADB OR 0022, 198 ff.: Protocol of Meclis-i Mebusan debate on railways, March 31, 1911, based on *Stamboul*, March 30, 1911, in French.

64 Murat Özyüksel, *The Berlin–Baghdad Railway and the Ottoman Empire: Industrialization, Imperial Germany and the Middle East* (London: I. B. Tauris, 2016), 62–66.

65 Kieser *Talaat*, 79, 80; HADB OR 0022, 100, 101.

political and entrepreneurial support, which proved effective at times such as the parliamentary debate of 1911 and in the long run in securing the German-Ottoman alliance during World War I.[66]

Conclusion

Infrastructure, especially regarding transport, is a key issue around which publics articulate their interests, desires, hopes, and fears. Since the dawn of the modern era, concepts of a good life and good mobility remain intertwined, both in the immediate present and the imagined future.[67]. As is common for rhetorical tropes that have a potential impact on the public, statements on infrastructure—and their effect on a ruler's legitimacy—have changed little from the mid-nineteenth century to the present. Their basic content tends to legitimize or delegitimize a ruler, either through acts of boasting about their infrastructural accomplishments or through shaming them for their failures. When large-scale infrastructure goes awry, these legitimizing and delegitimizing tropes can take the form of smokescreening or scapegoating.

Victor Turner has demonstrated that for large-scale political mobilization to succeed—especially without a state propaganda apparatus at one's disposal—the tropes used must evoke images resonating with a wide public.[68] None of the nineteenth-, twentieth-, or twenty-first-century characters cited likely ever consciously pondered whether their rhetoric resembled that of Charles VI. And yet, the images of connectivity and prosperity the Austrian emperor used in 1724 have been repeated at regular intervals ever since, within the framework of ideologies as disparate as enlightened absolutism, liberalism, nationalism, communism, anti-communism, and Islamism. In this way, they remain within the repository of public memory and can be utilized for political mobilization.

Nevertheless, one should not confuse outer similarity with uniformity of purpose. While all the political agents cited above evoke the same general matrix of connectivity, happiness, and legitimacy, their intent varies according to both the zeitgeist and the particular situation. To demonstrate this,

66 Kieser observes that the two people mentioned most often besides Talât in Cavid's diary are Huguenin and Helfferich; Kieser, *Talaat*, 88–90.

67 Hannah Appel, Nikhil Anand, and Akhil Gupta, "Introduction: Temporality, Politics, and the Promise of Infrastructure," in *The Promise of Infrastructure*, ed. Hannah Appel, Nikhil Anand, and Akhil Gupta (Durham: Duke University Press, 2018), 1–38.

68 Victor Turner, *Dramas, Fields, and Metaphors: Symbolic Action in Human Society* (Ithaca: Cornell University Press, 1987).

let us compare some of the examples mentioned above where corruption is insinuated. The Istanbul mayor successfully gained an advantage in the 2024 election by making his political competitors appear corrupt in the eyes of the voters. The "wastefulness" paradigm used careful wording to convey the message without giving the AKP politicians an opening for legal action—even though, in 2025, the judiciary resorted to trumped-up charges to legitimize the mayor's arrest. While Andreas Mordtmann sought to gradually erode international and domestic support for leading Ottoman bureaucrats, and the *Balkanska Tribuna* writer likewise sought to tarnish the Bulgarian prince's reputation, the East German rock band of 1989 and the Serbian protesters of 2024–2025 confronted Ceauşescu and Vučić head-on in the effort to weaken or ideally topple their respective regimes. When directed toward companies—such as the Smyrna–Aydın Railway, the Anatolian Railway, or the *Compagnie des Eaux de Constantinople*— these claims are most often intended to alert the authorities to wrongdoing and prompt the state to either withdraw support or take legal action.

The fact that connectivity and legitimacy are closely related is both a blessing and a curse for rulers, administrators, and private companies. Their esteem can profit greatly from it, or it can ruin their election results, profits, or careers. What is striking is that discussions regarding infrastructure politics often continue uninhibited even under autocratic regimes, such as those of Sultan Abdülaziz (1861–1876), Erich Honecker (1971–1989), or Aleksandar Vučić and Recep Tayyip Erdoğan today. Infrastructural matters oscillate between the private (as one can be immediately affected by them), the commercial (as the companies involved are seen to be in need of control or restraint in their impositions on others), and the political (as officeholders can ultimately be blamed for infrastructural failure). It is these overlapping dimensions that make them so difficult to categorize and control, while consequently rendering them a fascinating object of study.

Bibliography

Alpkökin, Doç. Dr. Pelin (@alpkokin_pelin). "Proje eksiklikleri ve uygunsuzluklara ragmen…" X, March 22, 2024. https://x.com/alpkokin_pelin/status/1771046411951366425.

Appel, Hannah, Nikhil Anand, and Akhil Gupta. "Introduction: Temporality, Politics, and the Promise of Infrastructure." In *The Promise of Infrastructure*, edited by Hannah Appel, Nikhil Anand, and Akhil Gupta, 1–38. Durham: Duke University Press, 2018.

Ardeleanu, Constantin. *Steamboat Modernity: Travel, Transport, and Social Transformation on the Lower Danube, 1830–1860*. Budapest: Central European University Press, 2024.

Baletic, Katerina. "Serbian Prosecution Charges 13 over Railway Station Disaster." *Balkan Insight*, December 30, 2024. https://balkaninsight.com/2024/12/30/serbian-prosecution-charges–13-over-railway-station-disaster/bi/.

Bektaş, Yakup. "The British Technological Crusade to Post-Crimean Turkey: Electric Telegraphy, Railways, Naval Shipbuilding and Armament Technologies." PhD diss., University of Kent at Canterbury, 1995.

Birol, Nurettin. *Halil Rıfat Paşa dönemi ve icraatı (1827–1901)*. Ankara: Cedit Neşriyat, 2009.

Bronštejn, L. A, and B. N. Budrin. *Planirane i otčitane na avtomobilnija transport*. Translated by B. Janovski. Sofia: Nauka i Izkustvo, 1950.

Bundesarchiv (BA), Berlin. Dossier AA R 901/39576.

Çetin, Emrah. *Tanzimat'tan II: Meşrutiyet'e Anadolu'da Karayolu Ulaşımı*. Ankara: Türk Tarih Kurumu, 2017.

Cobb, Elvan. "Railway Crossings: Encounters in Ottoman Lands." PhD diss., Cornell University, 2018.

Çolak, Filiz. "Atatürk Döneminde Türkiye Cumhuriyeti'nin Ulaşım Politikasına Genel Bir Bakıs." *Turkish Studies* 8, no. 2 (2013): 345–364.

Danyıldız, Meral. "İmamoğlu AKP'nin saltanatını sergiledi: Hangi hakla bu araçlara bindiniz!." *BirGün*, September 5, 2019. https://www.birgun.net/haber/imamoglu-akp-nin-saltanatini-sergiledi-hangi-hakla-bu-araclara-bindiniz-267465.

Deb, Debal. *Beyond Developmentality: Constructing Inclusive Freedom and Sustainability*. London: Earthscan, 2009.

Farge, Arlette. *Le Goût de l'archive*. Paris: Seuil, 1989.

Fuhrmann, Malte. *Port Cities of the Eastern Mediterranean: Urban Culture in the Late Ottoman Empire*, Cambridge: Cambridge University Press 2020.

Fuhrmann, Malte. "The 'Veins and Arteries of the Country': Imagined and Actual Exclusion from Railway Connectivity in Bulgaria, 1878–1908." *Journal of Transport History* 45, no. 3 (2024): 646–670. https://doi.org/10.1177/00225266241254764.

Geyikdağı, Vesile Necla. *Foreign Investment in the Ottoman Empire: International Trade and Relations 1854–1914*. London: I. B. Tauris, 2011.

General Assembly of the Ottoman Empire. Chamber of Deputies. *Meclis-i Mebusan Zabıt Ceridesi* [Journal of the Chamber of Deputies]. Protocol of 29th session, 17 mart 1327 (March 30, 1911). https://www.tbmm.gov.tr/tutanaklar/TUTANAK/MECMEB/mmbd01ic03c004/mmbd01ic03c004ink069.pdf.

Guldi, Jo. *Roads to Power: Britain Invents the Infrastructure State*. Cambridge: Harvard University Press, 2012.

Helmedach, Andreas. "Infrastrukturpolitische Grundsatzentscheidungen des 18. Jahrhunderts am Beispiel des Landverkehrswesens: Großbritannien, Frankreich, Habsburgermonarchie." *Comparativ* 6, no. 2 (1996): 41–62.

Helmedach, Andreas. *Das Verkehrssystem als Modernisierungsfaktor: Straßen, Post, Fuhrwesen und Reisen nach Triest und Fiume vom Beginn des 18. Jahrhunderts bis zum Eisenbahnzeitalter*, Munich: Oldenbourg, 2002.

"Herbst in Peking—Immortality | 'The Autobiography of Nicolae Ceaușescu' a film by Andrei Ujica." DieAnderenBands, November 27, 2010. YouTube video, 3:28. https://www.youtube.com/watch?v=BS9fURbBTn4.

Historisches Archiv der Deutschen Bank (HADB), Frankfurt. Orient (OR) Dossiers 022, 254, 603.

Kanitz, Felix. *Donau-Bulgarien und der Balkan: Historisch-geographisch-ethnographische Reisstudien aus den Jahren 1860–1875*. Vol. 2. Leipzig: Hermann Fries, 1877.

Kieser, Hans-Lukas. *Talaat Pasha: Father of Modern Turkey, Architect of Genocide*. Princeton: Princeton University Press, 2018.

Lehmann, Matthias B. *The Baron: Maurice de Hirsch and the Jewish Nineteenth Century*. Stanford: Stanford University Press, 2022.

Lie, Jon Harald Sande. *Developmentality: An Ethnography of the World Bank-Uganda Partnership*. New York: Berghahn, 2015.

Lüdtke, Alf. "Einleitung: Herrschaft als soziale Praxis." In *Herrschaft als soziale Praxis: Historische und sozial-anthropologische Studien*, edited by Alf Lüdtke. Göttingen: Vandenhoeck & Ruprecht, 1991.

Mansel, Philip. *Levant: Splendour and Catastrophe on the Mediterranean*. London: John Murray, 2010.

Mardin, Şerif. *The Genesis of Young Ottoman Thought: A Study in the Modernization of Turkish Political Ideas*. Princeton: Princeton University Press, 1962.

Mishra, Pankaj. *Age of Anger: A History of the Present*. New York: Farrar, Straus and Giroux, 2017.

Mitchell, Timothy. *Rule of Experts: Egypt, Techno-Politics, Modernity*. Oakland: University of California Press, 2002.

Mordtmann sen., Andreas D. *Anatolien: Skizzen und Reisebriefe aus Klein-Asien, 1850–1859*. Edited by Franz Babinger. Hanover: Lafaire, 1925.

Neumann, Christoph K. "'Ein Osmane' und die Osmanen: Andreas David Mordtmann d.Ä. als Beobachter des geistigen Lebens im Osmanischen Reich seiner Zeit." In *Daheim in Konstantinopel: Deutsche Spuren am Bosporus ab 1850 / Memleketimiz Dersaadet: 1850'den itibaren Boğaziçi'ndeki Alman izleri*. Edited by Erald Pauw. Istanbul: Pagma, 2014.

Opitz, Florian, dir. *Capital B: Wem gehört Berlin?* Episode 5, "Die Stadt als Beute." Arte video, November 6, 2023. https://www.arte.tv/de/videos/087960-005-A/capital-b-wem-gehoert-berlin-5-5/.

Owen, Roger. *The Middle East in the World Economy, 1800–1914*. London: Methuen, 1981.
Özkan, Fulya. "Winding Road to Modernization: Trabzon–Erzurum–Bayezid Road in the Late Ottoman World." *Comparative Studies of South Asia, Africa and the Middle East* 34, no. 1 (2014): 191–205. https://muse.jhu.edu/article/546323.
Özyüksel, Murat. *The Berlin–Baghdad Railway and the Ottoman Empire: Industrialization, Imperial Germany and the Middle East*. London: I. B. Tauris, 2016.
Petrov, Milen. "Tanzimat for the Countryside: Midhat Paşa and the Vilayet of Danube, 1864–1868." PhD diss., Princeton University, 2006.
Pichul, Vasili. "Little Vera (part 1/9) eng subs." GiaCancer, March 31, 2013. YouTube video, 14:59. https://www.youtube.com/watch?v=Ab9XnGaA46M.
Progrès d'Orient (Constantinople). August 15, 22, and 27, 1874.
PVP. "Kada će krenuti brzi voz od Novog Sada do Subotice?" *N1*, December 9, 2024. https://n1info.rs/vesti/kada-ce-krenuti-brzi-voz-od-novog-sada-do-subotice/.
"The Rescript of Gülhane—Gülhane Hatt-ı Hümayunu (3 November 1839)." Website of Turkish Constitutional Law, accessed April 2, 2026. https://www.anayasa.gen.tr/gulhane.htm.
"Rescript of Reform—Islahat Fermanı (18 February 1856)." Website of Turkish Constitutional Law, accessed April 2, 2026. http://www.anayasa.gen.tr/reform.htm.
Riedler, Florian. "Crossroads Edirne: Building Modern Infrastructures on Ancient Routes." In *The Heritage of Edirne in Ottoman and Turkish Times: Continuities, Disruptions and Reconnections*, edited by Florian Riedler and Birgit Krawietz. 151–174. Berlin: De Gruyter, 2020.
Riedler, Florian. "'Ottoman Infrastructure Governors' as Agents of Technological and Economic Modernization." In *Transforming Southeast Europe during the Long 19th Century: Persons and Personalities as Agents of Modernization in the Ottoman and Post-Ottoman Space*, edited by Boriana Antonova-Goleva and Ivelina Masheva, 323–346. Paderborn: Brill Schöningh, 2024.
Smith, Helena. "Rage in Greece as Second Anniversary of Train Disaster Prompts Mass Protests," *The Guardian*, February 28, 2025. https://www.theguardian.com/world/2025/feb/28/rage-in-greece-as-second-anniversary-of-train-disaster-prompts-mass-protests.
Smyrna Mail (İzmir), September 23, 1862.
Stanišev, Hristo. *Istorija na stroežite i săobštenijata v Bălgarija ot osvobošdenieto do kraja na 1939g*. Sofia: Kompas, 1948.
Tekeli, İlhan, and Selim İlkin. *Cumhuriyetin harcı*. Vol. 3, *Modernitenin altyapısı oluşurken*. Istanbul: İstanbul Bilgi Üniversitesi Yayınları, 2004.

Turkey. Ministry of Transport and Infrastructure. Transport Safety Investigation Center. “Demiryolu Kaza İnceleme Raporları,” accessed April 3, 2026. https://ulasimemniyeti.uab.gov.tr/demiryolu.

Turkey. Ministry of Transport and Infrastructure. “İstanbul’a 4,5 kilometrelik yeni metro hattı,” March 21, 2025. https://www.uab.gov.tr/haberler/istanbul-a–4-5-kilometrelik-yeni-metro-hatti.

Turner, Victor. *Dramas, Fields, and Metaphors: Symbolic Action in Human Society.* Ithaca: Cornell University Press, 1987.

Üstel, Aziz. “İstanbul ‘İstemezük’ Havalimanı.” *Star,* June 8, 2014. https://m.star.com.tr/yazar/istanbul-istemezuk-havalimani-yazi-893293/.

Zandi-Sayek, Sibel. *Ottoman Izmir: The Rise of a Cosmopolitan Port, 1840–1880.* Minneapolis: University of Minnesota Press, 2012.

About the Author

Malte Fuhrmann is a Berlin-based historian of the modern Eastern Mediterranean, Turkey, and Southeastern Europe. He has taught at universities in Istanbul, Bochum, and Konstanz and has been a research fellow in Berlin, Istanbul, Bucharest, and Sofia. His recent publications include *Port Cities of the Eastern Mediterranean: Urban Culture in the Late Ottoman Empire* (Cambridge University Press, 2020).

2. Transportation Infrastructure and Corruption Scandals

Silvia Marton

Abstract: From a constructivist standpoint, this conceptual chapter outlines scholarship on public corruption charges that transportation infrastructure development brought under the spotlight in the nineteenth century. It discusses, first, the powerful performative capacity that infrastructure polemics had in shaping the public, allowing for society-wide disputes on the overlapping factors conducive to public corruption. These included financial-economic or moral and political factors, ranging from individual corruption to systemic decay. Second, the chapter examines how infrastructure controversies unraveled traditional conceptions of corruption and questioned public norms and ideals. Finally, it discusses the impact that infrastructure construction had on Othering corruption in the nineteenth century as an era of liberal nationalism, when infrastructure development intertwined with nation-building, the politics of identity, and ideas of Othering.

Keywords: Public opinion, Public interest, Infrastructure polemics, Othering, Liberal nationalism, Constructivist perspective

In 1875, Dimitrie A. Sturdza (1833–1914), the former Liberal minister of public works, published a long article in the Liberal Party's mouthpiece in which he denounced systematic and rampant political corruption in Romania. His accusations were so harsh that the high-profile statesman published his pamphlet under a pseudonym.[1] Sturdza unambiguously blamed corruption

1 Erdmann de Hahn, "Germania, România şi principele Carol de Hohenzollern," Part 1, *Românul* 19, October 8, 1875, 907; Part 2, *Românul* 19, October 9, 1875, 910–911.

Marton, Silvia, and Andrei-Dan Sorescu, eds. *Infrastructural Development, Corruption, Xenophobia, and Colonization in Central and Southeastern Europe*. Amsterdam: Amsterdam University Press/Central European University Press, 2026.
 https://doi.org/10.5117/9789048577279-3

on railway concessions; financial mismanagement, public deficits and debts, and heavy taxation were all proof of "bad" and corrupt governance. Though a major advocate of the railway concessions when he was a minister, Sturdza now criticized them as excessively costly and built through contracts concluded "with crooks," thus draining the country not only of its financial but also of its political and moral resources.[2]

More specifically, Sturdza pointed his finger at the notorious Strousberg concession.[3] In 1870, while the construction of some 915 kilometers of railways was well underway, the concession turned into a major national and international scandal when the concessionaire—the Prussian-Jewish banker Bethel Henry Strousberg, dubbed the "Railway King"[4]—declared the consortium's bankruptcy (see fig. 2). Politicians (with the Liberals in opposition now at the forefront) and the public alike circulated a variety of accusations regarding the hidden motives of decision-makers, who had allegedly allowed themselves to be corrupted by Prussia and the Habsburg Empire, by "foreigners," or by Jewish capitalists. German presence and influence were, in fact, often framed as a form of "colonization" in the partisan press, eager to stoke fears of national peril or demise; Sturdza's intervention is best understood against this rhetorical background of implausible yet dramatic claims.[5] The Strousberg scandal reached its climax at the Congress of Berlin in 1878, when the Romanian state was forced to solve the financial intricacies the concession had generated and ultimately nationalize the Strousberg-built lines under a resented settlement imposed by Germany as a condition for the recognition of the country's independence.

In his 1875 article, Sturdza already accused Germany of having "humiliated" Romania. Worse still, the Liberal statesman believed that the Prussian-born ruling prince, Carol I of Hohenzollern-Sigmaringen (1839–1914; r. 1866–1914), was its accomplice. Sturdza alleged that Germany's policy "stemmed from petty profiteering interests," while Carol "treated his subjects with the same contempt and disregard as Berlin's foreign authority, providing

2 De Hahn, *Românul* 19, October 9, 1875, 910.

3 Silvia Marton, "Shades of Dependency and the Discourse on Corruption: Railway Concessions in Romania in the Nineteenth Century," *East European Politics and Societies* 39, no. 3 (2025): 691–716.

4 Ralf Roth, "Difficulties of International Railway Investments in Germany: The Example of the 'Railway King' Bethel Henry Strousberg, 1855–1875," in *Across the Borders: Financing the World's Railways in the Nineteenth and Twentieth Centuries*, ed. Ralf Roth and Günter Dinhobl (Abingdon: Routledge, 2008), 33–47.

5 Andrei Sorescu, "The Perils (and Promise) of German Colonization: Civilizational Hierarchies and Anxieties in Nineteenth Century Romania." *East European Politics and Societies* 39, no. 3 (2025): 717–740.

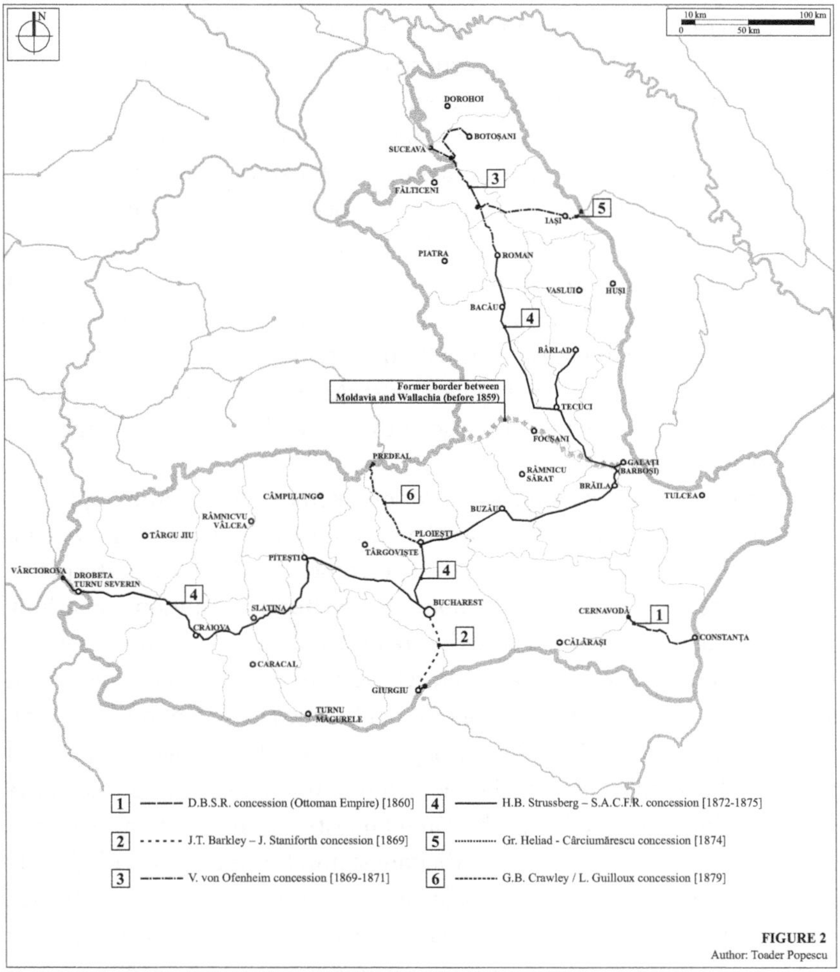

Figure 2. Railway concessions in Romania (1879)

his powerful protection to Prussian railway speculators."[6] To Sturdza, this amounted to an immense "disappointment" with German influence: "The bad habits of both German politics and Prince Carol" made "Germany more hated than even Russia was before the Crimean War."[7] Sturdza conveyed the Romanian nation- and state-builders' painful awareness that they had engaged in an immense effort to build the railways but had to do so with

6 De Hahn, *Românul* 19, October 8, 1875, 907.
7 De Hahn, *Românul* 19, October 9, 1875, 910.

foreign capital—all in the name of an emerging nation-state eager to have its sovereignty acknowledged by the international community.

Sturdza connected the Strousberg debacle to another controversial concession: the company represented by the British entrepreneur George B. Crawley, which was tasked with building the junction with the Habsburg network[8] (see fig. 2). The former Liberal minister enumerated the members of parliament from the Conservative majority who were but "obedient instruments," acting as members of the railway company "commissioned by their master to buy [other MPs'] votes." Allegedly, this was a case of "blatant and shameless bribery." Moreover, Sturdza also accused the government of forcing Parliament, busy as it was with discussing the Crawley concession, to quickly accept a trade agreement with Austria-Hungary without any real debate. He was firmly convinced that "Crawley's money bought the vote" on this treaty.[9]

To Sturdza, railway concession scandals made visible and legible all forms of corruption and their areas of manifestation. Railway money bought individual consciences, corrupted the government, and tarnished the polity. The pamphleteer went beyond denouncing electoral corruption, something which at that time usually caused the greatest and indeed unanimous public outrage.[10] Sturdza identified an overarching causality and hierarchy behind systemic corruption: "If public money is not handled conscientiously, government cannot be under the influence of probity."[11] And he claimed that political and moral corruption trickled down from Carol's entourage. State affairs were not characterized by "probity and honesty," Sturdza wrote; instead, a camarilla, made up of "the most dishonest and tainted men," ruled in its own selfish interest so as "to shamelessly exploit the state" and "every honest citizen" alike, while the prince was its prisoner or, worse yet, its accomplice.[12]

Sturdza's indictment shows just how important transportation infrastructure was for political, societal, economic, and territorial transformation in the nineteenth century.[13] Then as now, infrastructure and transport

8 Romania eventually annulled the Crawley concession, see Toader Popescu, *Proiectul feroviar românesc (1842–1916)* (Bucharest: Simetria, 2014), 53–55.

9 De Hahn, *Românul* 19, October 9, 1875, 911.

10 Silvia Marton, "La 'corrupción' electoral en Rumania: Los comienzos titubeantes de la democracia," *Ayer: Revista de Historia Contemporánea* 115, no. 3 (2019): 77–104.

11 De Hahn, *Românul* 19, October 9, 1875, 910.

12 De Hahn, *Românul* 19, October 9, 1875, 910.

13 For a historiographical discussion of infrastructure's fragmentation, its conflictual and segregation effects, and of its integrative power, see Jens Ivo Engels, "Infrastructure and Fragmentation: The Limits of the Integration Paradigm," in *Linking Networks: The Formation of*

technologies functioned as a terrain of power and contestation; they are critical and complex socio-technical systems integrated into social, aesthetic, material, and political configurations. Infrastructure may be understood as "critical" in more ways than one, as Jens Ivo Engels explains in his chapter, not least in terms of how imagining its importance is itself a "critical" act on the part of historical actors.[14] Small wonder, then, that debates on it proved a breeding ground for many controversies, including corruption accusations and scandals with outsize visibility and public impact. As the controversy over railway construction in Romania illustrates, infrastructure construction allowed for society-wide disputes on the overlapping factors conducive to public corruption, ranging from financial-economic factors to moral and political ones, from individual corruption to systemic decay.[15] Romania was a remarkable case in this regard. Within less than fifteen years, and through a huge financial effort involving exclusively foreign capital, several concessions built the main railway network and junctions with the neighboring Habsburg, Ottoman, and Russian empires. Around 1,400 kilometers of track were either completed or under construction between the mid–1860s and 1879 (see fig. 2).

For this chapter, Sturdza's pamphlet serves as a starting point for discussing the powerful performative capacity that infrastructure polemics had: first, in shaping the public; second, in unravelling conceptions of "corruption" and questioning public norms and ideals; and finally, in ascribing blame and Othering corruption in the nineteenth-century era of liberal nationalism, when transportation infrastructure construction intertwined with nation-building, the politics of identity, and ideas of Othering.

Common Standards and Visions for Infrastructure Development, ed. Hans-Liudger Dienel and Martin Schiefelbusch (Abingdon: Routledge, 2018), 19–33.

14 See Jens Ivo Engels, "Writing the History of Critical Infrastructures: Some Conceptual Remarks" in this volume. See also Hannah Appel, Nikhil Anand, and Akhil Gupta, "Introduction: Temporality, Politics, and the Promise of Infrastructure," in *The Promise of Infrastructure*, ed. Nikhil Anand, Akhil Gupta, and Hannah Appel (Durham: Duke University Press, 2018), 1–38. For the material dimension of transportation infrastructure and its key role in techno-political modernization and territorial transformation in Romania and Belgium see, respectively, Constantin Ardeleanu, "'Commercial and Political Needs Demand the Establishment of a Black Sea Port': Infrastructure Development, Opportunities and Anxieties in an Eastern European Periphery (1860s–1870s)," *The Journal of Transport History* 44, no. 3 (2023): 436–456; Greet De Block, "The Material Politics of Infrastructure Networks Infrastructure Design and Territorial Transformation in Belgium, 1830–40s," *Social Science History* 45, no. 2 (2021): 341–362.

15 The construction of port facilities in the Black Sea port of Constanța led to wide-range corruption accusations in the 1890s, see Constantin Ardeleanu, "The Hallier Affair, Political Corruption and National Dignity in Romania in Late Nineteenth and Early Twentieth Centuries," *East European Politics and Societies* 39, no. 3 (2025): 741–761.

The Public

Building major infrastructure projects meant competition for (state) resources and the disturbance of local, regional, or even international politics and society, while also contributing to the increased visibility and scrutiny—even surveillance—of political power by the public.[16] Within the broader transformations resulting from the political and industrial revolutions of the nineteenth century, public outrage over major affairs such as the Panama scandal—a quasi-foundational scandal with its European and indeed global ripples[17]—or the Strousberg scandal, with its continental echoes, stigmatized corruption and exposed it in an increasingly interconnected transnational mediatic arena. Public opinion was thus increasingly becoming a tribunal, not only for political corruption but also for moral scandals,[18] at the onset of a veritable "age of scandals."[19] Legitimizing and delegitimizing rhetorical framings of transportation infrastructure controversies—boasting, shaming, smokescreening, and scapegoating—had a major public and political impact, as explained in Malte Furhmann's contribution to this volume.[20]

The performative dynamics of corruption denunciation in the realm of major transportation infrastructure projects contributed decisively to making and widening the "public" that mattered (both sociologically and as a normative ideal). During the second half of the nineteenth century, the public was not yet understood as a universal critical authority, and its (in)

16 Pierre Rosanvallon, *Counter-Democracy: Politics in an Age of Distrust*, trans. Arthur Goldhammer (Cambridge: Cambridge University Press, 2008). For local and national power hierarchies and accusations of corruption in the railway urban planning process in Romania, see Toader Popescu, "Hoping and Coping: Local Anxieties and Railway Development in Romania" in this volume.

17 Jean-Yves Mollier, *Le Scandale de Panama* (Fayard, 1991); Christophe Portalez, "*La Revue socialiste* face à la corruption politique: du scandale de Panama à l'affaire Rochette (1892–1914)," *Cahiers Jaurès* 209, no. 3 (2013): 15–32; Christophe Portalez and Anna Rothfuss, "Panama and the Opposition: The Perception of French and German Socialists of the Panama Scandal," in *Scandales et corruption à l'époque contemporaine*, ed. Olivier Dard, Jens Ivo Engels, Andreas Fahrmeir, and Frédéric Monier (Paris: Armand Colin, 2014), 181–194.

18 For an annotated bibliography, see Dard, Engels, Fahrmeir, and Monier, eds., *Scandales et corruption*, 279–296; Frédéric Monier, "La corruption politique: une histoire européenne," *Cahiers Jaurès* 209, no. 3 (2013): 3–13.

19 John B. Thompson, *Political Scandal: Power and Visibility in the Media Age* (Cambridge: Polity Press, 2000). On the sociology of scandals, see also Damien de Blic and Cyril Lemieux, "The Scandal as Test: Elements of Pragmatic Sociology," *Politix* 71, no. 3 (2005): 9–38.

20 Malte Furhmann, "On Hegemony and Connectivity: A Short History of Infrastructure-Political Struggles and their Discourses" in this volume.

capacity for political judgment was seen as the source of legitimate restrictions on participation.[21] But societal and political developments—such as increased literacy, the gradual extension of suffrage, media culture, and print technology—contributed, in their profound interdependence, to creating the conditions for scandals and for an increasingly demanding public.[22]

While repeatedly invoking in his pamphlet the entire nation—in fact, the Liberals' middle-class urban constituency, which he dubbed the "true nation"—Sturdza was convinced that the Strousberg railway-concession scandal raised wider awareness. "Discontent grows," he emphatically warned.[23] The very use of the pamphlet as a form and vector for his discourse, and more broadly of the press as a medium of publicization, shows that statesmen like Sturdza knew that political language—including that of (anti)corruption—was becoming ever more open to wider social strata.[24] More generally, and beyond the Romanian controversy, the very fact that (infrastructure) corruption scandals occurred proves that "corruption" had become a key concept with not only partisan and ideological connotations but also wider, trans-partisan significance.[25] Its denunciation translated the new, democratic-liberal demand for distrust and resistance to power,[26] cultures of vigilance,[27] and a concern with the pursuit of "good government." (Anti)corruption rhetoric thus contributed to steering the language of politics in new directions.

Moreover, Sturdza and his contemporaries realized that long-term financial and political decision-making for the construction of railways—involving subnational, national, imperial, transregional, or transimperial

21 Jodi Dean, "Publicity's Secret," *Political Theory* 29, no. 5 (2001): 624–650; Alan S. Kahan, *Liberalism in Nineteenth-Century Europe: The Political Culture of Limited Suffrage* (London: Palgrave Macmillan, 2003).

22 On the relation between infrastructures and the production of publics, see Appel, Anand, and Gupta, "Introduction: Temporality, Politics, and the Promise of Infrastructure"; Antina von Schnitzler, "Infrastructure, Apartheid Technopolitics, and Temporalities of 'Transition,'" in *The Promise of Infrastructure*, ed. Anand, Gupta, and Appel, 133–154.

23 De Hahn, *Românul* 19, October 9, 1875, 911.

24 Reinhart Koselleck and Michaela Richter, "Introduction and Prefaces to the *Geschichtliche Grundbegriffe* (Basic Concepts in History: A Historical Dictionary of Political and Social Language in Germany)," *Contributions to the History of Concepts* 6, no. 1 (2011): 10–11, 14.

25 Koselleck and Richter, "Introduction and Prefaces," 1–37.

26 Rosanvallon, *Counter-Democracy.*

27 As conceptualized by members of the collaborative research center based at the University of Munich, see Arndt Brendecke, "Foreword: A Commentary on Corruption and Control from the Perspective of the Cultures of Vigilance," in *Corruption, Anti-Corruption, Vigilance, and State Building from Early to Late Modern Times*, ed. Ricard Torra-Prat, Joan Pubill-Brugués, and Arndt Brendecke (Abingdon: Routledge, 2024), x–xvii.

actors in intricate asymmetrical relations—could not be made without the government being held politically accountable, even in this era of limited census franchise and exclusionary social hierarchies and publics. Mihai Chiper's chapter in this volume shows how railway decision-making increased developmental inequalities and urban decline in Moldavia and how it pushed Moldavian members of parliament and notables to accuse their counterparts in the capital city, Bucharest, of abusing their power and using the new infrastructure for their own private interests.[28] The principle of politicians' accountability was frail yet increasingly important, part and parcel of the processes of internal restructuring of states and empires alike. This was the case both for emerging nation-states engaged in nationalizing policies and for multiethnic empires engaged in reforms and/or nationalizing endeavors, but also in an international arena where multiple actors interacted at various scales, ranging from private entrepreneurs and bankers to middle- or top-level government officials and cross-border leagues or associations. As Alex R. Tipei shows in this volume, accountability for infrastructural development *qua* proof of good rule and modernity was sometimes presumed to be of greater importance to an international public than to domestic stakeholders, depending on the political juncture and the relative (un)importance of accountability on a more local scale.[29] Likewise, the railway and harbor constructed in Ottoman Dobrudja in the 1860s were the subject of considerable diplomatic negotiations in London and Istanbul between authorities, investors, and capitalists but also gained significant public exposure, with different understandings of "private" and "public" interests at stake.[30] After the turn of the century, universal suffrage pushed demands for accountability toward antiliberal "populism," which contested (infrastructural) modernization and pitted an allegedly corrupt and alienated ruling class against an idealized people, composed notably of peasants.[31]

An increased concern with voters was a sign of advancing politicization, which also brought with it an increased demand for openness in decision-making and, to a certain extent, for the public's right to be informed, even if

28 Mihai Chiper, "Regional Competition, Economic Interests, and Military Stumbling-Blocks in the Making of the Moldavian Railway Network" in this volume.

29 Alex R. Tipei, "Accusations of Tyranny: Mobilizing the Ottoman Imperial Past in Service of West European Informal Empire in the Early Greek State" in this volume.

30 Constantin Ardeleanu, "British Investors and the 'Kustendjie Harbor Dues Issue'" in this volume.

31 Raul Cârstocea, "Railway Infrastructure and Student Anti-Semitism in Interwar Romania" in this volume.

the "public" and "voters" did not yet wholly overlap as categories.[32] Implicit in Sturdza's denunciation of the negative influence that the camarilla allegedly had on Prince Carol was that the latter and his acolytes were making hidden political maneuvers and thus were illegitimately keeping the (limited) public in the dark. Denouncing occult political maneuvering was consubstantial with political corruption scandals and was a key trope in Romanian politics at the time.[33] Implicit in many nineteenth-century political scandals was that publicity was gradually becoming a value and that concealment was slowly losing ground to demands for disclosure, though secrecy was still seen as inherent to publicity at the beginning of democratic politics.[34]

Conceptions of Corruption: Party Politics and Public Interest

While Sturdza rhetorically referred to the entire nation in order to legitimize his denunciation, his polemical positions were unambiguously partisan. Through the political instrumentalization of a corruption scandal and its mediatization, partisan rivalry acted as a magnifying glass, exposing political power relations, the pathologies of political systems, hierarchies, and ideological clashes in a language bristling with intended exaggerations and discursive violence.[35]

32 Jens Ivo Engels and Frédéric Monier, "Researching the History of Transparency: Introduction," in *History of Transparency in Politics and Society,* ed. Jens Ivo Engels and Frédéric Monier (Göttingen: V&R unipress, 2020), 7–20; Christopher Hood, "Transparency in Historical Perspective," in *Transparency: The Key to Better Governance?*, ed. Christopher Hood and David Heald (Oxford: Oxford University Press / British Academy, 2006), 2–23; John Pitseys, "Publicité et transparence: Le statut de la représentation et de la visibilité politique chez Kelsen et Schmitt," *Revue française de science politique* 66, no. 1 (2016): 117–135.

33 Silvia Marton and Andrei-Dan Sorescu, "Política 'oculta': Publicidad, secretismo, transparencia e inteligibilidad en la Rumanía de finales del siglo xix," in *Las sombras de la transparencia. Secreto, corrupción y "Estado profundo" en la Europa contemporánea: Secreto, corrupción y "Esatdo profundo" en la Europa contemporánea*, ed. Frédéric Monier, Lluís Ferran Toledano, Joan Pubill, and Gemma Rubí (Granada: Comares, 2022), 21–43.

34 Eva Horn, "Logics of Political Secrecy," *Theory, Culture & Society* 28, nos. 7–8 (2011): 103–122; Mark Fenster, "Transparency in Search of a Theory," *European Journal of Social Theory* 18, no. 2 (2015): 150–167; Frédéric Monier, Lluís Ferran Toledano, Joan Pubill, and Gemma Rubí, "Introducción: Una mirada desde la Historia sobre la Transparencia, el Secreto y el Estado profundo en la Europa contemporánea," in *Las sombras de la transparencia*, ed. Monier, Toledano, Pubill, and Rubí, IX-XX.

35 Nicolas Pitsos, "Presse, mensonges et chemins de fer: L'affaire Homs-Bagdad à l'ombre de la question d'Orient," in *Scandales et corruption*, ed. Dard, Engels, Fahrmeir, and Monier, 80–95; Frank Bösch, "Politics with Scandals: Germany and Britain in Transnational Perspective (1880–1914)," *Historia y Política* 39 (2018): 53–77; Jens Ivo Engels and Anna Rothfuss, "Les usages de

Beyond the occasional hyperbole, partisan accusations and the political weaponization of corruption charges stirred debates on the meaning of general interest and the public conduct of officials. From a historical perspective, corruption scandals simultaneously reveal *and* reconfigure conflicts of values, public expectations, transgressions of social and political norms, and how all of these were constructed and disputed by historical actors themselves.[36] Our constructivist positioning on "corruption" focuses particularly on scandals and historically situated debates as gateways to a better understanding not only of public values and their reinterpretations but also of other social and political techniques of power.[37] In short, there was no (public) corruption unless contemporaries—such as the statesman, politician, and pamphleteer Sturdza—talked about it, criticized it, evaluated it, or judged it from a moral, political, or legal angle.[38] Together, this scholarship argues that "corruption" needs to be understood as an expression of historical actors' judgment and their ever-changing critical discourses denouncing political power, rulers, or institutions, and of their reflections on the techniques of influence and interests they considered immoral or

la politique du scandale: Le SPD et les débats sur la corruption politique pendant le Kaiserreich (1873–1913)," *Cahiers Jaurès* 209, no. 3 (2013): 33–51.

36 As illustrated by Tipei, "Accusations of Tyranny" in this volume. See also Callie Wilkinson, "Scandal and Secrecy in the History of the Nineteenth-Century British Empire," *The Historical Journal* 65, no. 2 (2022): 545–569. See also Damien de Blic, "Moraliser l'argent: Ce que Panama a changé dans la société française (1889–1897)," *Politix* 71, no. 3 (2005): 61–82; Anubha Anushree, "Corrupt Traditions and Traditions of Corruption: Caste, Colonialism, and Corruption in Late Nineteenth-Century India," *Itinerario* 48, no. 2 (2024): 152–168.

37 For a discussion of scholarship on historical corruption and of ways of rethinking the relation between corruption and modernity, see Ricard Torra-Prat and Joan Pubill-Brugués, "Introduction: Corruption and the Modernisation of the State: A Short Introduction to a Long-term Process (1600–1900)," in *Corruption, Anti-Corruption, Vigilance*, ed. Torra-Prat, Pubill-Brugués, and Brendecke, 1–20.

38 James Kennedy, Pieter Wagenaar, Mark Rutgers, and Joris van Eijnatten, "Corruption and Public Values in Historical and Comparative Perspective: An Introduction," *Public Voices* 10, no. 2 (2008): 3–6; Jens Ivo Engels, "Corruption as a Political Issue in Modern Societies: France, Great Britain and the United States in the Long 19th Century," *Public Voices* 10, no. 2 (2008): 68–86; Toon Kerkhoff, Ronald Kroeze, and Pieter Wagenaar, "Corruption and the Rise of Modern Politics," *Journal of Modern European History* 11, no. 1 (2013); Gjalt de Graaf, Pieter Wagenaar, and Michel Hoenderboom, "Constructing Corruption," in *The Good Cause. Theoretical Perspectives on Corruption*, ed. Gjalt de Graaf, Patrick von Maravić, and Pieter Wagenaar (Barbara Budrich, 2010), 98–114; Jens Ivo Engels, "La nueva historia de la corrupción: Algunas reflexiones sobre la historiografía de la corrupción política en los siglos XIX y XX," *Ayer: Revista de Historia Contemporánea* 115, no. 3 (2019): 23–49; Françoise Dreyfus, *Sociologie de la corruption* (Paris: La Découverte, 2022).

contrary to the general interest. All of these, then, contributed to shaping and creating political reality.

Sturdza pointed out that "probity"—along with "conscientiousness, seriousness, morality, and honesty"—was the measure of good public conduct and government, the right attitude and values ensuring that the general interest prevailed. Also implicit in Sturdza's denunciation was that no corruption in the modern and contemporary sense was possible without the public-private divide, a separation doomed to remain ambiguous.[39] Sturdza assumed the posture of a morally superior outsider—hiding his identity under a pseudonym and staging the publication of his article in the foreign press. The intended audience of any polemic is naturally important for its framing, and Sturdza's case is interesting, if not altogether original: *Românul* explained that the piece had been initially published in a Prague-based German-language newspaper, *Politik*.[40] Adopting the knowledgeable and politically neutral posture of a benevolent outsider, Sturdza thus claimed to speak to *a different* public, even if his primary target was literate Romanians. As is also highlighted by Alex R. Tipei's contribution to this volume, geographic scales and the national–international divide are crucial to such polemics, yet they are also readily manipulated elements of their rhetoric.[41]

Additionally, Sturdza's accusations illustrate how the (re)negotiation of the shifting boundary between the public and the private spheres was at the core of nineteenth-century corruption scandals, with railway infrastructures as privileged sites of such contestations among state agencies, politicians, populations, notabilities, and transnational investors.[42] Sturdza mixed moral arguments and admonitions with considerations of liberal-elitist state action from above:

> If only Prince Carol of Hohenzollern had acquired a serious sense of right and duty; if only he had scorned illegality and wickedness; if only he had had a more open ear for conscientiousness, seriousness, morality

39 Peter Bratsis, "The Construction of Corruption, or Rules of Separation and Illusions of Purity in Bourgeois Societies," *Social Text* 21, no. 4 (2003): 9–33.

40 Other like-minded Liberal papers published the article. The pamphlet was also published as a booklet in 1875, see Nicoleta Stan-Țurcanu, *Lumea lui Dimitrie A. Sturdza* (Bucharest: Vremea, 2023), 104, footnote 297.

41 Tipei, "Accusations of Tyranny" in this volume.

42 Dard, Engels, Fahrmeir, and Monier, eds., *Scandales et corruption*, 17–18. On the complex entanglements between business and politics, and private and public interests, see Gábor Egry, "Steel and Wood: Railway Construction and Rent-Seeking" in this volume, and Chiper, "Regional Competition, Economic Interests" in this volume.

> and honesty, than for flattery; if only he had not turned aside all the greatest capacities in the whole country, only to surround himself with an unworthy camarilla; if he had united consideration with circumspection in all affairs of state; finally, if it had occurred to him that it would be a more beautiful course action to work toward the uplifting of a people, than to labor to ruin its future for the sake of foreign interest, even if that was a German interest—*then* Romania would be altogether different.[43]

While questioning public ethics, infrastructure-corruption controversies were also catalysts for the historical construction of corruption-related categories—such as immoral behavior of public officials, misconduct or improbity in public office, and abuse of influence—thus implicitly renegotiating the public-private boundary and leading to institutional and administrative change.[44] In this regard, corruption controversies were a significant part of state-building and of the history of administrative capacity-building in Central, South, and Eastern Europe,[45] as well as in overseas imperial rule.[46] Railway concessions and the corruption debates they generated were crucial moments in the transformation of administration in Romania and its regulatory standards and practices, notably regarding conflicts of interest in public and elected office. In the early 1880s, following the Strousberg debacle, the Romanian state sought to prevent similar controversies by shifting toward the construction of new lines

43 De Hahn, *Românul* 19, October 9, 1875, 911.

44 See also Ardeleanu, "British Investors and the 'Kustendjie Harbor Dues Issue'" in this volume.

45 Benjamin Landais, "Du micropolitique à la corruption: Joseph II et l'idéal de l'État de police dans le Banat," in *Moralité du pouvoir et corruption en France et en Roumanie XVIIIe–XXe siècle*, ed. Silvia Marton, Frédéric Monier, and Olivier Dard (Paris: PUPS, 2017), 37–49; Alexandra Iancu, "Le pouvoir à l'abri des sanctions: Les débuts du contrôle parlementaire en Roumanie," in *Moralité du pouvoir*, ed. Marton, Monier, and Dard, 137–149; Mihai Olaru, "From Disloyalty to Law-breaking: The Emergence of Administrative and Judicial Malpractice in Eighteenth-Century Wallachia," *Südost-Forschungen* 77, no. 1 (2018): 10–29; Ovidiu Olar, "Corruption and Anticorruption in the Romanian Principalities: Rules of Governance, Exceptions, and Networks, Seventeenth to the Nineteenth Century," in *Anticorruption in History: From Antiquity to the Modern Era*, ed. Ronald Kroeze, André Vitória, and Guy Geltner (Oxford: Oxford University Press, 2018), 225–238; Klaus Buchenau, "Corruption in Southeastern Europe and Latin America: An Introduction," *Comparative Southeast European Studies* 66, no. 3 (2018), 293–298; Klaus Buchenau, "Historicizing Corruption: An Outline on Serbia (19th–21st Century)," in *Klientelismus in Südosteuropa, 54. Internationale Hochschulwoche der Südosteuropa-Gesellschaft in Tutzing, 5.- 9. Oktober 2015*, ed. Klaus Roth and Ioannis Zelepos (Berlin: Peter Lang Verlag, 2018), 109–132; Gábor Kármán, ed., *Tributaries and Peripheries of the Ottoman Empire* (Leiden: Brill, 2020).

46 Ronald Kroeze, Pol Dalmau, and Frédéric Monier, eds., *Corruption, Empire and Colonialism in the Modern Era* (London: Palgrave Macmillan, 2021).

through public credit—rather than through companies and concessions voted on directly by parliament—and gradually nationalized the main railways. As Gábor Egry shows in his chapter, railway concessions in the Ottoman Empire and in Transylvania—in Eastern Hungary—allowed local political and business elites to manifest their agency and profit from railway construction by employing various schemes of rent-seeking and informal tools, thus leveraging more powerful partners.[47]

In Romania, as elsewhere, the often colossal financial effort involved in nineteenth-century infrastructure construction—be it via concessions or with public money—along with the genuine and inherent difficulties of controlling expenditure, provided fertile ground for accusations of economic and financial individual corruption, notably bribery, embezzlement, personal enrichment, or favoritism.[48]The old moral-religious denunciation of the sin of greed, meanwhile, was always lurking behind such charges. Sturdza's pamphlet listed the names of allegedly corrupt members of parliament. During the 1870s and 1880s, the Romanian parliament set up several commissions of inquiry to ascertain the potential culpability of members of parliament and whether civil servants had been bribed in connection with railway concessions. In Vienna, the trial of businessman Viktor von Ofenheim, the sometime chief director of the Lemberg-Cernowitz-Jassy Railway Company that had built and operated a railway section in Romania, also involved high-profile Austrian politicians on suspicion of bribery and financial corruption.[49]

Ascribing Blame and Othering Corruption

Finally—yet no less importantly—the question of the categories and the sources of corruption was as important in infrastructure controversies as the question of who was corrupt. Identifying and describing the corrupting

47 Egry, "Steel and Wood."

48 Christian Ebhardt, "In Search of a Political Office: Railway Directors and Electoral Corruption in Britain and France, 1820–1870," *Journal of Modern European History* 11, no. 1 (2013): 72–87; Stephan Ruderer, "Crisis and Corruption: The Anglo-Argentine Scandal Surrounding the Privatisation of the Buenos-Aires Sanitary Works 1888–1891," in *Scandales et corruption*, ed. Dard, Engels, Fahrmeir, and Monier, 66–79. See also Toader Popescu, "Hoping and Coping: Local Anxieties and Railway Development in Romania" in this volume.

49 Pieter M. Judson, *Exclusive Revolutionaries: Liberal Politics, Social Experience, and National Identity in the Austrian Empire, 1848–1914* (Ann Arbor: University of Michigan Press, 1997), 182–183, and 182 footnotes 42, 43.

agent(s) and ascribing blame were at the core of corruption scandals; unfailingly, these could only be one's political opponents, as well as various local or foreign groups or individuals. Corruption accusations, with their performative function, were not just a critique of governance; they also actively constituted political antagonisms and identities (internally and internationally) and allowed for Othering corruption in the process. From this angle, infrastructure-corruption scandals were also intimately connected not only to state-building but equally to nation-building and the process of political democratization. Nation-builders used the state-centric language of (anti)corruption to create the nation's internal and external Others, often building ascriptive hierarchies in which corrupting ethnoreligious groups were placed.

Accusing one's political opponents of being corruptible and in cahoots with the foreign Other in blatant disregard of the national interest was something repeatedly mirrored by both the Liberals and the Conservatives in Romania. They believed collusion with foreign interests was the most scandalous matter of all in the Strousberg debacle. As Andrei-Dan Sorescu notes in his chapter in the present volume, infrastructure scandals and Othering converged in anxieties over the potential "colonization" of the country, and the Strousberg scandal became something of a template for later panics in decades to come.[50] In his pamphlet, Sturdza blamed German officials and entrepreneurs, who were allegedly ready to engage in shady business, speculation, embezzlement, and fraud and to infiltrate local decision-making to gain accomplices and to corrupt from within. He explicitly accused Prince Carol and his government of defending foreign and private interests and speculators instead of standing for the entire political body. Sturdza also feared the trickle-down effect of corruption, which he described as a kind of "moral contagion" that would ultimately spread to the entire social and political body. The very worst kind of corruption, then, was that induced and spread by a powerful, occult group, closely tied to the highest decision-makers and to a foreign great power and its interests. Such accusations casted doubt on parliamentarism in France, Germany, Spain, and Britain at that time.[51] Other Romanian Liberals even more explicitly accused local politicians of being in the pay of foreign interests and, worse

50 Andrei-Dan Sorescu, "The Infrastructural Power of the 'Colonial' as a Concept" in this volume.

51 Frédéric Monier, *Un pouvoir malhonnête: La corruption publique en France, 18e–20e siècles* (Ceyzérieu: Champ Vallon, 2025), 176–192; Jens Ivo Engels, *A korrupció története: A korai újkortól a 20. századig*, trans. Kurdi Imre (Budapest: Corvina, 2016), 257–328.

yet, of Jewish interests. Their strong antidynasticism coalesced with their perception that the country had become a "colony," a mere extension of the political and economic reach of Prussia on the Lower Danube and in the Orient.[52]

Moreover, the internal politics of the scandal quickly became international by engaging in controversies over Romania's foreign policy. Even more than merely denouncing Prince Carol's collusion with the Conservatives and with Prussia, the Liberals professed republican leanings,[53] indulged in explicit Teutophobia,[54] and strongly questioned the Conservatives' foreign policy toward the Habsburg Empire and Germany by accusing them of promoting an antinational agenda. And when Chancellor Bismarck interfered at the Congress of Berlin in 1878 to defend the German bankers and investors and conditioned the acknowledgment of Romania's sovereignty on the emancipation of its Jews and the nationalization of the Strousberg-built railway lines, Romanian politicians' xenophobic and anti-Semitic accusations thrived even more acutely.[55]

Identifying the foreigner as an agent of corruption, with or without local complicity, allowed for widening the scale of corruption—from the strictly individual (such as bribery, personal profit, abuse of power, and negligence) to more systemic corruption and the diffuse degeneration of collective practices and behaviors (such as the Jews' alleged pernicious and occult influence). Moreover, pinpointing the foreigner from within as an agent of corruption—notably the Jews—contributed decisively to the consolidation of anti-Semitism and extremist right-wing, nationalist antiparliamentarism, with infrastructure controversies acting as conduits for this nexus of accusations well into the interwar period, as shown by Raul Cârstocea in this

52 Andrei Sorescu, "The Perils (and Promise) of German Colonization."

53 Silvia Marton, *"Republica de la Ploiești" și începuturile parlamentarismului în România* (Bucharest: Humanitas, 2016).

54 Andrei-Dan Sorescu, "The Political Language of Moral Panics: Teutophobia and the Image of a Besieged Romania in 1871," in *Penser le XIXe siècle: nouveaux chantiers de recherche*, ed. Silvia Marton and Constanţa Vintilă-Ghiţulescu (Iași: Editura Universităţii "Alexandru Ioan Cuza" din Iași, 2013), 153–173.

55 Marton, "Shades of Dependency"; Fritz Stern, *Gold and Iron: Bismarck, Bleichröder, and the Building of the German Empire* (New York: Alfred Knopf, 1977), chapter 14; Constantin Iordachi, *Liberalism, Constitutional Nationalism, and Minorities: The Making of Romanian Citizenship, c. 1750–1918* (Leiden: Brill, 2019), 306–351. On anti-Semitism's connection to the national liberal movement in Romania and Algeria, see the historiographical essay by Lisa Moses Leff, "Liberalism and Antisemitism: A Reassessment from the Peripheries," in *Jews, Liberalism, Antisemitism: A Global History*, ed. Abigail Green and Simon Levis Sullam (London: Palgrave Macmillan, 2020), 23–45.

volume.[56] Corruption was thus seen as a social pathology, yet one difficult to separate from individual acts that were classified as such, which almost always included a moral judgment on social and political pollution and on immoral economic exploitation.[57] The accusations of corruption induced by financial capitalism and speculation allowed for disputes over both individual transgressions and systemic social, political, and moral decay.[58]

The focus on corruption controversies originating in transportation infrastructure adds a further layer to our understanding of state development in the polities of Central, South, and Eastern Europe on the imperial peripheries and interstices of the Russian, Ottoman, and Habsburg empires. Here, transport infrastructure was simultaneously part of an emerging process of state-building[59] while allowing for new transimperial and cross-regional

56 Raul Cârstocea, "Conduits of Violence: Railway Infrastructure and Student Anti-Semitism in Interwar Romania" in this volume. See also Olivier Dard, "La corruption dans la France des années 1930: historiographie et perspectives de recherche," in *La politique vue d'en bas: Pratiques privées et débats publics, XIXe–XXe siècles*, ed. Jens Ivo Engels, Frédéric Monier, and Natalie Petiteau (Paris: Armand Colin, 2012), 209–222; Olivier Dard, "Le moment Barrès: nationalisme et critique de la corruption," *Cahiers Jaurès* 209 (2013): 93–111; Raul Cârstocea, "Anti-Semitism at the Intersection of Corruption and Colonialism: Continuities of Political Rhetoric from the Nineteenth Century to the Interwar Period," *East European Politics and Societies* 39, no. 3 (2025): 762–783.

57 Marton, "Shades of Dependency." For such debates in early nineteenth-century Britain, see Ian Cawood, "The Black Book and the Reform of Public Life in Early Nineteenth-Century Britain," *Journal of Liberal History* 113 (2022): 6–17.

58 On the moralizing and anti-Semitic discourse on money, finance, and capitalism, see De Blic, "Moraliser l'argent"; Emmanuel Kreis, *Quis ut Deus? Antijudéo-maçonnisme et occultisme en France sous la IIIe République* (Paris: Les Belles Lettres, 2017); Nina Valbousquet, *Catholique et antisémite: Le réseau de Mgr Benigni: Rome, Europe, États-Unis, 1918–1934* (Paris: CNRS éditions, 2020); Stephanie Chasin, *Anticapitalism and the Emergence of Antisemitism* (New York: Peter Lang, 2020). For the history of transnational credit and speculative capitalism and of their role in the realization of colonial, national, and statehood ambitions in the nineteenth century, see Christopher Clay, *Gold for the Sultan: Western Bankers and Ottoman Finance, 1856–1881: A Contribution to Ottoman and to International Financial History* (London: I. B. Tauris, 2000); Ralf Roth and Günter Dinhobl, eds., *Across the Borders: Financing the World's Railways in the Nineteenth and Twentieth Centuries* (Abingdon: Routledge, 2008); Glenda Sluga, "'Who Hold the Balance of the World?' Bankers at the Congress of Vienna, and in International History," *The American Historical Review* 122, no. 5 (2017): 1403–1430; Darina Martykánová and Juan Pan-Montojo, eds., *Misioneros del capitalismo: Aventureros, hombres de negocios y expertos transnacionales en el siglo XIX* (Granada: Comares, 2023).

59 For the contribution of transport infrastructure to state-building, see Jo Guldi, *Roads to Power: Britain Invents the Infrastructure State* (Cambridge: Harvard University Press, 2012); Jordi Martí-Henneberg, "From State-Building to European Integration: The Role of the Railway Network in the Territorial Integration of Europe, 1850–2020," *Social Science History* 45, no. 2 (2021): 221–231. For the role of railways for the territorial integration of the Portuguese nation-state, see Hugo Silveira Pereira, "Appropriation, Integration, and Nation Building: Portuguese Railways

connectivity, which meant both cooperation and competition.[60] Such polities were, we argue, central to the history of mobilities and connectivities.[61] The integrative effects of transportation infrastructure on what would otherwise have remained peripheral regions proved crucial both for emerging national territories and for large imperial spaces. Yet railway construction had the opposite effects as well, in that it fostered developmental inequalities and peripheralization and prevented regional integration, as Mihai Chiper's chapter in this volume documents for the province of Moldavia in Romania.[62] Additionally, Gábor Egry shows in his chapter that railway concessions also allowed for semicolonial situations in Transylvania, a peripheral region of dualist Hungary, and in the territories still dependent on the late Ottoman Empire, while they also permitted local political and business elites to profit from railway construction.[63]

Moreover, connectivity meant inherent tension: railways may have integrated politically bounded territories internally, but they simultaneously transcended any politically defined territory, be it a nation-state or an empire.[64] The communicative function of railways clashed with nationalist claims in the era of liberal nationalism. Conflicting forces underpinned railway-network construction: the forces of transnational globalization, on the one hand, and the forces of rising ethnic nationalism, on the other.[65] For Sturdza and like-minded nationalist Liberals, concessions built with German

in the Second Half of the Nineteenth and Early Years of the Twentieth Century," *Social Science History* 45, no. 2 (2021): 391–416.

60 Frithjof Benjamin Schenk, "Die Transsibirische Eisenbahn als Brücke zwischen Europa und Asien," *Religion & Gesellschaft in Ost und West* 45, no. 12 (2017): 14–16; L. Pozharliev, F. Riedler, and S. Rohdewald, "Introduction to the Special Issue: Transottoman Infrastructures and Networks across the Black Sea," *Journal of Balkan and Black Sea Studies* 3, no. 5 (2020): 13–18; Florian Riedler, "Around the Black Sea in Forty-Five Days: Transottoman Space, Time, and Infrastructure," in *Transottoman Matters: Objects Moving through Time, Space, and Meaning*, ed. Arkadiusz Blaszczyk, Robert Born, and Florian Riedler (Göttingen: V&R unipress, 2021), 27–60.

61 An overview of the scholarship can be found in Arkadiusz Blaszczyk, Robert Born, and Florian Riedler, "Introduction: Movable Objects," in *Transottoman Matters*, ed. Blaszczyk, Born, and Riedler, 15–20.

62 Chiper, "Regional Competition, Economic Interests."

63 Egry, "Steel and Wood."

64 For a historical and comparative analysis of infrastructure networks that helped international integration of the globalized economy, markets, and politics, see the contributions to Hans-Liudger Dienel and Martin Schiefelbusch, eds., *Linking Networks: The Formation of Common Standards and Visions for Infrastructure Development* (Abingdon: Routledge, 2018).

65 Felix Jeschke, "Dracula on Rails: The Pressburgerbahn between Imperial Space and National Body, 1867–1935," *Central Europe* 10, no. 1 (2012): 1–17; Jan Musekamp, *Shifting Lines, Entangled Borderlands: Mobilities and Migration along the Prussian Eastern Railroad* (Bloomington: Indiana University Press, 2024) (particularly chapter 2); Malte Fuhrmann, "The 'Veins and Arteries of the

and other foreign capital posed a danger to precisely what railways were meant to achieve—by facilitating connectivity beyond the national territory, they allowed neighbors and/or more distant potential enemies to reach and ultimately expand their political, economic, or cultural dominion over the home country. Foreign investors were thus seen as bridgeheads for their respective governments' interests, if not for outright colonial interference and economic extraction. After the Crimean War, fear of the Great Powers' involvement in internal affairs by way of major infrastructure concessions was widespread.[66] In sum, while nation-builders in the region welcomed the benefits of railways as a means of integration and connectivity beyond the nation-state, they also aimed at preserving territorial, national, and ethnic disparities.

Conclusion

Sturdza's pamphlet was an altogether typical intervention in an age of both nation and empire, an age of steam, (corruption) scandals, and slippery slopes of xenophobic anxiety. As the present volume even more amply demonstrates, the language of (anti)corruption and the Othering of corruption proved empowering discursive devices in the context of a global capitalist imperialism that advanced (both figuratively and literally) through the expansion of infrastructure. What makes Sturdza's intervention representative is not only its ultimate banality, however, but also the way in which the various scales at which corruption was felt to be at play intersected in the narrative. Sturdza was at home when providing minute anecdotal detail and local color, yet equally so when discussing the far-reaching implications of international power politics. By choosing a German *nom de plume*, he articulated his criticism from what was intended to be the position of a foreign observer who, though part of the reviled out-group, had managed to rise above his own national interest to defend the "truth," "probity," and the

Country': Imagined and Actual Exclusion from Railway Connectivity in Bulgaria, 1878–1908," *The Journal of Transport History* 45, no. 3 (2024): 646–670.

66 J. H. Jensen and Gerhard Rosegger, "British Railway Builders along the Lower Danube, 1856–1869," *The Slavonic and East European Review* 46, no. 106 (1968): 105–128; Boriana Antonova-Goleva, "'Top-Down' or 'Bottom-Up' Modernisation: Local Railway Entrepreneurs in the Ottoman Empire in the Second Half of the 19th Century," *Turkish Historical Review* 14, no. 1 (2023): 26–57. For a historical and comparative study of the integration challenges of railways, see Martin Schiefelbusch, "Visions of Rail Development—The 'Internationality of Railways' Revisited," in *Linking Networks*, ed. Dienel and Schiefelbusch, 73–94.

honor of a small nation. This was, of course, hardly a convincing disguise, even if the Liberal press reprinted his pamphlet in translation by playing along. But this is nevertheless symptomatic of how the international nature of the corruption scandals connecting Southeast Europe to a great power influenced the political imagination. This, then, will also prove to be one of the key recurring themes of our volume: it was not simply infrastructure that connected the regions but the corruption controversies that came with it.

Bibliography

Antonova-Goleva, Boriana. "'Top-Down' or 'Bottom-Up' Modernisation: Local Railway Entrepreneurs in the Ottoman Empire in the Second Half of the 19th Century." *Turkish Historical Review* 14, no. 1 (2023): 26–57. https://doi.org/10.1163/18775462-bja10007.

Anushree, Anubha. "Corrupt Traditions and Traditions of Corruption: Caste, Colonialism, and Corruption in Late Nineteenth-Century India." *Itinerario* 48, no. 2 (2024): 152–168. https://doi.org/10.1017/S0165115324000226.

Appel, Hannah, Nikhil Anand, and Akhil Gupta. "Introduction: Temporality, Politics, and the Promise of Infrastructure." In *The Promise of Infrastructure*, edited by Nikhil Anand, Akhil Gupta, and Hannah Appel, 1–38. Durham: Duke University Press, 2018.

Ardeleanu, Constantin. "'Commercial and Political Needs Demand the Establishment of a Black Sea Port': Infrastructure Development, Opportunities and Anxieties in an Eastern European Periphery (1860s–1870s)." *The Journal of Transport History* 44, no. 3 (2023): 436–456. https://doi.org/10.1177/00225266231180284.

Ardeleanu, Constantin. "The Hallier Affair, Political Corruption and National Dignity in Romania in Late Nineteenth and Early Twentieth Centuries." *East European Politics and Societies* 39, no. 3 (2025): 741–761. https://doi.org/10.1177/08883254251352115.

Blaszczyk, Arkadiusz, Robert Born, and Florian Riedler. "Introduction: Movable Objects." In *Transottoman Matters: Objects Moving through Time, Space, and Meaning*, edited by Arkadiusz Blaszczyk, Robert Born, and Florian Riedler, 9–26. Göttingen: V&R unipress, 2021.

De Blic, Damien. "Moraliser l'argent: Ce que Panama a changé dans la société française (1889–1897)." *Politix* 71, no. 3 (2005): 61–82. https://doi.org/10.3917/pox.071.0061.

De Blic, Damien, and Cyril Lemieux. "The Scandal as Test: Elements of Pragmatic Sociology." *Politix* 71, no. 3 (2005): 9–38. https://shs.cairn.info/journal-politix–2005-3-page-9?lang=en.

De Block, Greet. "The Material Politics of Infrastructure Networks: Infrastructure Design and Territorial Transformation in Belgium, 1830–40s." *Social Science History* 45, no. 2 (2021): 341–362. https://doi.org/10.1017/ssh.2021.5.

Bösch, Frank. "Politics with Scandals: Germany and Britain in Transnational Perspective (1880–1914)." *Historia y Política* 39 (2018): 53–77. https://doi.org/10.18042/hp.39.03.

Bratsis, Peter. "The Construction of Corruption, or Rules of Separation and Illusions of Purity in Bourgeois Societies." *Social Text* 21, no. 4 (2003): 9–33. https://doi.org/10.1215/01642472-21-4_77-9.

Brendecke, Arndt. "Foreword: A Commentary on Corruption and Control from the Perspective of the Cultures of Vigilance." In *Corruption, Anti-Corruption, Vigilance, and State Building from Early to Late Modern Times*, edited by Ricard Torra-Prat, Joan Pubill-Brugués, and Arndt Brendecke, x–xvii. Abingdon: Routledge, 2024.

Buchenau, Klaus. "Corruption in Southeastern Europe and Latin America: An Introduction." *Comparative Southeast European Studies* 66, no. 3 (2018), 293–298. https://doi.org/10.1515/soeu–2018-0025.

Buchenau, Klaus. "Historicizing Corruption: An Outline on Serbia (19th–21st Century)." In *Klientelismus in Südosteuropa, 54. Internationale Hochschulwoche der Südosteuropa-Gesellschaft in Tutzing, 5.-9. Oktober 2015*, edited by Klaus Roth and Ioannis Zelepos, 101–120. Berlin: Peter Lang, 2018.

Cârstocea, Raul. "Anti-Semitism at the Intersection of Corruption and Colonialism: Continuities of Political Rhetoric from the Nineteenth Century to the Interwar Period." *East European Politics and Societies* 39, no. 3 (2025): 762–783, https://doi.org/10.1177/08883254251352116.

Cawood, Ian. "The Black Book and the Reform of Public Life in Early Nineteenth-Century Britain." *Journal of Liberal History* 113 (2022): 6–17.

Chasin, Stephanie. *Anticapitalism and the Emergence of Antisemitism*. New York: Peter Lang, 2020.

Clay, Christopher. *Gold for the Sultan: Western Bankers and Ottoman Finance, 1856–1881: A Contribution to Ottoman and to International Financial History.* London: I. B. Tauris, 2000.

Dard, Olivier. "La corruption dans la France des années 1930: historiographie et perspectives de recherche." In *La politique vue d'en bas: Pratiques privées et débats publics, XIXe–XXe siècles,* edited by Jens Ivo Engels, Frédéric Monier, and Natalie Petiteau, 209–222. Paris: Armand Colin, 2012.

Dard, Olivier. "Le moment Barrès: nationalisme et critique de la corruption." *Cahiers Jaurès* 209 (2013): 93–111, https://doi.org/10.3917/cj.209.0093.

Dean, Jodi. "Publicity's Secret." *Political Theory* 29, no. 5 (2001): 624–650, https://www.jstor.org/stable/3072532.

Dienel, Hans-Liudger, and Martin Schiefelbusch, eds. *Linking Networks: The Formation of Common Standards and Visions for Infrastructure Development.* Abingdon: Routledge, 2018.

Dreyfus, Françoise. *Sociologie de la corruption.* Paris: La Découverte, 2022.

Ebhardt, Christian. "In Search of a Political Office: Railway Directors and Electoral Corruption in Britain and France, 1820–1870." *Journal of Modern European History* 11, no. 1 (2013): 72–87. https://doi.org/10.17104/1611-8944_2013_1_72.

Engels, Jens Ivo. "Corruption as a Political Issue in Modern Societies: France, Great Britain and the United States in the Long 19th Century." *Public Voices* 10, no. 2 (2008): 68–86.

Engels, Jens Ivo. "Infrastructure and Fragmentation: The Limits of the Integration Paradigm." In *Linking Networks: The Formation of Common Standards and Visions for Infrastructure Development*, edited by Hans-Liudger Dienel and Martin Schiefelbusch, 19–34. Abingdon: Routledge, 2018.

Engels, Jens Ivo. *A korrupció története: A korai újkortól a 20. századig.* Translated by Kurdi Imre. Budapest: Corvina, 2016.

Engels, Jens Ivo. "La nueva historia de la corrupción: Algunas reflexiones sobre la historiografía de la corrupción política en los siglos XIX y XX." *Ayer: Revista de Historia Contemporánea* 115, no. 3 (2019): 23–49, https://doi.org/10.55509/ayer/115-2019-02.

Engels, Jens Ivo, and Frédéric Monier. "Researching the History of Transparency: Introduction." In *History of Transparency in Politics and Society*, edited by Jens Ivo Engels and Frédéric Monier, 7–20. Göttingen: V&R unipress, 2020.

Engels, Jens Ivo, and Anna Rothfuss. "Les usages de la politique du scandale: Le SPD et les débats sur la corruption politique pendant le Kaiserreich (1873–1913)." *Cahiers Jaurès* 209, no. 3 (2013): 33–51.

Fenster, Mark. "Transparency in Search of a Theory." *European Journal of Social Theory* 18, no. 2 (2015): 150–167, https://doi.org/10.1177/1368431014555257.

Fuhrmann, Malte. "The 'Veins and Arteries of the Country': Imagined and Actual Exclusion from Railway Connectivity in Bulgaria, 1878–1908." *The Journal of Transport History* 45, no. 3 (2024): 646–670. https://doi.org/10.1177/00225266241254764.

De Graaf, Gjalt, Pieter Wagenaar, and Michael Hoenderboom. "Constructing Corruption." In *The Good Cause: Theoretical Perspectives on Corruption*, edited by Gjalt de Graaf, Patrick von Maravić, and Pieter Wagenaar, 98–114. Opladen: Barbara Budrich, 2010.

Guldi, Jo. *Roads to Power: Britain Invents the Infrastructure State.* Cambridge: Harvard University Press, 2012.

De Hahn, Erdmann. "Germania, România și principele Carol de Hohenzollern," Part 1. *Românul* 19, October 8, 1875; Part 2, *Românul* 19, October 9, 1875.

Hood, Christopher, "Transparency in Historical Perspective." In *Transparency: The Key to Better Governance?*, edited by Christopher Hood and David Heald, 2–23. Oxford: Oxford University Press, 2006.

Horn, Eva. "Logics of Political Secrecy." *Theory, Culture & Society* 28, nos. 7–8 (2011): 103–122, https://doi.org/10.1177/0263276411424583.

Iancu, Alexandra. "Le pouvoir à l'abri des sanctions: Les débuts du contrôle parlementaire en Roumanie." In *Moralité du pouvoir et corruption en France et en Roumanie XVIIIe–XXe siècle*, edited by Silvia Marton, Frédéric Monier, and Olivier Dard, 137–149. Paris: PUPS, 2017.

Iordachi, Constantin. *Liberalism, Constitutional Nationalism, and Minorities: The Making of Romanian Citizenship, c. 1750–1918.* Leiden: Brill, 2019.

Jensen, J. H., and Gerhard Rosegger. "British Railway Builders along the Lower Danube, 1856–1869." *The Slavonic and East European Review* 46, no. 106 (1968): 105–128.

Jeschke, Felix. "Dracula on Rails: The Pressburgerbahn between Imperial Space and National Body, 1867–1935." *Central Europe* 10, no. 1 (2012): 1–17, https://doi.org/10.1179/1479096312Z.00000000001.

Judson, Pieter M. *Exclusive Revolutionaries: Liberal Politics, Social Experience, and National Identity in the Austrian Empire, 1848–1914*. Ann Arbor: University of Michigan Press, 1997.

Kahan, Alan S. *Liberalism in Nineteenth-Century Europe: The Political Culture of Limited Suffrage.* London: Palgrave Macmillan, 2003.

Kármán, Gábor, ed. *Tributaries and Peripheries of the Ottoman Empire*. Leiden: Brill, 2020.

Kennedy, James, Pieter Wagenaar, Mark Rutgers, and Joris van Eijnatten. "Corruption and Public Values in Historical and Comparative Perspective: An Introduction." *Public Voices* 10, no. 2 (2008): 3–6.

Kerkhoff, Toon, Ronald Kroeze, and Pieter Wagenaar. "Corruption and the Rise of Modern Politics." *Journal of Modern European History* 11, no. 1 (2013): 19–30.

Koselleck, Reinhart, and Michaela Richter. "Introduction and Prefaces to the *Geschichtliche Grundbegriffe*: (Basic Concepts in History: A Historical Dictionary of Political and Social Language in Germany)." *Contributions to the History of Concepts* 6, no. 1 (2011): 1–37.

Kreis, Emmanuel. *Quis ut Deus? Antijudéo-maçonnisme et occultisme en France sous la IIIe République*. Paris: Les Belles Lettres, 2017.

Kroeze, Ronald, Pol Dalmau, and Frédéric Monier, eds. *Corruption, Empire and Colonialism in the Modern Era: A Global Perspective.* London: Palgrave Macmillan, 2021.

Landais, Benjamin. "Du micropolitique à la corruption: Joseph II et l'idéal de l'État de police dans le Banat." In *Moralité du pouvoir et corruption en France et*

en Roumanie XVIIIe–XXe siècle, edited by Silvia Marton, Frédéric Monier, and Olivier Dard, 37–49. Paris: PUPS, 2017.

Leff, Lisa Moses. "Liberalism and Antisemitism: A Reassessment from the Peripheries." In *Jews, Liberalism, Antisemitism. A Global History*, edited by Abigail Green and Simon Levis Sullam, 23–45. London: Palgrave Macmillan, 2020.

Martí-Henneberg, Jordi. "From State-Building to European Integration: The Role of the Railway Network in the Territorial Integration of Europe, 1850–2020." *Social Science History* 45, no. 2 (2021): 221–31. https://doi.org/10.1017/ssh.2021.7.

Marton, Silvia. "La 'corrupción' electoral en Rumania: Los comienzos titubeantes de la democracia." *Ayer: Revista de Historia Contemporánea* 115, no. 3 (2019): 77–104, https://www.jstor.org/stable/26773638.

Marton, Silvia. *"Republica de la Ploiești" și începuturile parlamentarismului în România*. Bucharest: Humanitas, 2016.

Marton, Silvia. "Shades of Dependency and the Discourse on 'Corruption': Railway Concessions in Romania in the Nineteenth Century." *East European Politics and Societies* 39, no. 3 (2025): 691–716, https://doi.org/10.1177/08883254251352113.

Marton, Silvia, and Andrei-Dan Sorescu. "Política 'oculta': Publicidad, secretismo, transparencia e inteligibilidad en la Rumanía de finales del siglo XIX." In *Las sombras de la transparencia. Secreto, corrupción y «Estado profundo» en la Europa contemporánea*, edited by Frédéric Monier, Lluís Ferran Toledano, Joan Pubill, and Gemma Rubí, 21–43. Granada: Comares, 2022.

Martykánová, Darina, and Juan Pan-Montojo, eds. *Misioneros del capitalismo: Aventureros, hombres de negocios y expertos transnacionales en el siglo XIX*. Granada: Comares, 2023.

Mollier, Jean-Yves. *Le Scandale de Panama*. Paris: Fayard, 1991.

Monier, Frédéric. "La corruption politique: une histoire européenne." *Cahiers Jaurès* 209, no. 3 (2013): 3–13, https://doi.org/10.3917/cj.209.0003.

Monier, Frédéric. *Un pouvoir malhonnête: La corruption publique en France, 18e–20e siècles*. Ceyzérieu: Champ Vallon, 2025.

Monier, Frédéric, Lluís Ferran Toledano, Joan Pubill, and Gemma Rubí. "Introducción: Una mirada desde la Historia sobre la Transparencia, el Secreto y el Estado profundo en la Europa contemporánea." In *Las sombras de la transparencia: Secreto, corrupción y "Esatdo profundo" en la Europa contemporánea*, edited by Frédéric Monier, Lluís Ferran Toledano, Joan Pubill, and Gemma Rubí, IX–XX. Granada: Comares, 2022.

Musekamp, Jan. *Shifting Lines, Entangled Borderlands: Mobilities and Migration along the Prussian Eastern Railroad*. Bloomington: Indiana University Press, 2024.

Olar, Ovidiu. "Corruption and Anticorruption in the Romanian Principalities: Rules of Governance, Exceptions, and Networks, Seventeenth to the Nineteenth

Century." In *Anticorruption in History: From Antiquity to the Modern Era*, edited by Ronald Kroeze, André Vitória, and Guy Geltner, 225–238. Oxford: Oxford University Press, 2018.

Olaru, Mihai. "From Disloyalty to Law-breaking: The Emergence of Administrative and Judicial Malpractice in Eighteenth-Century Wallachia." *Südost-Forschungen* 77, no. 1 (2018): 10–29. https://doi.org/10.1515/sofo-2018-770105.

Pereira, Hugo Silveira. "Appropriation, Integration, and Nation Building: Portuguese Railways in the Second Half of the Nineteenth and Early Years of the Twentieth Century." *Social Science History* 45, no. 2 (2021): 391–416. https://doi.org/10.1017/ssh.2021.4.

Pitseys, John. "Publicité et transparence: Le statut de la représentation et de la visibilité politique chez Kelsen et Schmitt." *Revue française de science politique* 66, no. 1 (2016): 117–135. https://doi.org/10.3917/rfsp.661.0117.

Pitsos, Nicolas. "Presse, mensonges et chemins de fer: L'affaire Homs-Bagdad à l'ombre de la question d'Orient." In *Scandales et corruption à l'époque contemporaine*, edited by Olivier Dard, Jens Ivo Engels, Andreas Fahrmeir, and Frédéric Monier, 80–95. Paris: Armand Colin, 2014.

Popescu, Toader. *Proiectul feroviar românesc (1842–1916)*. Bucharest: Simetria, 2014.

Portalez, Christophe. "*La Revue socialiste* face à la corruption politique: du scandale de Panama à l'affaire Rochette (1892–1914)." *Cahiers Jaurès* 209, no. 3 (2013): 15–32. https://doi.org/10.3917/cj.209.0015.

Portalez, Christophe, and Anna Rothfuss. "Panama and the Opposition: The Perception of French and German Socialists of the Panama Scandal." In *Scandales et corruption à l'époque contemporaine*, edited by Olivier Dard, Jens Ivo Engels, Andreas Fahrmeir, and Frédéric Monier, 181–194. Paris: Armand Colin, 2014.

Pozharliev, L., F. Riedler, and S. Rohdewald. "Introduction to the Special Issue: Transottoman Infrastructures and Networks across the Black Sea." *Journal of Balkan and Black Sea Studies* 3, no. 5 (2020): 13–18.

Riedler, Florian. "Around the Black Sea in Forty-Five Days: Transottoman Space, Time, and Infrastructure." In *Transottoman Matters: Objects Moving through Time, Space, and Meaning*, edited by Arkadiusz Blaszczyk, Robert Born, and Florian Riedler, 27–60. Göttingen: V&R unipress, 2021.

Rosanvallon, Pierre. *Counter-Democracy: Politics in an Age of Distrust*. Translated by Arthur Goldhammer. Cambridge: Cambridge University Press, 2008.

Roth, Ralf. "Difficulties of International Railway Investments in Germany: The Example of the 'Railway King' Bethel Henry Strousberg, 1855–1875." In *Across the Borders: Financing the World's Railways in the Nineteenth and Twentieth Centuries*, edited by Ralf Roth and Günter Dinhobl, 33–47. Abingdon: Routledge, 2008.

Roth, Ralf, and Günter Dinhobl, eds. *Across the Borders: Financing the World's Railways in the Nineteenth and Twentieth Centuries*. Abingdon: Routledge, 2008.

Ruderer, Stephan. "Crisis and Corruption: The Anglo-Argentine Scandal Surrounding the Privatisation of the Buenos-Aires Sanitary Works 1888–1891." In *Scandales et corruption à l'époque contemporaine*, edited by Olivier Dard, Jens Ivo Engels, Andreas Fahrmeir, and Frédéric Monier, 66–79. Paris: Armand Colin, 2014.

Schenk, Frithjof Benjamin. "Die Transsibirische Eisenbahn als Brücke zwischen Europa und Asien." *Religion & Gesellschaft in Ost und West* 45, no. 12 (2017): 14–16.

Schiefelbusch, Martin. "Visions of Rail Development—The 'Internationality of Railways' Revisited." In *Linking Networks: The Formation of Common Standards and Visions for Infrastructure Development*, edited by Hans-Liudger Dienel and Martin Schiefelbusch, 73–94. Abingdon: Routledge, 2018.

Sluga, Glenda. "'Who Hold the Balance of the World?' Bankers at the Congress of Vienna, and in International History." *The American Historical Review* 122, no. 5 (2017): 1403–1430, https://doi.org/10.1093/ahr/122.5.1403.

Sorescu, Andrei. "The Perils (and Promise) of German Colonization: Civilizational Hierarchies and Anxieties in Nineteenth Century Romania." *East European Politics and Societies* 39, no. 3 (2025): 741–761. https://doi.org/10.1177/08883254251355736.

Sorescu, Andrei-Dan. "The Political Language of Moral Panics: Teutophobia and the Image of a Besieged Romania in 1871." In *Penser le XIXe siècle: nouveaux chantiers de recherche*, edited by Silvia Marton and Constanţa Vintilă-Ghiţulescu, 153–173. Iași: Editura Universității "Alexandru Ioan Cuza" din Iași, 2013.

Stan-Țurcanu, Nicoleta. *Lumea lui Dimitrie A. Sturdza*. Bucharest: Vremea, 2023.

Stern, Fritz. *Gold and Iron: Bismarck, Bleichröder, and the Building of the German Empire*. New York: Alfred Knopf, 1977.

Thompson, John B. *Political Scandal: Power and Visibility in the Media Age*. Cambridge: Polity Press, 2000.

Torra-Prat, Ricard and Joan Pubill-Brugués. "Introduction: Corruption and the Modernisation of the State: A Short Introduction to a Long-term Process (1600–1900)." In *Corruption, Anti-Corruption, Vigilance, and State Building from Early to Late Modern Times*, edited by Ricard Torra-Prat, Joan Pubill-Brugués, and Arndt Brendecke, 1–20. Abingdon: Routledge, 2024.

Valbousquet, Nina. *Catholique et antisémite: Le réseau de Mgr Benigni: Rome, Europe, États-Unis, 1918–1934*. Paris: CNRS Éditions, 2020.

Von Schnitzler, Antina. "Infrastructure, Apartheid Technopolitics, and Temporalities of 'Transition.'" In *The Promise of Infrastructure*, edited by Nikhil Anand, Akhil Gupta, and Hannah Appel, 133–154. Durham: Duke University Press, 2018.

Wilkinson, Callie. "Scandal and Secrecy in the History of the Nineteenth-Century British Empire." *The Historical Journal* 65, no. 2 (2022): 545–569. https://doi.org/10.1017/S0018246X21000066.

About the Author

Silvia Marton is associate professor at the Faculty of Political Science, University of Bucharest, and senior researcher at the New Europe College—Institute for Advanced Study, Bucharest. Her scientific interests include the history and sociology of political corruption, and nation-state building in Central-South-East Europe. She is principal investigator of "Transnational histories of 'corruption' in Central-South-East Europe (1750–1850)" (ERC-2022-AdG no. 101098095).

3. Writing the History of Critical Infrastructures: Some Conceptual Remarks

Jens Ivo Engels

Abstract: The chapter outlines several conceptual reflections based on the terminology of the social sciences and technical infrastructure research, and explores their use for historians. First, it argues for a narrow definition of "infrastructure" in terms of technologically based networks, as described through the concept of graph theory. Their core function is the movement of goods, people, and information. Second, the concept of "criticality" can be used both in technical-functional terms and in terms of social attribution. When used in a technical-functional way, a positive concept should prevail, highlighting the connecting capacities of a system rather than failure-based criticality. Criticality is a relational term useful for researching the importance of infrastructure and asking questions such as "under which conditions?," "when?," and "to whom?"

Keywords: Criticality, Socio-technical systems, Graph theory, Networks, Organizations, History

The aim of this chapter is to introduce several concepts from interdisciplinary infrastructure research and determine their usefulness for historical research. The point of departure is the term "critical infrastructures." The chapter will address the two components of this term and discuss their background and implications.

Why should historians write about "critical infrastructure"? Obviously, it is not a term that comes from the historical sources. The concept of infrastructure only gradually gained acceptance from the end of the nineteenth

Marton, Silvia, and Andrei-Dan Sorescu, eds. *Infrastructural Development, Corruption, Xenophobia, and Colonization in Central and Southeastern Europe*. Amsterdam: Amsterdam University Press/Central European University Press, 2026.
 https://doi.org/10.5117/9789048577279-4

century; the word was first used in France in 1875 to denote the fixed installations of the railway system. It was only around 1950 that the term came to be used broadly in the way we still use it today.[1] The combination "critical infrastructure" is even more recent. The term came into use in the 1990s, initially in the United States and shortly thereafter in Europe. Public debate on critical infrastructure received a significant boost following the attacks of September 11, 2001, and the subsequent threat of international terrorism. Hurricane Katrina, again in the United States, proved in 2005 how vulnerable technical systems are to natural disasters. After these events, political observers and decision-makers became aware of the precariousness of technical systems and their fundamental importance for the stability of industrialized societies. In English, the term "critical" is well suited to this dual description: infrastructures are critical because modern life would not be conceivable without them; however, they are also critical because they are threatened by various damaging effects and are prone to failure. In quick succession, individual states and then the EU developed binding strategies for the protection of critical infrastructures. The United States adopted the first version of its Critical Infrastructure Protection Program as early as 1998 and updated it in 2003.[2] The European Union followed in 2006, with Germany adopting its national strategy in 2009 and Romania in 2011.[3]

The concept of critical infrastructure is thus quite young, and it is ahistorical insofar as it simply does not appear in sources prior to around 1990. Nevertheless, I argue that it is worth engaging with. There are two reasons for this: first, the concept is indispensable in current public and academic debates on infrastructure. From a pragmatic point of view, it makes sense for us as historians to comment on this and clarify the historical aspect of infrastructure development. For example, we can ask whether critical infrastructures existed before 1900 and what their criticality was. This leads directly to the second, much more important argument for using critical infrastructure in historical research: it is an analytically interesting and fruitful concept. For example, it is possible to discuss whether critical infrastructures are limited to industrialized societies or whether they also exist in other cultures in their capacity as "critical"—that is, both precarious and essential at the same time.

1 Dirk van Laak, "Infra-Strukturgeschichte," *Geschichte und Gesellschaft* 27, no. 3 (2001): 370.

2 United States Government, *The National Strategy for the Physical Protection of Critical Infrastructures and Key Assets* (Washington, DC: White House, 2003).

3 For an overview on the debate in the German language, see Jörn Birkmann et al., *State of the Art der Forschung zur Verwundbarkeit Kritischer Infrastrukturen am Beispiel Strom/Stromausfall, Schriftenreihe Sicherheit 2* (Berlin: Forschungsforum Öffentliche Sicherheit, Freie Universität Berlin, 2010).

What are infrastructures?

At first glance, this question seems easy to answer. If we look at it more deeply, however, there are many unresolved issues to address. The definition given by the German National Strategy for the Protection of Critical Infrastructure reads: "Critical infrastructures are organizations and facilities that are of great importance to the public community (*staatliches Gemeinwesen*), the failure or impairment of which would result in lasting supply shortages, significant disruptions to public safety, or other dramatic consequences."[4]

The definition I prefer is slightly different: infrastructures are networked socio-technical systems that supply physical resources and goods by moving them. One of the main problems with the federal German definition is the broad concept of infrastructure underpinning it. If one lists organizations and facilities, then countless institutions can be included in the notion. According to this strategy, hospitals, schools, and the entire education system are examples of critical infrastructure. Even cultural sites with a central significance for the identity of the population are included. In this sense, the Brandenburg Gate in Berlin is also infrastructure. In more recent developments of the German strategy, there is a tendency to focus more on the services that infrastructures provide to society, meaning that the actual systems take a back seat. This reflects a shift "from infrastructure protection to service continuity."[5] Such a shift makes sense in the context of political governance, where efforts are made to act as comprehensively as possible. However, while governance endeavors to encompass as many areas as possible, scientific research requires concepts that are as clear and definable as possible.

Therefore, we should conceive of infrastructure based on two elements: first, function, and, second, shape. Regarding function, infrastructures are systems that provide for the physical resources, such as energy, transportation, communication transmission, water provision, and waste(-water) disposal. These needs pertain not only to industrial society but also to nearly every human community. Provision means, conversely, that there is a movement of these resources toward, away from, or between users. German historian Dirk van Laak defines infrastructure as "all that is stable and

4 Germany, Federal Ministry of the Interior, *Nationale Strategie zum Schutz Kritischer Infrastrukturen (KRITIS-Strategie)* (Berlin: Bundesministerium des Innern, 2009), 4.

5 Susanne Krings, ed., *10 Jahre KRITIS-Strategie: Einblicke in die Umsetzung der nationalen Strategie zum Schutz Kritischer Infrastrukturen*. Praxis im Bevölkerungsschutz Band 21 (Bonn: Bundesamt für Bevölkerungsschutz und Katastrophenhilfe, 2020), 25. "Vom Schutz der Infrastruktur zur Aufrechterhaltung der Dienstleistung."

needed to enable mobility and the exchange of people, goods and ideas."[6] However, the question remains open as to how "stable" infrastructures really are, or if they are highly dynamic instead.[7] In any case, we can establish that infrastructures move energy, information, goods, and other resources. Second, as concerns form, we should think of infrastructures as technical networks. These networks can be extremely diverse, including highly integrated technical systems—such as electricity and sewer networks—as we have known them since the early twentieth century. However, technical infrastructures can also include much more rudimentary objects, such as the medieval road network, or even networks that are only physically manifested at specific points, such as overseas shipping lines materialized in ports and navigation routes. The components of even these rudimentary networks are interconnected; in this sense, an isolated well does not constitute a networked infrastructure, but a spring that has been tapped and whose water is piped to different users is such a system.

If we want to understand this concept in more abstract terms, we can use graph theory (or network analysis). Graph theory represents networks in the simplest possible way through the use of nodes and edges. These two elements can be used to describe adequately almost any network-based infrastructure.[8] The usefulness of graph theory in historical analysis was recently demonstrated in a study of the history of urban pneumatic tube systems in the first half of the twentieth century in Berlin and Hamburg. The author used it to examine the centrality of different post offices, showing that certain neighborhoods were clearly favored over others. She also used graph theory to demonstrate the difficulties involved when the

6 Dirk van Laak, *Alles im Fluss: Die Lebensadern unserer Gesellschaft—Geschichte und Zukunft der Infrastruktur* (Frankfurt a.M.: S. Fischer, 2018), 13. "Alles Stabile, das notwendig ist, um Mobilität und einen Austausch von Menschen, Gütern und Ideen zu ermöglichen."

7 This is what the author of this chapter maintains together with his colleagues in the TU Darmstadt-based research training group "KRITIS," cf. Jens Ivo Engels et al., *Transformation, Zirkulation, System of Systems Für ein dynamisches Verständnis netzgebundener Infrastrukturen* (Darmstadt: TUprints, 2021), https://doi.org/10.26083/tuprints-00017923.

8 See Christian Stegbauer and Roger Häußling, eds., *Handbuch Netzwerkforschung* (Wiesbaden: Springer VS, 2010), chap. 5.1; Gerrit J. Schenk, "Knoten in Netzen? Überlegungen zur Hafenstadt als Kritische Infrastruktur," in *Flusstäler, Flussschifffahrt, Flusshäfen: Befunde aus Antike und Mittelalter*, ed. Peter Ettel and Achim T. Hack (Mainz: Verlag des Römisch-Germanischen Zentralmuseums, 2019); Stefan Höhne and Michael Naumann, "Infrastruktur. Zur Analyse sozio-technischer Netzwerke zwischen altem und neuem Materialismus," in *Raumproduktionen: Theorie under gesellschaftliche Praxis*, ed. Bernd Belina et al. (Münster: Westfälisches Dampfboot, 2018); Marcus Dombois, *Multilayer Graph Networks for Modeling and Analyzing Exercise Scenarios in Civil Protection and Disaster Response*, ed. Uwe Rüppel (Düren: Shaker Verlag, 2020).

system was developed from a polygonal to a radial shape.[9] The advantage of graph theory is that the nodes can be scaled as desired: one can define the tens of thousands of taps in a municipal freshwater system as nodes, or the substations in an electricity system. One can even further generalize and think of entire infrastructure systems as nodes if they are coupled, such as energy, information and communication systems, and transport in modern urban settings. Thinking of infrastructures as networks enables us to model their (historically increasing) complexity and the interrelatedness of individual components as well as integral systems. When we research the railway system of the late nineteenth century, for instance, we must not underestimate the accompanying challenges and complexities of energy provision. In a certain sense, thinking of infrastructures as complex interconnected systems is the first step toward writing infrastructure history as *histoire totale*.

When it comes to infrastructure, we often talk about the technical system. Our usage of "system" as a concept should be as pragmatic as possible. A "system" refers to two characteristics: the boundary and the interaction. First, a system is defined by its boundaries according to a binary scheme: there are elements and components that belong to the system and those that belong to the environment. The decision as to what lies within the system boundaries is made on the basis of functional affiliation and interaction. All those elements involved in the transport of goods, people, energy, or information belong to the corresponding infrastructure system; all others lie outside the system boundaries. This concept is just as flexible as graph theory: the functional determination can be carried out differently depending on the research project. One can look at the pneumatic tube system of a city or at the postal system of an entire country (into which the urban pneumatic tube system is integrated). In this case, the boundaries are drawn spatially. One can also look at the inner-city communication system, in which the pneumatic tube system, the telephone, the telegraph, and the parcel post form individual subsystems. In this case, the boundaries would have been drawn according to sub-functions. There is a significant overlap between systems theory and graph or network theory, as has been demonstrated by Hartmut Böhme.[10]

9 Cleopatra Schuhmacher, *Die Stadtrohrpost: Geschichte einer Kritischen Infrastruktur in Berlin und Hamburg im 19. und 20. Jahrhundert* (Baden-Baden: Nomos, 2023).

10 Hartmut Böhme, "Einführung: Netzwerke: Zur Theorie und Geschichte einer Konstruktion," in *Netzwerke: Eine Kulturtechnik der Moderne*, ed. Jürgen Barkhoff, Hartmut Böhme, and Jeanne Riou (Cologne: Böhlau, 2004).

Up to this point, my argument has been focused on technology. However, it should be noted that infrastructures should be thought of as socio-technical networks. The concept of socio-technical systems has been developed in science and technology studies since the 1990s and is well established in interdisciplinary technology research.[11] Infrastructures are not only technically complex, but they also consist of governance structures, legal systems, and financing models. They involve practices and have symbolic meaning. Infrastructures embody power and include or exclude social groups. Infrastructures shape sociocultural reality and are shaped by sociocultural influences in turn.

A simple example of the importance of social factors for infrastructure building is the following: there is no "natural" or "objective" solution for the route of a railway line. In very rare cases, certain landscape hurdles—such as mountains or seas—cannot be overcome, making a certain connection objectively impossible; however, in nearly every case, there are many different options. A balancing of different influences, therefore, is responsible for the actual shape of a route. What did the engineers learn in their education (knowledge and training)? Was there a possibility to buy or expropriate land (legal system)? Which route is more cost-effective (economic consideration)? Are political influences decisive for connecting a certain city (power relations)? While this argument is probably self-evident for historians, another aspect of the sociotechnical effects of infrastructures often receives less attention: infrastructures have a facilitation function, but they also have restrictive effects. This means that they can influence social action technically. For example, modern rail transport has allowed the intensive exchange of goods since the mid-nineteenth century and has brought about market integration in large parts of Europe. However, regions that were not connected remained excluded from this integration, leading to a relative loss of prosperity. This exclusion put peripheral regions in a bind: there was hardly any option left but to pursue railway construction. Once decided, railway construction in turn offered potential for conflict because—for instance—the necessary land had to be provided and because

11 Paul N. Edwards, "Infrastructure and Modernity: Force, Time, and Social Organization in the History of Sociotechnical Systems," in *Modernity and Technology*, ed. Thomas J. Misa, Philip Brey, and Andrew Feenberg (Cambridge, 2003); Laur Kanger and Johan Schot, "Deep Transitions: Theorizing the Long-Term Patterns of Socio-Technical Change," *Environmental Innovation and Societal Transitions* 32 (2019): 7–21; Ulrike Felt et al., eds., *The Handbook of Science and Technology Studies*, 4th ed. (Cambridge: MIT Press, 2017); Samuel A. Markolf et al., "Interdependent Infrastructure as Linked Social, Ecological, and Technological Systems (SETSs) to Address Lock-in and Enhance Resilience," *Earth's Future* 6, no. 12 (2018).

the required capital and know-how sometimes had to be imported from Europe's industrial centers. Furthermore, the installed network cemented the preferential settlement of industry near the stations. In this respect, the sociotechnical system of the railway developed a momentum of its own.

Criticality

Criticality is one of the concepts used in current infrastructure policy. Other concepts that often arise in this context are vulnerability, preparedness, and, above all, resilience. All these terms are used both in public debate and in academia. I would like to advocate the application of the concept of criticality in historical research as well. One important aim here is to strip criticality of its strongly normative use.

Before I discuss criticality in more abstract terms below, I would first like to shed light on the historical background to the concept's rise. We saw at the beginning that the protection of critical infrastructures has been put on the agenda in recent decades primarily by governments. It is definitively a state concern, as infrastructures are critical from the point of view of public authorities. This idea is hardly new. States have been involved in infrastructure building since time immemorial. We all know Wittfogel's famous (albeit somewhat outdated) thesis that complex irrigation infrastructures correlate with the development of early statehood, bureaucracy, and power concentration.[12] Yet even beyond such strong arguments, there is enough evidence of authorities' interest in infrastructure construction. Large structures such as the Roman military roads or the early modern Canal du Midi in France were planned or at least used by rulers to increase their fame and reputation[13]—something that offered, to borrow a term from Malte Fuhrmann's contribution to this volume, grounds for infrastructural "boasting" as a public strategy for defending good rule.[14] In our context,

12 Karl August Wittfogel, *Die orientalische Despotie: Eine vergleichende Untersuchung totaler Macht* (Cologne: Kiepenheuer & Witsch, 1962).

13 Chandra Mukerji, "Cartography, Entrepreneurialism, and Power in the Reign of Louis XIV: The Case of the Canal du Midi," in *Merchants & Marvels: Commerce, Science, and Art in Early Modern Europe*, ed. Pamela Smith and Paula Findlen (New York: Routledge, 2002); Jacques Morand, *Le canal du Midi et Pierre-Paul Riquet: Histoire du Canal royal en Languedoc* (Aix-en-Provence: Édisud, 1993).

14 Malte Furhmann, "On Hegemony and Connectivity: A Short History of Infrastructure-Political Struggles and their Discourses" in this volume.

however, we are primarily interested in modern states and their relationship to technical infrastructures.

The early twentieth century was crucial for the present-day relationship between the state and infrastructure. It was then that experts in public law, such as the German Ernst Forsthoff, invented the principle of the state providing public services (*Daseinsvorsorge*). The idea is that the state has a responsibility for the provision of a range of basic services to society—in particular, technical infrastructure.[15] Forsthoff coined the term, but the idea behind it is about half a century older. As early as the decades around 1900, city governments in many European and North American metropolises saw it as their responsibility to guarantee certain technical standards within their jurisdictions. In addition to public transport and energy, modern systems for the supply and disposal of fresh water and wastewater were developed during this period. Public authorities have repeatedly outsourced the construction of infrastructure to private actors, but they still claimed infrastructure development as a public responsibility sustained with two objectives: improving the (hygienic and economic) condition of the population and shaping that population.[16] On the one hand, this meant integrating the population according to what has been termed the "modern infrastructure ideal" or "municipal socialism."[17] This was largely about giving socially weaker classes access to modern infrastructure, which was financed collectively, and about the spatial integration of cities so that, ideally, all neighborhoods would be provided the same standard of technical infrastructure. However, what was often meant to integrate actually turned out to fragment or to divide.[18] On the other hand, infrastructure was built with social engineering aims: even a prosaic undertaking such as the New York subway aimed to influence the everyday behavior of its users and

15 Van Laak, *Alles im Fluss*, chap. 2. Incidentally, it should not be forgotten that Forsthoff's concept was developed under the conditions of Nazi rule. Forsthoff saw in the citizens' claim to *Daseinsvorsorge* vis-à-vis the state a kind of counterpart to the loss of political rights.

16 Dirk van Laak, "Technological Infrastructure, Concepts and Consequences," *ICON: Journal of the International Committee for the History of Technology* 10 (2004): 53–64.

17 Olivier Coutard, "Placing Splintering Urbanism: Introduction," *Geoforum* 39, no. 6 (2008); Timothy Moss, "'Cold Spots' of Urban Infrastructure: 'Shrinking' Processes in Eastern Germany and the Modern Infrastructural Ideal," *International Journal of Urban and Regional Research* 32, no. 2 (2008).

18 Jens Ivo Engels, "Infrastructure and Fragmentation: The Limits of the Integration Paradigm," in *Linking Networks: The Formation of Common Standards and Visions for Infrastructure Development*, ed. Martin Schiefelbusch and Hans-Liudger Dienel (Farnham: Ashgate, 2014).

thus educate a "civilized" urban population.[19] This civilizing mission of infrastructure was by no means limited to the Western Hemisphere; it was also a very important motivation for building railways in the late Tsarist Empire. Frithjof Schenk, a historian of Eastern Europe, speaks of a "project of socio-spatial ordering." Some reformers even hoped that building the railways would help combat alcohol abuse and corruption. Of course, this utopia failed.[20] Similar patterns can be found in the colonial context. In the French colony of Senegal, the introduction of European time measurement using modern communication infrastructure—the telegraph—was also intended to lead to a European "civilized" and, above all, "disciplined" life for all inhabitants.[21] Since the mid-twentieth century, disciplinary fantasies in Western societies have eroded. It remained the responsibility of the public sector to create the conditions for economic development and a high standard of living for the population by constructing infrastructure. In recent decades, additional responsibility has been added, such as providing society with climate-neutral energy or environmentally friendly transportation. These remarks are intended only to explain why state actors speak of critical infrastructure at all. It illustrates why infrastructure is also "critical" to the state itself: if it fails or is threatened, part of the state's *raison d'être* is called into question—again, to borrow from Fuhrmann, something that might translate into infrastructural "shaming" on the part of the public as a form of contestation.[22] We will now turn to the term as a (scientifically usable) category of analysis.[23]

Let us start, again, from the official definitions given by various states. The 2003 US strategy for the protection of infrastructure cites the Patriot Act in stating that critical infrastructures are "systems ... so vital to the United

19 Stefan Höhne, *New York City Subway: Die Erfindung des urbanen Passagiers* (Cologne: Böhlau, 2017).

20 Frithjof Benjamin Schenk, "Im Kampf um Recht und Ordnung: Zivilisatorische Mission und Chaos auf den Eisenbahnen im Zarenreich," in *Neue Wege in ein neues Europa: Geschichte und Verkehr im 20. Jahrhundert*, ed. Ralf Roth and Karl Schlögel (Frankfurt a.M.: Campus, 2009), 200.

21 Sebastian-Manès Sprute, "U(H)Reigene Zeiten: Grenzen der Implementierung von europäischen Zeitnormen in Senegal, 1890–1930," in *Der Faktor Zeit: Perspektiven kulturwissenschaftlicher Zeitforschung*, ed. Katja Patzel-Mattern and Albrecht Franz (Stuttgart: Franz Steiner Verlag, 2015), 77–105.

22 Fuhrmann, "On Hegemony and Connectivity."

23 In the paragraph on criticality, I am relying mainly on two publications: Jens Ivo Engels, "Relevante Beziehungen: Vom Nutzen des Kritikalitätskonzepts für Geisteswissenschaftler," in *Was heißt Kritikalität? Zu einem Schlüsselbegriff der Debatte um Kritische Infrastrukturen*, ed. Jens Ivo Engels and Alfred Nordmann (Bielefeld: transcript, 2018); Kristof Lukitsch, Marcel Müller, and Chris Stahlhut, "Criticality," in *Key Concepts for Critical Infrastructure Research*, ed. Jens Ivo Engels (Wiesbaden: Springer VS, 2018).

States that the incapacity or destruction of such systems and assets would have a debilitating impact."[24] The German strategy, which I have already quoted above, reads very similarly: "Criticality is a relative measure of the importance of infrastructure in terms of the consequences of disruption or failure for the safety and security of the supply of essential goods and services to society."[25] However, there has been important criticism of the state-dominated concepts of criticality, as put forth by Sara Bouchon:

> The given criticality criteria leave problematic room for interpretation. They are limited to the standpoint of the authorities, while the criticality of an infrastructure might be different for the owner, insurer, or other stakeholder. They refer as well to a national scale, while the criticality might be different at regional or local level. Finally, they are valid at a certain time, but they might evolve, as political concerns evolve themselves.[26]

From all this, we can derive several features of criticality.

First, criticality is a property, usually of a system or an infrastructure.

Second, criticality is a relational property. It is measured in relation to the surrounding society or state, for which the infrastructure fulfills a function. At this point, however, we need to make an important distinction. In the scholarly literature, we speak of two different dimensions of relational criticality: system-based and consequence-based criticality. These represent the internal and external effects of an infrastructure.[27] System-based criticality means that certain components of a system are indispensable to the system itself. Consequence-based criticality aims to describe the importance of the system to its environment; it is therefore concerned with the significance of system boundaries. A switch is critical to the operation of a particular railway line. However, this individual switch is not critical to the economy of the cities connected by the line; what is critical for the economy of the cities is the transport system as a whole.

The third finding is that criticality is constructed. And this constructive character has two different dimensions as well. On the one hand, criticality

24 United States Government, *The National Strategy*, 6.

25 Bundesministerium des Innern, *Nationale Strategie*, 7.

26 Sara Bouchon, *The Vulnerability of Interdependent Critical Infrastructures Systems: Epistemological and Conceptual State-of-the-Art* (Ispra: Institute for the Protection and Security of the Citizen, 2006), 43.

27 Cf. Lukitsch, Müller, and Stahlhut, "Criticality"; Matthew J. Egan, "Anticipating Future Vulnerability: Defining Characteristics of Increasingly Critical Infrastructure-Like Systems," *Journal of Contingencies and Crisis Management* 15, no. 1 (2007): 4–17.

is constructed in the sense of culture; it is ascribed to a system by the government, by its users, or by public debate.[28] On the latter point—to refer to the introductory remarks of Silvia Marton in the present volume—a public itself may be said to be constructed around and through these very controversies.[29] Its size and rhetoric reflect the values and expectations of the society in question, or at least of those sections that possess the political capital and leverage to make it matter, either through the pressure they may apply upon stakeholders or, at the very least, as the object of the stakeholders' rhetorical invocations. The concept of criticality can be used for historical research, since users, rulers, or societies have always ascribed meaning to infrastructures in the past. However, in addition to the cultural construction of criticality, there is a technological construction. It is also man-made but on a different level: technical systems are always built on their components. Or, consider the coupling of systems: if a railway system relies entirely on electric locomotives, then the electricity supply is critical to the operation of the railway, regardless of what journalists ascribe to it.

The fourth dimension is that criticality is relative. Criticality is not an absolute value; rather, it is the result of comparisons. This is part of the relational nature of criticality, since it is possible to assign a higher criticality to one system and a lower criticality to another. This also depends on the functions performed by a system. In terms of consequence-based criticality, waterways and railways were more critical than lorries for the transport of manufactured goods in Central Europe around 1900. In the case of system-related criticality, it is possible to determine which components are more important to the stability of the system than others. Graph theory is useful here: central nodes are more critical than peripheral ones; a large power station is more critical than a local substation or a single power line.

As historians, we can draw a fifth lesson from this: the criticality of a system changes over time, depending on social or economic demand and on technological change. The criticality of canal transport for the industrial societies of Central Europe decreased with the expansion of the road system, but also as a result of the tertiarization of the economy as a whole. On the one hand, new technical possibilities, such as motorways, have shifted the transport of goods from canals to roads. On the other hand, services have become more important

28 Andreas Folkers, "Was Ist kritisch an Kritischer Infrastruktur? Kriegswichtigkeit, Lebenswichtigkeit, Systemwichtigkeit und die Infrastrukturen der Kritik," in *Was heißt Kritikalität? Zu einem Schlüsselbegriff der Debatte um Kritische Infrastrukturen*, ed. Jens Ivo Engels and Alfred Nordmann (Bielefeld: transcript, 2018), 124–125.

29 Silvia Marton, "Transportation Infrastructure and Corruption Scandals" in this volume.

than the production of goods, so the transport of heavy industrial products has lost relative importance. As some of these goods are now imported from overseas, the criticality of sea transport has increased. This was demonstrated during the COVID-19 pandemic when global supply chains were disrupted. The criticality of container shipping between Asia and Europe for the supply of medicines to the "Old Continent" is a recent phenomenon.

In the context of time, a consideration put forward Alexander Fekete is relevant. He identifies several criticality factors, including time, proportion, and quality.[30] At this point, I would like to expand on the topic of time only: the consequences of an infrastructure failure differ depending on when it occurs. This can be modeled in the short term: gas supply is more critical on a January day than in July because nobody needs to turn on their heating in July. But it can also be applied to different historical periods; in a digitized society, the failure of communication infrastructure has a different impact than it did for societies around 1900. For the purpose of assessing the variability of criticality over time, the concept of "dynamic criticality" has been coined.[31] The concept attempts to show that all elements of an infrastructure are constantly changing in their relationships and, therefore, in their respective criticality to each other. The same applies to the relationship between infrastructure and society; underlying this statement is the realization that infrastructure develops dynamically.[32] In this respect, Dirk van Laak's statement quoted above is not entirely accurate. Infrastructure may appear relatively stable at first sight, but on closer inspection, it turns out to be constantly changing. Another important factor is the influence of temporal social contexts on criticality as an attribution. Different societies ascribe different criticality to the same infrastructure, and this can change over time. The idea of comprehensive public services provided by the state did not yet exist in the early nineteenth century; therefore, infrastructures were critical in a different sense than they are today.

Some critiques of policy strategies on criticality have pointed to the fact that they are often overly focused on infrastructure's potential for failure, as they tend to define criticality in terms of the estimated resulting

30 Alexander Fekete, "Common Criteria for the Assessment of Critical Infrastructures," *International Journal of Disaster Risk Science* 2 (2011), 18.

31 Leon Hempel, Benjamin D. Kraff, and Robert Pelzer, "Dynamic Interdependencies: Problematising Criticality Assessment in the Light of Cascading Effects," *International Journal of Disaster Risk Reduction* 30, Part B (2018). Cf. also George Stergiopoulos et al., "Time-Based Critical Infrastructure Dependency Analysis for Large-Scale and Cross-Sectoral Failures," *International Journal of Critical Infrastructure Protection* 21 (2016).

32 Engels et al., *Transformation, Zirkulation, System.*

damage. This could be described as a negative attribution of criticality. The dominance of this negative view of criticality is likely due to the fact that, in the years after 2001, concepts of protection were often developed against the background of the terrorist threat. Consequently, a security mentality was responsible for the concept of criticality at that time.[33] However, this negative conception can easily be juxtaposed with a positive notion of criticality. If we think of infrastructures and their importance to society in terms of graph theory, then the centrality of a component or subsystem determines its criticality. The more connections a component has, the more critical it is to the system. The greater the number of "relevant relations" a component has— whether within a technical system or between a system and its environment—the higher its criticality.[34]

The positive conception of criticality is particularly important for historians. If criticality is defined in terms of the consequences of infrastructure failure, then the level of criticality attributed to the infrastructure in question is confirmed with certainty only when a disaster occurs. Only when the infrastructure actually fails do affected people or organizations produce relevant sources that show the extent to which the infrastructure is critical to society. In most cases, however, the infrastructure does not fail completely, either because the failure is anticipated and adequate protection strategies are developed, or because the effects are mitigated. If a positive approach of identifying relevant relationships is taken, the problem does not arise. However, these limitations apply in particular to the aspect of the technological construction of criticality; the study of the social attribution of criticality remains independent of this.

Some recent empirical studies have shown that it is worthwhile to use criticality in historiography. These include studies of industrialized societies as well as research on the preindustrial period. For example, Benedikt Vianden has highlighted the criticality of transport infrastructure for Protestant Christian settlers in Palestine at the end of the nineteenth century. From the perspective of Protestant immigrants from Europe, modern transport was supposed to contribute to the "advancement" of local civilization. At the same time, transport services operating on this infrastructure were an important source of income for parts of the community.[35] Medieval studies have also

33 Andreas Folkers, "Kritische Infrastruktur," in *Ortsregister: Ein Glossar zu Räumen der Gegenwart*, ed. Nadine Marquardt and Verena Schreiber (Bielefeld: transcript, 2012).

34 This is the main argument in Engels, "Relevante Beziehungen."

35 Benedikt Friedhelm Vianden, *Ebnet in der Steppe eine Straße für unseren Gott! Das Infrastrukturnetz der Tempelkolonien im osmanischen Palästina* (Darmstadt: wbg Academic, 2022).

made profitable use of the concept. Gerrit Schenk has used criticality to examine the significance of local and regional water infrastructures in the Upper Rhine region in the late Middle Ages. As suggested here, he has applied the distinction between attributed and functional criticality, as well as the distinction between system-based and consequence-based criticality. With this perspective, he was able to describe the systemic character of the extensive canal systems in this region and to present a nuanced picture of their importance to the societies located there, including significant dependencies on these systems.[36]

The chapter has aimed to show that several concepts from the social sciences and technical infrastructure research can be useful for historians. It argues for a narrow definition of "infrastructure" in terms of technologically based networks—as described through the concept of graph theory—whose core function is the movement of goods, people, and information. The concept of criticality is also useful, offering the advantage that it can be used both in technical-functional terms and in terms of social attribution. When used in a technical-functional way, a positive concept—determining the number and importance of an infrastructure's relevant connections—should be employed instead of a negative concept of criticality—one which focuses on the consequences of failure. This approach avoids the danger of counterfactual historiography and the tendency to securitize historical infrastructure research. Moreover, criticality is a relational term useful for answering questions such as: Under what conditions is an infrastructure important? When is it important? And to whom?

Bibliography

Birkmann, Jörn, Claudia Bach, Silvie Guhl, Maximilian Witting, Torsten Welle, and Miron Schmude. *State of the Art der Forschung zur Verwundbarkeit Kritischer Infrastrukturen am Beispiel Strom/Stromausfall.* Schriftenreihe Sicherheit 2. Berlin: Forschungsforum Öffentliche Sicherheit, Freie Universität Berlin, 2010.

36 Gerrit J. Schenk, "Der Straßen halb uff der Yll. Wasserwege im Elsaß als Kritische Infrastruktur für Wirtschaft und Gesellschaft (1350–1550)," in *Weg und Steg: Aspekte des Verkehrswesens von der Spätantike bis zum Ende des Alten Reiches*, ed. Kurt Andermann and Nina Gallion (Ostfildern: Thorbecke, 2018); Schenk, "Knoten in Netzen"; Gerrit J. Schenk, "Das Baby in der Wiege: Hochwasser zwischen Alltag und Katastrophe (ca. 1250–1550): Beobachtungen zur 'fluvialen Anthroposphäre' Im Kraichgau," in *Wasser: Ressource—Gefahr—Leben*, ed. Kurt Andermann and Gerrit J. Schenk (Ostfildern: Thorbecke, 2020).

Böhme, Hartmut. "Einführung: Netzwerke: Zur Theorie und Geschichte einer Konstruktion." In *Netzwerke: Eine Kulturtechnik der Moderne*, edited by Jürgen Barkhoff, Hartmut Böhme, and Jeanne Riou, 17–36. Cologne: Böhlau, 2004.

Bouchon, Sara. *The Vulnerability of Interdependent Critical Infrastructures Systems: Epistemological and Conceptual State-of-the-Art*. Ispra: Institute for the Protection and Security of the Citizen, 2006.

Bundesministerium des Innern. *Nationale Strategie zum Schutz Kritischer Infrastrukturen (KRITIS-Strategie)*. Berlin: Bundesministerium des Innern, 2009.

Coutard, Olivier. "Placing Splintering Urbanism: Introduction." *Geoforum* 39, no. 6 (2008): 1815–1820.

Dombois, Marcus. *Multilayer Graph Networks for Modeling and Analyzing Exercise Scenarios in Civil Protection and Disaster Response*. Edited by Uwe Rüppel. Düren: Shaker Verlag, 2020.

Edwards, Paul N. "Infrastructure and Modernity: Force, Time, and Social Organization in the History of Sociotechnical Systems." In *Modernity and Technology*, edited by Thomas J. Misa, Philip Brey, and Andrew Feenberg, 185–225. Cambridge, 2003.

Egan, Matthew Jude. "Anticipating Future Vulnerability: Defining Characteristics of Increasingly Critical Infrastructure-Like Systems." *Journal of Contingencies and Crisis Management* 15, no. 1 (2007): 4–17. https://doi.org/10.1111/j.1468-5973.2007.00500.x.

Engels, Jens Ivo. "Infrastructure and Fragmentation: The Limits of the Integration Paradigm." In *Linking Networks: The Formation of Common Standards and Visions for Infrastructure Development*, edited by Martin Schiefelbusch and Hans-Liudger Dienel, 19–33. Farnham: Ashgate, 2014.

Engels, Jens Ivo. "Relevante Beziehungen: Vom Nutzen des Kritikalitätskonzepts für Geisteswissenschaftler." In *Was heißt Kritikalität? Zu einem Schlüsselbegriff der Debatte um Kritische Infrastrukturen*, edited by Jens Ivo Engels and Alfred Nordmann, 17–45. Bielefeld: transcript, 2018.

Engels, Jens Ivo, Sybille Frank, Iryna Gurevych, Martina Heßler, Michèle Knodt, Jochen Monstadt, Alfred Nordmann et al. *Transformation, Zirkulation, System of Systems: Für ein dynamisches Verständnis netzgebundener Infrastrukturen*. Darmstadt: TUprints, 2021. https://doi.org/10.26083/tuprints-00017923.

Fekete, Alexander. "Common Criteria for the Assessment of Critical Infrastructures." *International Journal of Disaster Risk Science* 2 (2011): 15–24. https://doi.org/10.1007/s13753-011-0002-y.

Felt, Ulrike, Rayvon Fouché, Clark A. Miller, and Laurel Smith-Doerr, eds. *The Handbook of Science and Technology Studies*. 4th ed. Cambridge: MIT Press, 2017.

Folkers, Andreas. "Kritische Infrastruktur." In *Ortsregister: Ein Glossar zu Räumen der Gegenwart*, edited by Nadine Marquardt and Verena Schreiber, 154–159. Bielefeld: transcript, 2012.

Folkers, Andreas. "Was ist kritisch an Kritischer Infrastruktur? Kriegswichtigkeit, Lebenswichtigkeit, Systemwichtigkeit und die Infrastrukturen der Kritik." In *Was heißt Kritikalität? Zu einem Schlüsselbegriff der Debatte um Kritische Infrastrukturen*, edited by Jens Ivo Engels and Alfred Nordmann, 123–154. Bielefeld: transcript, 2018.

Hempel, Leon, Benjamin D. Kraff, and Robert Pelzer. "Dynamic Interdependencies: Problematising Criticality Assessment in the Light of Cascading Effects." *International Journal of Disaster Risk Reduction* 30, Part B (2018): 257–268. https://doi.org/10.1016/j.ijdrr.2018.04.011.

Höhne, Stefan. *New York City Subway: Die Erfindung des urbanen Passagiers.* Cologne: Böhlau, 2017.

Höhne, Stefan, and Michael Naumann. "Infrastruktur: Zur Analyse sozio-technischer Netzwerke zwischen altem und neuem Materialismus." In *Raumproduktionen: Theorie under gesellschaftliche Praxis*, edited by Bernd Belina, Johanna Hoerning, Henrik Lebuhn, and Boris Michel, 34–52. Münster: Westfälisches Dampfboot, 2018.

Kanger, Laur, and Johan Schot. "Deep Transitions: Theorizing the Long-Term Patterns of Socio-Technical Change." *Environmental Innovation and Societal Transitions* 32 (2019): 7–21. https://doi.org/10.1016/j.eist.2018.07.006.

Krings, Susanne, ed. *10 Jahre KRITIS-Strategie: Einblicke in die Umsetzung der nationalen Strategie zum Schutz Kritischer Infrastrukturen.* Praxis im Bevölkerungsschutz Band 21. Bonn: Bundesamt für Bevölkerungsschutz und Katastrophenhilfe, 2020.

Lukitsch, Kristof, Marcel Müller, and Chris Stahlhut. "Criticality." In *Key Concepts for Critical Infrastructure Research*, edited by Jens Ivo Engels, 11–20. Wiesbaden: Springer VS, 2018.

Markolf, Samuel A., Mikhail V. Chester, Daniel A. Eisenberg, David M. Iwaniec, Cliff I. Davidson, Rae Zimmerman, Thaddeus R. Miller, Benjamin L. Ruddell, and Heejun Chang. "Interdependent Infrastructure as Linked Social, Ecological, and Technological Systems (SETSs) to Address Lock-in and Enhance Resilience." *Earth's Future* 6, no. 12 (2018): 1638–1659.

Morand, Jacques. *Le canal du Midi et Pierre-Paul Riquet: Histoire du Canal royal en Languedoc.* Aix-en-Provence: Édisud, 1993.

Moss, Timothy. "'Cold Spots' of Urban Infrastructure: 'Shrinking' Processes in Eastern Germany and the Modern Infrastructural Ideal." *International Journal of Urban and Regional Research* 32, no. 2 (2008): 436–451. https://doi.org/10.1111/j.1468–2427.2008.00790.x.

Mukerji, Chandra. "Cartography, Entrepreneurialism, and Power in the Reign of Louis XIV: The Case of the Canal du Midi." In *Merchants & Marvels: Commerce, Science, and Art in Early Modern Europe*, edited by Pamela H. Smith and Paula Findlen, 248–276. New York: Routledge, 2002.

Schenk, Frithjof Benjamin. "Im Kampf um Recht und Ordnung: Zivilisatorische Mission und Chaos auf den Eisenbahnen im Zarenreich." In *Neue Wege in ein neues Europa: Geschichte und Verkehr im 20. Jahrhundert*, edited by Ralf Roth and Karl Schlögel, 197–221. Frankfurt a. M.: Campus, 2009.

Schenk, Gerrit J. "Das Baby in der Wiege: Hochwasser zwischen Alltag und Katastrophe (ca. 1250–1550): Beobachtungen zur 'fluvialen Anthroposphäre' im Kraichgau." In *Wasser: Ressource—Gefahr—Leben*, edited by Kurt Andermann and Gerrit J. Schenk, 79–113. Ostfildern: Thorbecke, 2020.

Schenk, Gerrit J. "Der Straßen halb uff der Yll. Wasserwege im Elsaß als 'Kritische Infrastruktur' für Wirtschaft und Gesellschaft (1350–1550)." In *Weg und Steg: Aspekte des Verkehrswesens von der Spätantike bis zum Ende des Alten Reiches*, edited by Kurt Andermann and Nina Gallion, 121–152. Ostfildern: Thorbecke, 2018.

Schenk, Gerrit J. "Knoten in Netzen? Überlegungen zur Hafenstadt als Kritische Infrastruktur." In *Flusstäler, Flussschifffahrt, Flusshäfen: Befunde aus Antike und Mittelalter*, edited by Peter Ettel and Achim T. Hack, 207–236. Mainz: Verlag des Römisch-Germanischen Zentralmuseums, 2019.

Schuhmacher, Cleopatra. *Die Stadtrohrpost: Geschichte einer Kritischen Infrastruktur in Berlin und Hamburg im 19. und 20. Jahrhundert*. Baden-Baden: Nomos, 2023.

Sprute, Sebastian-Manès. "U(H)Reigene Zeiten: Grenzen der Implementierung von europäischen Zeitnormen in Senegal, 1890–1930." In *Der Faktor Zeit: Perspektiven kulturwissenschaftlicher Zeitforschung*, edited by Katja Patzel-Mattern and Albrecht Franz, 77–105. Stuttgart: Franz Steiner Verlag, 2015.

Stegbauer, Christian, and Roger Häußling, eds. *Handbuch Netzwerkforschung*. Wiesbaden: Springer VS, 2010.

Stergiopoulos, George, Panayiotis Kotzanikolaou, Marianthi Theocharidou, Georgia Lykou, and Dimitris Gritzalis. "Time-Based Critical Infrastructure Dependency Analysis for Large-Scale and Cross-Sectoral Failures." *International Journal of Critical Infrastructure Protection* 21 (2016): 46–60. https://doi.org/10.1016/j.ijcip.2015.12.002.

United States Government. *The National Strategy for the Physical Protection of Critical Infrastructures and Key Assets*. Washington, DC: White House, 2003.

Van Laak, Dirk. *Alles im Fluss: Die Lebensadern unserer Gesellschaft—Geschichte und Zukunft der Infrastruktur*. Frankfurt a. M.: S. Fischer, 2018.

Van Laak, Dirk. "Infra-Strukturgeschichte." *Geschichte und Gesellschaft* 27, no. 3 (2001): 367–393. https://doi.org/10.14765/zzf.dok-2215.

Van Laak, Dirk. "Technological Infrastructure, Concepts and Consequences." *ICON: Journal of the International Committee for the History of Technology* 10 (2004): 53–64.

Vianden, Benedikt Friedhelm. *Ebnet in der Steppe eine Straße für unseren Gott! Das Infrastrukturnetz der Tempelkolonien im osmanischen Palästina*. Darmstadt: wbg Academic, 2022.

Wittfogel, Karl August. *Die orientalische Despotie: Eine vergleichende Untersuchung totaler Macht*. Cologne: Kiepenheuer & Witsch, 1962.

About the Author

Jens Ivo Engels is full professor of modern and contemporary history at Technische Universität Darmstadt. His scientific interests include the history and theory of critical infrastructures, corruption in Europe since the eighteenth century, environmental history, and popular culture in the late twentieth century. From 2016 to 2025 he was the director of the interdisciplinary research training group "Kritis" at TU Darmstadt.

4. The Infrastructural Power of the "Colonial" as a Concept

Andrei-Dan Sorescu

Abstract: This chapter offers a critical reflection on how the "colonial" was a surprisingly crucial category for historical actors in nineteenth-century Central and Southeastern Europe. Looking at the infrastructural importance of "anti-colonial" discourse in Romanian nationalism, it seeks to tease out how that discourse differed from more typical definitions by showing that (1) it was a secondary development rather than a primary impetus born from actual colonial relations; (2) it was less burdened by the need to culturally disentangle from the influence of a metropole and more concerned with identifying one worthy of emulation; and (3) it was a process paradoxically amplified, rather than ended, by Romanian independence. In so doing, the chapter calls attention to the importance of temporality and timing for imagining the "(anti-)colonial" as a feature of nationalism.

Keywords: Anti-colonial nationalism, Colonialism, Xenophobia, Othering, Romania

Introduction: A Moment of Colonial Anxiety

For months on end, between 1894 and 1895, the Romanian reading public found itself in the grips of a heated debate over the introduction of a new law regulating mining. As one appeal for a grand public rally in Bucharest in late 1894 phrased it, at stake for the opponents of the bill was "not a mere organic law regulating work and production in our country, as its authors claim, or a new impulse for the growth of national wealth—but [...] *the colonization*

Marton, Silvia, and Andrei-Dan Sorescu, eds. *Infrastructural Development, Corruption, Xenophobia, and Colonization in Central and Southeastern Europe*. Amsterdam: Amsterdam University Press/Central European University Press, 2026.
 https://doi.org/10.5117/9789048577279-5

of the country with foreigners—an attack on property—rural estates handed over to foreigners—the inalienability of peasants' lands nullified" (italics in original).[1] This dramatic checklist of anti-constitutional provisions, as interpreted by the Liberal opposition, was the presumed consequence of granting foreign investors access to and ownership of resources, the mandated establishment of a system of welfare provisions for immigrant laborers, and the future collusion between the state and the foreign capitalists providing infrastructural capacity and know-how for extraction.

That drafting such a law had already repeatedly been deferred by that point was a sign of its sensitive nature. Nowhere was the rhetorical potential of this nexus of anxieties over alienating the nation's land and resources more artfully mobilized than in a pamphlet published anonymously by writer and lawyer Barbu Ştefănescu Delavrancea (1858–1917), a rising star of the parliamentary tribune. The crux of Delavrancea's plea was that the key constitutional provisions regulating land ownership and forbidding foreign colonization were the expression of an organic impulse: "a sentence passed by the people, a historical energy, a mystery of our national existence."[2] It was not a hatred of foreigners that motivated this, the author insisted, but that a national monopoly on the national territory was to be defended only "so as to be what we are meant to be, as our race, history, and place in the world demand it." Regardless of the wealth that might still otherwise lie hidden underground, "[this land] has been fertilized on its surface by the blood of our kin and is fortified from beneath by the bones of our ancestors."[3] Yet, just as dangerous as the potential theft of the nation's resources seemed "the undoing of a centuries-long prohibition—the creation of foreign citadels among our villages, citadels of hundreds and thousands of souls, as needed for the building of galleries, wells, basins, dikes, dams, conduits, roads, canals, bridges, railroads, and buildings for men, cattle, and machines."[4] And this, Delavrancea further added, meant that any "vagabond" could lay a claim to the land that former serfs had at long last received through the great Land Reform of 1864.[5] A list of infrastructural components could, depending on the framing, read as an ode to progress—or, conversely, as an ominous intimation of infrastructural violence.[6]

1 "Lupta cea mare," *Voinţa Naţională*, December 18, 1894, 1.

2 [Barbu Ştefănescu Delavrancea, attrib.], *Români...*, [n. p.] Bucharest, 1894, 6–7.

3 [Delavrancea, attrib.], *Români...*

4 [Delavrancea, attrib.], *Români...*

5 [Delavrancea, attrib.], *Români...*, 13–15.

6 On lists as "ontographies," see Ian Bogost, *Alien Phenomenology, or, What It's Like to Be a Thing* (Minneapolis: University of Minnesota Press, 2012).

All of the above resembled and echoed arguments first advanced in the 1860s and 1870s. Fears of German settler colonies taking over the countryside, or of Prussian and Austrian railway concessions spelling bankruptcy and subjugation, were very much within recent memory and provided a backdrop for the reflections on private capital and national development that mining legislation brought to the fore.[7] Now, however, the added twist was that infrastructure did not transport invading colonists but required "colonization" for its very construction, and that land would be taken out of the nation's hands for the profit of foreign capitalists. Small wonder, then, that Gr. Vulturescu (1839–1913), former head of the Statistical Service, saw fit to reprint a fragment from the 1861 *Coloniștii germani și România* [German Colonists and Romania], authored by his predecessor, D. P. Marțian (1829–1865), as a companion piece to a pamphlet of his own.[8] As one commentator in the country's main law review noted, what had given rise to these anxieties anew had been, in fact, an article of the legislative proposal that stipulated the establishment of welfare and pensions for laborers, which in and of itself implied a critical mass of contributions from a presumably immigrant workforce.[9]

What these debates captured was the discursive convergence of capital, civilization, colonization, and national interest when major infrastructural projects were at stake, as illustrated by an exchange in a Senate debate on February 6, 1895. On the one hand, former radical and anti-monarchist George Panu (1848–1910), now in the Conservative camp, found colonization anxieties exaggerated. After all, he wondered, how else but with the aid of trained foreign personnel could work begin? Besides, talk of "treason" against nation and constitution was cheap. Had France not recently set out on an expedition to enforce a protectorate over Madagascar and made use of British ships for logistics, public scandals notwithstanding? "What was France to do? Give up on her expedition to Madagascar, only to listen to the cries of the chauvinists? Obviously not." Panu's example was explicitly drawn from the experience of a country readily taken as a model by Romania, and the lesson here was that even an empire would sometimes find itself

7 See Andrei Sorescu, "The Perils (and Promise) of German Colonization: Civilizational Hierarchies and Anxieties in Nineteenth-Century Romania," *East European Politics and Societies* 39, no. 3 (2025): 717–740.

8 Gr. Vulturescu, *Legea minelor: Respuns D-lui G. G. Disescu. Publicat în ziarul* Voinţa Naţională *din 1895 cu un Apendice coprindend parte din lucrarea lui D. P. Marţian. Coloniştii Germani şi România* (Bucharest: Tip. Gutenberg, Joseph Göbl, 1895).

9 Alexander, "Două cuvinte asupra legii minelor (urmare și fine)," *Dreptul* 24, no. 11, February 5, 1895, 89–93.

embarrassed in its expansion by infrastructural constraints. The answer was to forge ahead and ignore accusations of "treason" when the greater good, understood as economic progress, was at stake. Panu then continued with a jab against the kind of arguments found in Delavrancea's pamphlet: "Take, for instance, the Kyrgyz ['ckirchijii,' sic!] or the Lapps, who of course show such piety toward their ancestors that they do not want their bones to be plowed over or their blood to be profaned—and this is why they do not have a mining law of their own. (Prolonged applause.)"[10] Clearly, for Panu there was little room for sympathy with peoples who retained a supposedly atavistic attachment to the land of their ancestors, failing to see what lay beneath it as pragmatic resources for development. Romanian fears of colonization could, paradoxically, doom the country to colonial backwardness by denying it the blessings of infrastructural expansion.

To this, one rebuttal came from Liberal economist P. S. Aurelian (1833–1909), no stranger to the discourse of colonial anxiety in the 1870s. Aurelian, too, turned to the experience of others, yet for the purpose of drawing opposing conclusions. In Switzerland, he pleaded, even relocating between cantons required local approval—why, then, should Romania be cavalier about importing a workforce? And, in the United States, did the Homestead Act not come with a legal guarantee for the land transferred under the ownership of settlers' families—why, then, uproot Romanians for the sake of eminent domain? The mining law, Aurelian pleaded, threatened to bring an invasion of "foreign microbes" who, even if dispersed throughout the country rather than concentrated in mining colonies, could only bring contagion. And besides, he further warned, the experience of Jewish immigration—readily connoted as corrupting and depleting for Romania—was proof enough that the country could not afford to open its gates to foreigners on grounds of economic self-preservation.[11] The capillary reach of infrastructure across the spatialized body politic carried potential dangers.

The crux of the issue, as understood by historical actors themselves, was a tension between the imperative of economic development via infrastructural expansion, on the one hand, and the enforcement of a monopoly on the nation's territory and resources, on the other. This, as already pointed out, was not a new discursive configuration for the Romanian literate public by the late nineteenth century: xenophobic fears connoting German or Jewish presence as "colonial" provided easily understood talking points.

10 G. Panu, February 6, 1895, *Desbaterile Senatului. Sesiunea ordinară 1894-1895* (Bucharest: Imprimeria Statului, 1895), 383.

11 P. S. Aurelian, February 6, 1895, *Desbaterile Senatului*, 385–386.

Rather, this opening vignette allows us to home in on how xenophobia(s) and colonial anxieties can be co-constitutive. Moreover, it equally shows how fears over the unintended consequences of infrastructural development drew on a colonial discourse that was itself infrastructural in how it underpinned nationalist reflection on resource sovereignty and more. Let us note how both Delavrancea and Panu dismissed xenophobia as a pernicious attitude, yet they found themselves deploying essentialist and exclusionary arguments to strengthen their diverging rhetorical positions. At the same time, the broader archive of empire was a useful point of reference for both those who stoked fears of colonial encroachment and those who dismissed them, yet without giving much thought to *who* was dispossessed either by the invasion and ultimate colonial takeover of Madagascar or by America's westward push to fulfill "manifest destiny."

What contributed to the sense of urgency in our case was infrastructural expansion: infrastructure—broadly understood in both material and symbolic terms as a simultaneously constructive and disruptive presence—was (and still remains) an attractor for scandals and contestation. Infrastructure catalyzes debate, not just in terms of how resources for its construction or upkeep are (mis)allocated but also as a conduit for the extraction of resources and for the arrival of expertise, of human and financial capital, and the entanglements of credit and debt. That all of this could be seen to entail perceivedly "colonial" relationships and hierarchies of power, in turn, may be understood in its historical context as the echo of how imperial projects of infrastructural expansion strengthened the grip of domination elsewhere in more properly colonial settings. As the present volume examines, colonial anxieties were deeply intertwined with xenophobia in terms of how potentially colonizing Others came to be feared and stigmatized—but also with "corruption," as a category that is itself historicized, in terms of the effects that said Others and their subversive collaborators within the in-group were feared to have on state, economy, and mores. As such, the present chapter is intended to bring to the fore one particular element within a more complex nexus and unpack its workings: how did the "colonial" function within this discursive economy?

It must be stated here that colonial anxieties were not necessarily or strictly infrastructure-bound; more broadly, they proved "infrastructural" for xenophobias with (anti-) colonial overtones, providing a common denominator. Rather than attempting to trace a history of emotions in which "anxiety" would be understood as a label employed by actors themselves, its use is open-ended. "Anxieties" here describe concerns with the hypothetical effects of settler and/or extractive colonial subjection, manifesting themselves as projections of a future that troubled the present, rather than as the lingering aftereffect of a colonial past. This shift, in turn, helps explain the ways in which the recurring "anti-colonial" features of Romanian nineteenth-century nationalist discourse differ from what might be expected to fall under this label. As outlined further below, Romanian "anti-colonial" discourse distinguishes itself as (1) a secondary development of nationalism, rather than a primary impetus for it, born from actual colonial relations; (2) as less burdened by the need to culturally disentangle from the influence of a metropole and more concerned with *identifying* one worthy of emulation; and (3) as a process that was paradoxically amplified, rather than ended, by Romania winning its independence in the last Russo-Turkish war of 1877–1878. Common to all three features, then, was a temporal orientation toward the future, rather than against the past. Thus, underpinning all three was what we might term the infrastructural recurrence of the "colonial" as a common conceptual denominator across a range of public debates and contexts.

What we may hope to extrapolate from the Romanian case is thus laying the groundwork for potential comparisons and connections in the long nineteenth century and bringing the European periphery into the conversation. Currently, literature on colonialism in Europe is more concerned with processes of internal colonization or the colonial complicities of peripheral actors[12] than with how the "colonial" as a category was understood and gained prominence in countries that were neither colonized nor colonizers. By contrast, this chapter is interested in how these discourses recurred and led to connections that actors themselves drew, either between potential "colonizers" as Others or between the potential "colonization" of the nation and similar processes elsewhere.

The Shape and Limits of Anti-Colonial Xenophobia(s)

Romanian nationalism in the nineteenth century was not, on the whole, a primarily anti-colonial affair—nor could it have been so, strictly speaking, in the absence of actual colonial domination. Born of the union between the Principalities of Moldavia and Wallachia in 1859 and independent from the

12 The literature is by now vast and still growing. For a review, see Silvia Marton and Andrei Sorescu, "Introduction: Colonial Anxieties, Corruption Scandals, and Xenophobia in Nineteenth-Century Infrastructure Development in Romania," *East European Politics and Societies* 39, no. 3 (2025): 680–690.

Ottoman Empire as of its participation in the Russo-Turkish War of 1877–1878, modern Romania did, nevertheless, have a strong, if subsequently overlooked, anti-colonial streak as a feature of political discourse in the nineteenth century. To be sure, not every constitutive Other of Romanian nationalism was colonially coded or coded in the same way. The Roma, for instance, were racialized as subaltern and "backward" and, if anything, as occasional *subjects* of internal colonization proposals during the gradual abolition of slavery, rather than "colonizers" themselves.[13] Nor were Hungarians or Russians seen as posing a primary colonial threat for the "Old Kingdom" of Romania *per se*, but as colonial antagonists of Romanian kinfolk in their respective empires—though a Russian protectorate over the Principalities between the 1830s and 1850s was seen as carrying the latent danger of territorial assimilation.[14] More broadly, the imperial neighborhood appeared to nineteenth-century Romanian nation-state builders as a *colonial continuum*: a geographical space riddled with projects of internal colonization that sometimes threatened to spill over. So, we might then ask, in which instances were xenophobias most visibly "(anti-)colonial"?

Let us start by eliminating an obvious choice: the Ottomans. Perhaps surprisingly, in the Romanian case, the "post-" in "post-Ottoman" was not understood as the same "post-" in "post-colonial."[15] Even before achieving full independence from the Ottoman Empire in 1877–1878, there was a sense that, for the most part, whatever substantive sovereignty already existed was grounded in interpreting "ancient treaties" with the Porte. These were framed as proof of absent conquest, and their stipulations explicitly banned the settlement of Muslim Ottoman subjects in the Principalities. Equally true, however, the centuries-long position of the Romanian Principalities as Ottoman client states had meant the continued and problematic existence of extraterritorial consular jurisdiction.[16] Moreover, the Principalities' taxonomic co-inclusion with other peripheries of the Ottoman world by

13 Viorel Achim, *The Roma in Romanian History* (Budapest: Central European University Press, 1998), chapter 3 *passim*.

14 The most recent account of Russian domination in the early nineteenth-century Principalities is Viktor Taki, *Russia on the Danube: Empire, Elites, and Reform in Moldavia and Wallachia, 1812–1834* (Budapest: Central European University Press, 2021).

15 This is a nod to the questions of "post-ness" as articulated in classic interventions such as Kwame Anthony Appiah, "Is the Post- in Postmodernism the Post- in Postcolonial?," *Critical Inquiry* 17, no. 2 (1991): 336–357, and David Chioni Moore, "Is the Post- in Postcolonial the Post- in Post-Soviet? Toward a Global Postcolonial Critique," *PMLA* 116, no. 1 (2001): 111–128.

16 The right enjoyed by diplomatic personnel to deal with the judicial and fiscal affairs of individuals under foreign protection, exempting them from various obligations to local authorities. For more on this, see Holly Case, "'The Quiet Revolution,'" in *The Balkans as Europe*, ed.

diplomats and scholars of international law impelled reactions against "lateral Orientalism," understood as lumping together the Principalities as self-avowedly European polities with the likes of Egypt within a broader Ottoman zone of influence and sovereignty, to the outrage of would-be nation-state builders.[17] Looking at this from a more global perspective, however, an insistence on Europeanness as a standard of civilization and on the value of unequal treaties as proof of agency seems misplaced—a forlorn plea. After all, this came at a time when "empire by contract" was a source of inequality in international relations, from China and Japan to the expanding frontiers of Western colonialism the world over.[18] Such as it existed in the Romanian case, anti-Ottoman xenophobia drew more on stock Orientalist tropes of backwardness and decay yet did not translate into an abiding fixation in Romanian discourse, except when looking backward at a historical enemy that had never truly conquered the Principalities in the Middle Ages.[19] Rather, after the post-independence annexation of the Dobrudja from the Ottoman Empire, Romanian colonization policies aimed at altering the ethnic makeup of the province regarded local Muslim communities as "backward" targets for civilization and, ambivalently, as a population whose emigration was, if not desirable, then acceptable relative to the process of Romanianization by colonization.[20]

Perhaps a closer "post-" to that of the "post-colonial" was felt in drawing a chronological dividing line between the present of a nascent state in the service of the nation and the still-recent past of "Phanariot" domination. This, rather than Ottoman rule in and of itself, was seen as *the* difficult legacy that the present had to grapple with. Hellenophone elites acting as a comprador class of intermediaries for the Ottoman Empire, the Phanariotes

Timothy Snyder and Katherine Younger (Rochester, NY: University of Rochester Press, 2018), 110–138.

17 Andrei-Dan Sorescu, "National History as a History of Compacts: Jus Publicum Europaeum and Suzerainty in Romania in the mid-Nineteenth Century," *East Central Europe* 45, no. 1 (2018): 63–93.

18 See Saliha Belmessous, ed., *Empire by Treaty: Negotiating European Expansion, 1600–1900* (Oxford: Oxford University Press, 2014); Steven M. Harris, "Manufacturing International Law: Pre-Printed Treaties in the 'Scramble for Africa,'" *Journal of the History of International Law/ Revue d'histoire du droit international* 23, no. 3 (2020): 439–465; Jong-Pil Yoon, "Establishing Expansion as a Legal Right: An Analysis of French Colonial Discourse Surrounding Protectorate Treaties," *History of European Ideas* 46, no. 6 (2020): 811–826.

19 Sorescu, "National History."

20 Catalina Hunt, "Changing Identities at the Fringes of the Late Ottoman Empire: The Muslims of Dobruca, 1839–1914" (PhD diss., University of Ohio, 2015); Vladimir Hamed-Troyansky, "Abandoned Land: Muslim Refugees' Property in the Post-Ottoman Balkans," *Past & Present* (2025, ahead of print).

saw a century of domination in the Principalities ended by the local iteration of the Greek Revolution of 1821. Events were also pivotally shaped by a nativist (though not anti-Ottoman) movement led by Tudor Vladimirescu (1780–1821), canonized as a symbol of national regenerative agency around the middle of the century. Yet while colonial analogies such as describing Moldavia and Wallachia as the "Peru of the Greeks" were occasionally employed,[21] public discourse did not retrospectively connote the Phanariot age as a "colonial" one in a consistent manner. Rather, intimations of this potential framing sometimes surfaced when parallels were drawn between Phanariot presence in the past and Jewish presence in the present as a continuity in economic extraction by a foreign Other.[22] Rhetorical framings of continued Phanariot persistence, be it as a moral malaise of generalized "corruption" or through vilifying one's political opponents as "Phanariotes" in both ascendance and deeds, complicated the temporal claims to this past having been wholly done away with.[23] If anything, vestigial Phanariot "corruption" was ascribed, with xenophobic intent, to one's political opponents as a precondition for their collaboration with *other* Others, sometimes for the purpose of aiding in the "colonial" takeover of Romania.[24]

Already, then, we are closer to grasping both the limits and the importance of how and in which contexts the "colonial" undergirded Romanian nationalism. As alluded to, two targets of putatively "anti-colonial" discourse were the Germans and the Jews, resulting in two somewhat entangled but ultimately dissimilar kinds of xenophobia. For one thing, connoting German presence as potentially "colonizing" had, for the most part, less to do with the actually existing German community. It was more an anxiety relating to the *potential* of either German settlement as a consequence of a nebulously defined pan-German master plan of eastward expansion or German and Austrian economic predominance in conjunction with the infrastructural presence and financial fallout of railway development.[25] Still, in the longer run, this did not translate into an imagological projection of the "German" as a fundamental antagonist of Romania, nor into anything

21 Mihail Kogălniceanu, *Histoire de la Valachie, de la Moldavie et des Valaques transdanubiens*, vol. 1 (Berlin: B. Behr, 1837), 280.

22 See, for instance, Mina Savel, *Istoria judaismului și jidanii în România* (Iași: Tip. M. P. Popovici, 1902), 54–55, 75, 78, 140; Mina Savel, *Judaismul în România* (Iași: [n. p.], 1896).

23 Mihai Chioveanu, "The Constitutive Other: Topical and Tropical Phanariot in Modern Romania," *Studia Politica: Romanian Political Science Review* 9, no. 2 (2009): 213–227.

24 See also, on the connection with anti-Semitism, Cécile Folschweiller, "Les clichés anti-grecs chez les penseurs roumains de la seconde moitié du XIXe siècle," *Cahiers balkaniques* 42 (2014).

25 Sorescu, "The Perils (and Promise)."

other than rhetorical violence. The characterological essentialism ascribing a "conquering" nature to Germans as an ethnic group was always up against a (more or less begrudging) admiration for them as an embodiment of "civilization" and had more to do, in ultimately geopolitical terms, with a fear of Prussian, Austrian, or pan-German economic and political expansionism.

Contrast this, then, to how Jews were cast as "colonizers." One imagined link—owing both to the perceived similarity between Yiddish and German and to the protection afforded to some among the community by the neighboring Habsburgs—was that Jews could act as a vanguard of, and in conjunction with, "German" interests.[26] Yet another, predicated on the venerable *canard* of a "state within a state," was that continued Jewish immigration and demographic growth would turn either Moldavia in particular or Romania in general into a "new/second Palestine/Judea."[27] This sometimes carried with it the further implication that the likes of the *Alliance Israélite Universelle* were coordinating a demographic assault and underhanded political takeover by applying international pressure from without and corrupting an unscrupulous political class from within.[28] Occasional global comparisons with Chinese "coolies" reflected a focus on economic perniciousness and the impossibility of racial assimilation, a nod to anxieties of "reverse colonization" in an imperial metropole, scaled down to a Romanian level but argued to be identical in principle.[29] Still other tropes, of course, served to justify Jewish exclusion from citizenship and, by way of consequence, the acquisition of rural landed estates: that Jews would become the middle class that Romania was felt to lack, or that they would exert a morally (even physically) corrupting influence on the hapless peasantry, were also commonplace accusations. As such, Jewish estate administrators (*arendași*) became a focal point for xenophobia, even if their non-citizen status precluded ownership—extraction (to be precise, in the service of Romanian absentee landlords as part of a process

26 D. P. Marţian, *Coloniştii germani şi România. Ed. II: Cu mai multe anecse şi o introducere despre Economia Naţională*, ed. Al. Lupaşcu (Bucharest: Imprimeria Naţională, 1871); Silvia Marton, "Shades of Dependency and the Discourse on 'Corruption': Railway Concessions in Romania in the Nineteenth Century," *East European Politics and Societies* 39, no. 3 (2025): 691–716.

27 Andrei-Dan Sorescu and Raul Cârstocea, "The (Great) Numbers Game: Demographic Anxieties and Quotas in Nineteenth- and Twentieth-Century Romania and the Global Antisemitic Imaginary," in *Quotas: The "Jewish Question" and Higher Education in Central Europe, 1880–1945*, ed. Michael Miller and Judith Szapor (New York: Berghahn, 2024), 67–109.

28 N. R. Locusteanu, "Puterile occidentali și jidovii," *Trompetta Carpaților̆*, August 31, 1869.

29 Andrei-Dan Sorescu, "'The Same Causes Occasioning the Same Effects': The 'Jewish Question,' the 'Chinese Question' and the Global Precedents of Exclusion in Late Nineteenth Century Central Europe," *Journal of Global History* 19, no. 2 (2024): 281–300.

of internal colonialism) was taint enough. The colonial twist of Romanian anti-Semitism proved at once more long-lasting and more consequential than anti-German sentiment. While the inclusion of an article prohibiting the colonization of the country's territory with "peoples of foreign race" in the 1866 constitution—which marked the arrival of a German prince—was seen by contemporary commentators to have been triggered by the precedent of debates on the (un)desirability of German colonization, the provision readily lent itself to being invoked against Jewish presence, too.[30]

If Romanian nationalism was not purely anti-colonial, it was not purely anti-imperial, either. Without attempting here to tease apart the definitional differences between "colonialism" and "imperialism," we may nevertheless note that, for nineteenth-century Romanian nation-state builders, engaging with empire did not always mean engaging with the colonial.[31] In the first half of the century, neither the movement led by Vladimirescu in 1821 nor the Wallachian 1848ers were overtly anti-Ottoman. On the contrary, both revolutions attempted to describe themselves as fundamentally compatible with and respectful of Ottoman authority in their quest for internal reform, even when said reform was socially radical. Particularly after 1848, when a joint Russo-Ottoman invasion put an end to the Wallachian revolutionary experiment, the rejection of the Russian protectorate—as part of the exiled revolutionaries' discourse in the public European sphere and in line with the broader anti-Russian stance of the international revolutionary community—was, however, clear-cut. Though the institutional overhaul that the Russian occupation and protectorate had brought to the Principalities provided a launchpad for their early careers in the army or administration, these high-born founding fathers would, in their twilight years, feel particularly vindicated in their anti-Russian stance by the re-annexation of southern Bessarabia in 1878, even as Russia had been Romania's wartime ally against the Ottomans. Moreover, fears that Russian subversion undermined nation-building efforts and dynastic loyalty among the peasantry would linger until the Great War.[32]

Recurring suspicion did not necessarily amount to principled anti-imperialism. While by 1916 public discourse in the Old Kingdom had grown more overt about its hopes for a "Greater Romania" that would include some

30 Sorescu, "The Perils (and Promise)."

31 For a recent disambiguation, see Barbara Arneil, "Colonialism versus Imperialism," *Political Theory* 52, no. 1 (2023): 146–176.

32 Andrei Sorescu, "Peddlers, Peasants, Icons, Engravings: The Portrait of the Tsar and Romanian Nation-Building, 1888–1916," *New Europe College Ştefan Odobleja Program Yearbook 2019–2020* (2021): 209–246.

if not all of the kinfolk in neighboring empires, the very reality of empire and the strictures of diplomatic pragmatism usually meant denying any ambitions beyond fostering Romanian cultural unity across borders. We must not overestimate how frail, anachronistic, or moribund "empire" was perceived to be as a political form by historical actors, even as they otherwise were more than willing to criticize its impact on "unredeemed" kinfolk. The horizon of expectation for pan-Romanian political unity was ill-defined, and the limits to anti-imperial sentiment when decrying the fate of, for instance, Transylvanian Romanians in Austria-Hungary were also set by a belief in their agency and exemplary self-preservation.[33] Cooperating with empire—such as when nation-building was carried from afar into the Balkans—meant that "Macedonian Romanians" were framed as not only exemplary Romanians but also as loyal Ottoman subjects. This was not only a way of dispelling official suspicion on the ground but also meant to contrast the claims made on the same populations and region by competing Greek or Bulgarian nationalisms.[34] Living with empire, anxieties and animosities aside, was part and parcel of the experience of the long nineteenth century—one that did not always translate into a sustained ideological critique thereof.

Beyond Romania, Inside Out: What Was the "Anti-Colonial"?

Looking from the inside out, Romanian nationalism was, at least in part, anti-colonial. Anti-colonial nationalism has, in fact, been *the* most common form taken by nationalism historically and the basis for the map of the world as we know it today—yet, at the same time, it remains a still relatively undertheorized phenomenon, once distinguished from "secessionist" or "separatist" nationalisms.[35] A now-classic taxonomy of the relationships between nationalism and the state, for instance, places the Romanian

33 Andrei Sorescu, "The 'Is' at Home, the 'Ought' Abroad: Self-Comparison as Self-Criticism and the Transylvanian Model in Early Twentieth-Century Romania," *Comparative Studies in Society and History* 66, no. 1 (2024): 213–237.

34 Andrei-Dan Sorescu, "Inventing a Prosthetic Bourgeoisie: Romania and the Aromanians, 1848–1906," *Nationalities Papers* 50, no. 4 (2022): 661–683.

35 Julian Go and Jake Watson, "Anticolonial Nationalism: From Imagined Communities to Colonial Conflict," *European Journal of Sociology* 60, no. 1 (2019): 31–68. Note that, for instance, Greece is included in this study as a case of "anti-colonial" nationalism, which would perhaps imply that, vis-à-vis the Ottoman Empire, Romania would be classified in the same way as per its criteria.

case under the heading of "separatist" in relation to the Habsburg and Ottoman empires, though one may equally think of it as a "unification" nationalism after the establishment of Romania as a state; the "anti-colonial" remains a separate category as opposed to either.[36] Yet the problem with such classifications is that they overlook the possibility of anti-colonial visions as *one* key discursive feature of a given nationalism, rather than its overarching typology.

In early nineteenth-century South and Central America,[37] then globally in the wake of the "Wilsonian moment" and Lenin's proclamations on self-determination,[38] or after World War II, anti-colonial nationalisms amounted to a protean assemblage of state-seeking, often multiethnic, sometimes federalist projects.[39] We may think of decolonization as a globalizing process,[40] but one whereby the state model and popular sovereignty within a national framework were ultimately universalized,[41] even if the very visions of anti-colonial nationalists reached toward "worldmaking" and rethinking the global as a scalar or racial category.[42] Yet the difficulties

36 John Breuilly, *Nationalism and the State* (Manchester: Manchester University Press, 1993), 135–139.

37 Benedict Anderson, *Imagined Communities: Reflections on the Origin and Spread of Nationalism* (London: Verso, 1983).

38 Erez Manela, *The Wilsonian Moment: Self-Determination and the International Origins of Anticolonial Nationalism* (Oxford: Oxford University Press, 2007); Rita Augestad Knudsen, *The Fight over Freedom in 20th- and 21st-Century International Discourse: Moments of "Self-Determination"* (London: Palgrave Macmillan Cham, 2020); Borislav Chernev, "The Brest-Litovsk Moment: Self-Determination Discourse in Eastern Europe before Wilsonianism," *Diplomacy & Statecraft* 22, no. 3 (2011): 369–387; Michael Goebel, *Anti-Imperial Metropolis Interwar Paris and the Seeds of Third World Nationalism* (Cambridge: Cambridge University Press, 2015). Note, however, that "self-determination" was quickly co-opted by settler colonialism as a category and is therefore neither coterminous nor inherently part of anti-colonial nationalism; see Joseph Massad, "Against Self-Determination," *Humanity: An International Journal of Human Rights, Humanitarianism, and Development* 9, no. 2 (2018): 161–191.

39 For internationalism versus nationalism as visions of the anticolonial, see Manu Goswami, "Imaginary Futures and Colonial Internationalisms," *The American Historical Review* 117, no. 5 (2012): 1461–1485.

40 Martin Thomas and Andrew Thompson, "Empire and Globalization: From 'High Imperialism' to Decolonization," *The International History Review* 36, no. 1 (2014): 142–170.

41 Karuna Mantena, "Popular Sovereignty and Anti-Colonialism," in *Popular Sovereignty in Historical Perspective*, ed. Richard Bourke and Quentin Skinner (Cambridge: Cambridge University Press, 2016), 297–319.

42 Musab Younis, *On the Scale of the World: The Formation of Black Anticolonial Thought* (Oakland: University of California Press, 2022); Adom Getachew, *Worldmaking after Empire: The Rise and Fall of Self-Determination* (Princeton: Princeton University Press, 2019). For a skeptical take on the underlying tensions between "cosmopolitical demands" and those of anticolonial nationalism in terms of development, for instance, see Partha Chatterjee, "Nationalism, Internationalism,

of post-independence nation- and state-building complicate the picture of continuities and discontinuities before, during, and after the colonial moment.[43] The centrifugal or centripetal effects of colonial-era borders on the emergence of national movements and subsequent states are, of course, hard to downplay,[44] and tensions between top-down "modernizing" nationalisms and deeper forms of local patriotism branded as "tribal" remain.[45] Postcolonial nation-states have thus had to grapple with the creation of internal and external Others and to deal with the legacies of xenophobia left behind by imperially generated hierarchies of class and race as an integral part of the puzzle.

Perhaps all nation-building is somehow "colonial" in its attempts to turn peasants into whatever-men, and a tragedy of postcolonial nationalism has been that it is forced to replicate this, even in its revolutionary forms.[46] However, even when nationalism is on the rise in a given context, not all anti-colonial sentiment and praxis need be nationalist as such.[47] This, in turn, raises the issue of how nationalism is often normativized as a supposedly inevitable historical development both by the hindsight of scholarly interpretation and by anti-colonial nationalists themselves. The question that arises, then, is of its (un)originality as an ideology: what is there left to invent for the would-be non-European nation "discovering" it in its struggle against its European colonizers? One classic perspective is that, in reaction, emerging anti-colonial nationalist discourse carves out "the spiritual" as a domain of resistance and identity, as a counterpoint to the material leverage of progress *qua* domination.[48] Yet this is not to say that the economic is unimportant to the anticolonial nationalist: as we have seen in the Romanian case as well, economics provided a framework

and Cosmopolitanism: Some Observations from Modern Indian History," *Comparative Studies of South Asia, Africa and the Middle East* 36, no. 2 (2016): 332–333.

43 Frederick Cooper, "Possibility and Constraint: African Independence in Historical Perspective," *The Journal of African History* 49, no. 2 (2008): 167–196.

44 David E. F. Henley, "Ethnogeographic Integration and Exclusion in Anticolonial Nationalism: Indonesia and Indochina," *Comparative Studies in Society and History* 37, no. 2 (1995): 286–324.

45 John M. Lonsdale, "Anti-Colonial Nationalism and Patriotism in Sub-Saharan Africa," in *The Oxford Handbook of the History of Nationalism*, ed. John Breuilly (Oxford Academic, 2013).

46 Arif Dirlik, "Rethinking Colonialism: Globalization, Postcolonialism, and the Nation," *Interventions* 4, no. 3 (2002): 428–448.

47 Adam Guerin, "'Not a Drop for the Settlers': Reimagining Popular Protest and Anti-Colonial Nationalism in the Moroccan Protectorate," *The Journal of North African Studies* 20, no. 2 (2015): 225–246.

48 Partha Chatterjee, *The Nation and Its Fragments* (Princeton: Princeton University Press, 1993).

for thinking about certain relationships in "colonial" terms; and when reference was made to the blood of ancestors, it was in tandem with more substantively economic arguments about capacity-building and resistance.

These furtive reflections on the conundrums of anticolonial nationalism will now be doubled by a short comparison between some of its key features and those of Romanian nationalism in the long nineteenth century. A first feature would be the existence of an *actual* relationship of political dependence, subverted and critiqued by the rise of a self-assumedly "native" intelligentsia, invested in a nation-building project even before the end of colonial rule. The story of Romanian nationalism fits the bill only in part, but the mismatch is instructive. The foundational French discourse of "civilization" that reached the Principalities in the first half of the nineteenth century and contributed to the local growth of nationalism was diffused through the still-important Greek-language networks that did not die out with the events of 1821.[49] Anti-Phanariot discourse notwithstanding, the ethnicization of a "civilized" and European Romanian nation as a subject of discourse and praxis was not a sudden move but a gradual disentanglement, coextensive with the retrospective invention and increased politicization of a local 1821 as separate in aims and nature from the Greek Revolution. So it was, then, too, that anticolonial attitudes did not, *a priori,* exert an uncontested hegemony, either. Up until the 1860s, debates on whether encouraging the settlement of foreigners precisely as a means of hastening economic development and importing "civilization" were still open-ended. It was only when, in Weberian terms, the imperative of monopolistic closure was perceived as superior to that of development in the abstract that the tide turned in favor of outlawing "colonization," yet even then, it left enough ambiguity for the immigration of "kindred" Latin peoples to remain at least a theoretical possibility. The importance of "race" (*gintă, rasă*), while not yet biologically connoted, was felt when imagining a broader community of descent and, as detailed below, lent itself to politicization via xenophobia. The case of the Principalities/Romania points to the possibility of nationalism *becoming* "anticolonial" rather than having this as a core motivation from the very beginning.

Secondly, anticolonial nationalism is, in Andersonian terms, always-already haunted by the experience of "living comparatively," of managing the creation of national selfhood in and against the image of the imperial

49 Alex R. Tipei, *Unintended Nations: France's Empire of Civilization, Southeast Europe, and the Post-Napoleonic World* (Montreal: McGill-Queen's University Press, 2025).

metropole that lingers as a point of reference for development.[50] In the Romanian case, the extent to which the structural, scaled-down emulation of either the Porte or Petersburg in the first half of the century was perceived as a necessity remains debatable. Further research is needed to more fully ascertain whether the proto-constitutional Organic Regulations, though promulgated under Russian tutelage, were felt by local actors to also encapsulate their own agency and vision—or whether the Ottoman reforms of the Tanzimat era were credited as a model or impetus for local change.[51] Still, neither was a class of Russian or Ottoman imperial administrators confronted, conversely, with managing "mimicry" on the part of locals whose set role in perpetuating the structures of domination threatened to be overstepped.[52] What does crop up at various times, however, is the more or less explicit assumption that Romanians could self-identify as something of a "colonial" outpost of French influence on the edges of the Orient. From fleeting plans of French settler colonization to the petitions of Ion C. Brătianu (1821–1891) to Napoleon III in the 1850s,[53] the sense that a more or less metaphorical colonial position *vis-à-vis* France was seen as desirable by some Romanians is worth reflecting upon. At the same time, as early as the 1830s and 1840s, criticism of the seemingly hollow yet intense adoption of French language and habits by urban classes gained traction—though it, too, came from the very literati-cum-founding fathers who were otherwise avid Francophiles.[54] At stake was drawing a line between taking French culture, language, and "civilization" as models and stimuli for national regeneration and the dangers that excess might pose to the very same development of a native culture and language. This, in short, was a question of how French "soft power" was to be productively instrumentalized, rather than allowed

50 Benedict Anderson, *The Spectre of Comparisons: Nationalism, Southeast Asia, and the World* (London: Verso, 1998).

51 However, see Taki, *Russia on the Danube*, and, for a tentative research agenda, see Radu F. Florescu, "The Romanian Impact upon the Ottoman Tanzimat," *The Journal of Southeastern European Studies* 6 (1978): 227–238.

52 Homi Bhabha, "Of Mimicry and Man: The Ambivalence of Colonial Discourse," *October* 28 (1984): 125–33.

53 Ion C. Brătianu, "Memoriu asupra Românilor dat Împăratului Napoleon III," in *Ion C. Brătianu: Acte și cuvântări*, vol. 1.1, ed. G. Marinescu and C. Grecescu (Bucharest: Cartea Românească, 1938 [1853]), 27–32. First published in *Românulŭ*, December 6, 1861. For French remarks on the colonial potential of Romanian lands, see Ovidiu Muntean, *Imaginea românilor în Franța la mijlocul secolului al XIX-lea* (Cluj-Napoca: Napoca Star, 2005), 174–180.

54 One illuminating example being '48er poet and playwright Vasile Alecsandri. On French influence in his works, see Charles Drouhet, *Vasile Alecsandri și scriitorii francezi* (Bucharest: Cultura națională, 1924).

to be uncritically adopted at the expense of national specificity. Though at present largely absent from current studies of French "informal empire,"[55] a deeper engagement with the Romanian case may yield interesting reflections on how a metropole may be *chosen* as a model for emulation as part of a program of national "revival."

The racial aspect of nineteenth-century Romanian nationalism prevented *some* potentially colonial relationships from engendering xenophobias, and visions of pan-Latin kinship played a part in this, just as they arose in reaction to fears of pan-German or pan-Slavic designs within an imagined landscape of competing ethnic and historical forces. Interpretations of "self-colonizing cultures" as a regional phenomenon notwithstanding,[56] the transnational imaginary of Latin kinship in question here included both the occasionally political and an abiding (if, in fairness, not uncontested) cultural commitment. Hierarchies of kin-making impelled by geopolitical and cultural prestige meant keeping a closer eye on France than on Italy—to say nothing of the feebleness of Iberian connections or of the near absence of trans-Atlantic connections with "Latin America." Size and perceived comparability, conversely, made Belgium into a model for emulation, if not a *metropole*,[57] and the overall repertoire of national self-comparison remains an under-researched topic. But the thrust of the argument remains: "anti-colonial" nationalism in Romania could have, as a flipside, the imagining of *other*, desirable (para-)colonial relationships.

Thirdly, anticolonial nationalisms become postcolonial nationalisms when the odds are in their favor, but this seldom means that the ties that bind colony and metropole are well and truly severed, as neocolonial forms of extraction arise. In the Romanian case, the yearly tribute paid to the Ottoman Empire was, by the time of independence, not a hobbling financial burden, and the relative economic monopoly on extraction previously enjoyed by Istanbul had been replaced with the integration of the Principalities into the global capitalist market system as of the 1830s. The

55 With the notable recent exception of Tipei, *Unintended Nations*.

56 Alexander Kiossev, "Notes on Self-Colonising Cultures," in *Cultural Aspects of the Modernisation Process*, ed. Dimitŭr Ginev, Francis Sejersted, and Kostadinka Simeonova (Oslo: TMV-Senteret, 1995), 73–81.

57 Silvia Marton, "'La Belgique de l'Orient' et les Chemins de Fer—les Raisons d'une Comparaison: la construction politique de l'État-nation dans le Parlement roumain (1866–1871)," *Studia Politica: Romanian Political Science Review* 8, no. 1 (2008): 27–44; Raul Cârstocea and Maarten Van Ginderachter, "Smallness and the East-West Binary in Nationalism Studies: Belgium and Romania in the Long Nineteenth Century," in *The Politics of Smallness in Modern Europe: Size, Identity and International Relations since 1800*, ed. Samüel Kruizinga (London: Bloomsbury, 2022), 55–72.

diplomatic implications of non-sovereignty weighed more heavily, even if trade agreements were concluded without Ottoman oversight, but after independence there remained nothing in the way of lingering Ottoman dominion. However, between the 1850s and 1870s, as the Russian protectorate was replaced after the Crimean War with the collective stewardship of the European Great Powers, fears of existential frailty nevertheless persisted. The feeling that independence threatened to remain unachieved due to economic entanglements was pervasive. It is here once more that infrastructure makes its appearance. Flipping the usual script on its head, the "anticolonial" streak in Romanian nationalist discourse was more inflamed by the paradoxical threat of *becoming* something of a colony, even as the practical and symbolic threshold of emancipation from the "Turkish Yoke" was being reached. This, then, was part and parcel of the decade-long scandal that surrounded the so-called Strousberg affair in the 1870s. The bankruptcy of the major concession that was to build much of Romania's railway network had to be handled while maintaining the country's financial standing *and* navigating the pressure that the German Empire applied in order to protect the interests of the capitalists involved.[58] That the very recognition of Romanian independence, in fact, was conditioned by Germany on bailout and buyout was inflammatory—that this came with the further obligation of offering a legal framework for Jewish emancipation was doubly so. Though in this case Strousberg's Jewish origin was initially not flagged as such, we see here how two strands of "anticolonial" xenophobic discourse were braided in Romania, even if, as already established, their shelf life and intensity were decidedly different. Is this fundamentally part of the story of a "Balkan colony" coming to grips with its entry into the capitalist "world system"?[59] Perhaps. But that is not quite the same story as whether or why historical actors would also come to think of their predicament in somewhat different, historicizably "colonial" terms.

One way of summing up these three potential avenues of rethinking the "anticolonial" as a qualifier for Romanian nationalism is by looking at temporality and timing. On the one hand, colonial anxieties came *after* the process of imperial disentanglement was already underway, and local elites had both the agency and the public venues for debating the potential benefits or perils of colonization. On the other hand, colonization or (para-) colonial relationships, when advocated, were seen as a way of overcoming

58 Marton, "Shades of Dependency."

59 Daniel Chirot, *Social Change in a Peripheral Society: The Creation of a Balkan Colony* (New York: Academic Press, 1976).

lag and "backwardness" by importing "civilization"—and, when opposed, seen as a form of backsliding into a past that had not even existed, even as independence was increasingly within reach. It is by taking all of the above together, then, that we may return to the observation that thinking about the colonial performed an infrastructural function for ideology: a common frame of reference, even if manifesting itself across seemingly disparate contexts.

Conclusion

On February 7, 1895, the floor of the Senate saw the mining law being defended by P. P. Carp (1837–1919), the Conservative minister of estates (*Ministrul Domeniilor*) and a key influence in its drafting.[60] Carp was a towering figure of the Conservative establishment, yet cut from a different cloth when it came to his lukewarm but open support of the Jewish community. His reasoning was that the Romanians needed to fear competition less and bolster their work ethic to catch up with an industrious minority. In this speech, too, Carp was unwilling to compromise on his vision for national development and proceeded to dismantle the arguments of his opponents by drawing upon a vast array of legal and economic arguments.

There were no grounds for his opponents, Carp began, to claim that Article 2 of the Constitution, outlawing the cession of national territory; Article 19, preventing extralegal expropriation; or Article 7, restricting ownership of rural landed estates, were in danger of being violated. As mines were a *sui generis* type of property, Carp argued, peasants would be duly compensated rather than displaced.[61] The Liberals' concern for the politically disenfranchised and economically dependent peasant majority, in fact, seemed to him to be little but demagoguery: "I do think, gentlemen, that there will come a time when not only the peasantry but also the rest of our population, which is comprised of neither *ciocoi* [upstarts] nor Phanariotes, will begin to tire of tribunes who only speak words of love but, whenever in a position to contribute to the betterment of their lot, prove the most obstinate adversaries."[62]

60 On Carp's take on the "Jewish Question" see Silvia Marton, *La construction politique de la nation: La nation dans les débats du Parlement de la Roumanie, 1866–1871* (Iași: Institutul European, 2009), 302–303.

61 P. P. Carp, *Discursul Domnului P. P. Carp, ministru al domenielor, rostit asupra Legei minelor în ședința Senatului la 7 Februarie 1895* (Bucharest: Tip. Gutenberg, Joseph Göbl, 1895), 3–9.

62 Carp, *Discursul Domnului P. P. Carp*, 9.

The rhetorical specter of the Phanariot—here understood as a proxy for all non-peasants in the essentialist discursive binary pitting foreigner-filled town against purely Romanian countryside—was something that Carp could ventriloquize on behalf of his opponents. And with this, he turned to their "anti-foreigner-ism" (*anti-străinism*) as a more general issue. Whereas the Liberals seemed to suggest that "when we shall be more cultivated, stronger, with nothing to fear, then we shall be more open to foreigners, not showing them the hatred of today," Carp countered: "The weaker a country, the less cultivated it is, the more it is in need of strengthening, the more the cases in which it needs foreigners." Against a logic of potentially indefinite deferral, Carp counterposed one of urgency and offered a list of historical examples. Emulation—from that of the Phoenicians, Assyrians, and Egyptians by the Greeks to that of the Germans by the Russians—could be taken as proof of how modernization was always in motion, "and we shall see where the blanket exclusion that [the Russians] have for some time begun to apply to all foreign elements will get them."[63]

If "culture," here understood to include technological know-how, was the only true measure of a nation's worth, then Romania had to reckon with its still-incomplete westernization. Carp dared his opponents to deny that their very own ideas were Western imports, and imperfect ones at that. What better example of how the need for foreigners met a need for modernization could one offer in Romania than that of Anghel Saligny (1854–1925), an engineer of French origin who had designed the massive infrastructural centerpiece that was the railway bridge connecting Wallachia to the (colonial) Dobrudja across the Danube? So, too, with the Jews, Carp continued: even if prevented from acquiring citizenship *en masse* or owning rural land, capital and experience made the foreigner valuable rather than dangerous. As an "adolescent" people, still in a "phase of assimilation," the Romanians and their political class were, on Carp's reading, forced to abandon xenophobia if they were ever to take "the path of culture, civilization, and progress."[64]

After an excursus on the origin and nature of property, Carp finally arrived at the charge of "colonization." "You, Sirs, seem not to know what 'colonization' means"—he asserted, before proceeding to explain that a handful of foreign workers working on a clearly defined infrastructural project were not granted land to form a self-administering community, but remained under state control in terms of access to property and rights. "I do not know," Carp continued, "how you could have reached the conclusion that

63 Carp, *Discursul Domnului P. P. Carp*, 10–11.

64 Carp, *Discursul Domnului P. P. Carp*, 11–13.

this might mean colonization, or what gives you the right to utter such empty and bombastic phrases."[65] The very definition of "colonization," we might thus note, was something that was contested in the Romanian case, yet the rhetorical value of invoking the term was not, since it was acknowledged to have inflammatory potential. Finally, Carp came to the economic argument of how bolstering local resource extraction could help the development of Romanian industry, in due course pushing beyond dependence on foreign fuel and the mere assembly of prefabricated parts. This was a law for the future strengthening and independence of not just the country as a whole but of its laboring classes, beyond populist promises of political inclusion without material emancipation.[66]

Carp's arguments, or rather the comfortable majority support that the government could count on in parliament, won the day. The Senate moved to vote in favor of the bill a few days after his speech, and fierce debates arose again in the Chamber of Deputies come springtime. The law would prove crucial for the development of the Romanian economy. Petroleum (also covered by its stipulations) rose dramatically in importance as an export by World War I, and, perhaps ironically, the influx of foreign capital it brought only heightened the sense of economic dependency and entanglement.

Summing up Carp's intervention is also helpful for drawing a more general conclusion. By 1895, the consensus that land was to remain a monopoly of ethnically defined citizens seemed broadly unassailable, even if "colonization" appeared to some as merely the exaggerated description of an implausible danger. Yet the question of how best to fortify the Romanian economy and, more broadly, to bring "civilization" and development remained an open one. Xenophobia continued to be pitted against a need for capital, and this may help reinterpret well-known debates on protectionism and the (non-) industrialization of the Romanian economy.[67] Carp's surprisingly "liberal" discourse, especially if compared to that of the self-styled Liberals, may be taken as emblematic of how, in nineteenth-century Romania, protectionism was favored by the latter camp. Yet Carp also acknowledged the need for reaching some form of economic autarky after having amassed the requisite capital and experience from elsewhere, pointing to the transpartisan nature

65 Carp, *Discursul Domnului P. P. Carp*, 18–19.

66 Carp, *Discursul Domnului P. P. Carp*, 24–28.

67 John Michael Montias, "Notes on the Romanian Debate on Sheltered Industrialization: 1860–1906," in *Social Change in Romania, 1860–1940: A Debate on Development in a European Nation*, ed. Kenneth Jowitt (Berkeley: University of California Press, 1978), 53–71; Joseph L. Love, *Crafting the Third World: Theorizing Underdevelopment in Rumania and Brazil* (Stanford: Stanford University Press, 1996).

of economic nationalism as a shared framework. Carp is a welcome outlier, helping nuance the temporalities of development that were imagined at the time. For some, the future was anxiogenic *because* of the nation's perceived frailty and "backwardness" in the present, whereas for him, the future threatened to fail to arrive *because* imagined frailty prevented "civilization" from taking root due to xenophobia. For all the potential vagueness of the "colonial" as a category for our historical actors, the bottom line remained squarely material. The far-reaching and complex set of issues discussed in parliament and elsewhere in our particular case had the very concrete matter of resource extraction and infrastructural development at heart.

To return to the bigger picture one last time, colonial anxieties acted infrastructurally in the sense that they fueled xenophobias, just as infrastructure could itself be connoted as a gateway for colonial intrusion. This, in turn, points to the problematic tension between nation-building and state-building. The imperatives of the latter, including infrastructural development as part of a drive for "modernization," could come into conflict with those of the former if seen as entailing unintended and potentially pernicious consequences for the nation. Neither this nor the politics of grappling with "backwardness" is peculiar to the Romanian case, of course. Nor, then, should our reflections on how thinking about it against the grain of "anticolonial nationalism" be understood as strictly relevant to it alone. If we are to consider how the "anticolonial" could be a secondary, rather than primary feature of a nationalism, how a "metropole" is a less than fixed point of reference, and how the temporal politics of state-building map onto this, then the field lies open for other, comparative perspectives broadening our understanding even more. As this volume will further explore, the shifting configurations of infrastructure, colonialism, corruption, and xenophobia on the European periphery provide fertile ground for both empirical and theoretical reassessment.

Bibliography

Achim, Viorel. *The Roma in Romanian History.* Budapest: Central European University Press, 1998.

Alexander. "Două cuvinte asupra legii minelor (urmare și fine)." *Dreptul* 24, no. 11, February 5, 1895: 89–93.

Anderson, Benedict. *Imagined Communities: Reflections on the Origin and Spread of Nationalism.* London: Verso, 1983.

Anderson, Benedict. *The Specter of Comparisons: Nationalism, Southeast Asia, and the World*. London: Verso, 1998.

Appiah, Kwame Anthony. "Is the Post- in Postmodernism the Post- in Postcolonial?" *Critical Inquiry* 17, no. 2 (1991): 336–357.

Arneil, Barbara. "Colonialism versus Imperialism." *Political Theory* 52, no. 1 (2023): 146–176. https://doi.org/10.1177/00905917231193107.

Augestad Knudsen, Rita. *The Fight over Freedom in 20th- and 21st-Century International Discourse: Moments of "Self-Determination."* Cham: Palgrave Macmillan, 2020.

Belmessous, Saliha, ed. *Empire by Treaty: Negotiating European Expansion, 1600–1900*. Oxford: Oxford University Press, 2014.

Bhabha, Homi. "Of Mimicry and Man: The Ambivalence of Colonial Discourse." *October* 28 (1984): 125–133.

Bogost, Ian. *Alien Phenomenology, or, What It's Like to Be a Thing*. Minneapolis: University of Minnesota Press, 2012.

Brătianu, Ion C. "Memoriu asupra Românilor dat Împăratului Napoleon III." In *Ion C. Brătianu: Acte și cuvântări*. Vol. 1.1, edited by G. Marinescu and C. Grecescu, 27–32. Bucharest: Cartea Românească, 1938 [1853].

Breuilly, John. *Nationalism and the State*. Manchester: Manchester University Press, 1993.

Carp, P. P. *Discursul Domnului P. P. Carp, ministru al domenielor, rostit asupra Legei minelor în şedinţa Senatului la 7 Februarie 1895*. Bucharest: Tip. Gutenberg, Joseph Göbl, 1895.

Cârstocea, Raul, and Maarten Van Ginderachter. "Smallness and the East-West Binary in Nationalism Studies. Belgium and Romania in the Long Nineteenth Century." In *The Politics of Smallness in Modern Europe. Size, Identity and International Relations since 1800*, edited by Samüel Kruizinga, 55–72. London: Bloomsbury, 2022.

Case, Holly. "The Quiet Revolution: Consuls and the International System in the Nineteenth Century." In *The Balkans as Europe*, edited by Timothy Snyder and Katherine Younger, 110–138. Rochester, NY: University of Rochester Press, 2018.

Chatterjee, Partha. "Nationalism, Internationalism, and Cosmopolitanism: Some Observations from Modern Indian History." *Comparative Studies of South Asia, Africa and the Middle East* 36, no. 2 (2016): 332–334. https://doi.org/10.1215/1089201X-3603392.

Chernev, Borislav. "The Brest-Litovsk Moment: Self-Determination Discourse in Eastern Europe before Wilsonianism." *Diplomacy & Statecraft* 22, no. 3 (2011): 369–387. https://doi.org/10.1080/09592296.2011.599635.

Chioni Moore, David. "Is the Post- in Postcolonial the Post- in Post-Soviet? Toward a Global Postcolonial Critique." *PMLA* 116, no. 1 (2001): 111–128. https://doi.org/10.1632/pmla.2001.116.1.111.

Chioveanu, Mihai. "The Constitutive Other: Topical and Tropical Phanariot in Modern Romania." *Studia Politica: Romanian Political Science Review* 9, no. 2 (2009): 213–227.

Chirot, Daniel. *Social Change in a Peripheral Society: The Creation of a Balkan Colony.* New York: Academic Press, 1976.

Cooper, Frederick. "Possibility and Constraint: African Independence in Historical Perspective." *The Journal of African History* 49, no. 2 (2008): 167–196. https://doi.org/10.1017/S0021853708003915.

Delavrancea, Barbu Ștefănescu. *Români...* Bucharest: [n. p.], 1894.

Desbaterile Senatului. Sesiunea ordinară 1894-1895. Bucharest: Imprimeria Statului, 1895 (February 6, 1895).

Dirlik, Arif. "Rethinking Colonialism: Globalization, Postcolonialism, and the Nation." *Interventions* 4, no. 3 (2002): 428–448. https://doi.org/10.1080/136980 1022000013833.

Drouhet, Charles. *Vasile Alecsandri și scriitorii francezi.* Bucharest: Cultura națională, 1924.

Florescu, Radu F. "The Romanian Impact upon the Ottoman Tanzimat." *The Journal of Southeastern European Studies* 6 (1978): 227–238.

Folschweiller, Cécile. "Les clichés anti-grecs chez les penseurs roumains de la seconde moitié du XIXe siècle." *Cahiers balkaniques* 42 (2014): 1–17. https://doi.org/10.4000/ceb.4896.

Getachew, Adom. *Worldmaking after Empire: The Rise and Fall of Self-Determination.* Princeton: Princeton University Press, 2019.

Go, Julian, and Jake Watson. "Anticolonial Nationalism: From Imagined Communities to Colonial Conflict." *European Journal of Sociology* 60, no. 1 (2019): 31–68. https://doi.org/10.1017/S000397561900002X.

Goebel, Michael. *Anti-Imperial Metropolis: Interwar Paris and the Seeds of Third World Nationalism.* Cambridge: Cambridge University Press, 2015.

Goswami, Manu. "Imaginary Futures and Colonial Internationalisms." *The American Historical Review* 117, no. 5 (2012): 1461–1485. https://doi.org/10.1093/ahr/117.5.1461.

Guerin, Adam. "'Not a Drop for the Settlers': Reimagining Popular Protest and Anti-Colonial Nationalism in the Moroccan Protectorate." *The Journal of North African Studies* 20, no. 2 (2015): 225–246. https://doi.org/10.1080/13629387.2014.917586.

Hamed-Troyansky, Vladimir. "Abandoned Land: Muslim Refugees' Property in the Post-Ottoman Balkans." *Past & Present* (2025, ahead of print).

Harris, Steven M. "Manufacturing International Law: Pre-printed Treaties in the 'Scramble for Africa.'" *Journal of the History of International Law/ Revue d'histoire du droit international* 23, no. 3 (2020): 439–465. https://doi.org/10.1163/15718050-12340137.

Henley, David E. F. "Ethnogeographic Integration and Exclusion in Anticolonial Nationalism: Indonesia and Indochina." *Comparative Studies in Society and History* 37, no. 2 (1995): 286–324.

Hunt, Catalina. "Changing Identities at the Fringes of the Late Ottoman Empire: The Muslims of Dobruca, *1839–1914*." PhD diss., University of Ohio, 2015.

Kiossev, Alexander. "Notes on Self-Colonising Cultures." In *Cultural Aspects of the Modernisation Process*, edited by Dimitŭr Ginev, Francis Sejersted, and Kostadinka Simeonova, 73–81. Oslo: TMV-Senteret, 1995.

Kogălniceanu, Mihail. *Histoire de la Valachie, de la Moldavie et des Valaques transdanubiens*. Vol. 1. Berlin: B. Behr, 1837.

Locusteanu, N. R. "Puterile occidentali și jidovii." *Trompetta Carpaților̆*, August 31, 1869.

Lonsdale, John M. "Anti-Colonial Nationalism and Patriotism in Sub-Saharan Africa." In *The Oxford Handbook of the History of Nationalism*, edited by John Breuilly, 318–337. Oxford: Oxford Academic, 2013.

Love, Joseph L. *Crafting the Third World: Theorizing, Underdevelopment in Rumania and Brazil*. Stanford: Stanford University Press, 1996.

Manela, Erez. *The Wilsonian Moment: Self-Determination and the International Origins of Anticolonial Nationalism*. Oxford: Oxford University Press, 2007.

Mantena, Karuna. "Popular Sovereignty and Anti-Colonialism." In *Popular Sovereignty in Historical Perspective*, edited by Richard Bourke and Quentin Skinner, 297–319. Cambridge: Cambridge University Press, 2016.

Marțian, D. P. *Coloniştii germani şi România. Ed. II: Cu mai multe anecse şi o introducere despre Economia Naţională*, edited by Al. Lupaşcu. Bucharest: Imprimeria Naţională, 1871.

Marton, Silvia. "'La Belgique de l'Orient' et les Chemins de Fer—les Raisons d'une Comparaison: la construction politique de l'État-nation dans le Parlement roumain (1866–1871)." *Studia Politica: Romanian Political Science Review* 8, no. 1 (2008): 27–44.

Marton, Silvia. *La construction politique de la nation: La nation dans les débats du Parlement de la Roumanie, 1866–1871*. Iași: Institutul European, 2009.

Marton, Silvia. "Shades of Dependency and the Discourse on 'Corruption': Railway Concessions in Romania in the Nineteenth Century." *East European Politics and Societies* 39, no. 3 (2025): 691–716. https://doi.org/10.1177/08883254251352113.

Marton, Silvia, and Andrei Sorescu. "Introduction: Colonial Anxieties, Corruption Scandals, and Xenophobia in Nineteenth-Century Infrastructure Development in Romania." *East European Politics and Societies* 39, no. 3 (2025): 680–690. https://doi.org/10.1177/08883254251352114.

Massad, Joseph. "Against Self-Determination." *Humanity: An International Journal of Human Rights, Humanitarianism, and Development* 9, no. 2 (2018): 161–191. https://dx.doi.org/10.1353/hum.2018.0010.

Montias, John Michael. "Notes on the Romanian Debate on Sheltered Industrialization: 1860–1906." In *Social Change in Romania, 1860–1940: A Debate on Development in a European Nation*, edited by Kenneth Jowitt, 53–71. Berkeley: University of California, 1978.

Muntean, Ovidiu. *Imaginea românilor în Franța la mijlocul secolului al XIX-lea*. Cluj-Napoca: Napoca Star, 2005.

Savel, Mina. *Istoria judaismului și jidanii în România*. Iaşi: Tip. M. P. Popovici, 1902.

Savel, Mina. *Judaismul în România*. Iaşi: [n. p.], 1896.

Sorescu, Andrei. "The 'Is' at Home, the 'Ought' Abroad: Self-Comparison as Self-Criticism and the Transylvanian Model in Early Twentieth-Century Romania." *Comparative Studies in Society and History* 66, no. 1 (2024): 213–237. https://doi.org/10.1017/S001041752300035X.

Sorescu, Andrei. "Peddlers, Peasants, Icons, Engravings: The Portrait of the Tsar and Romanian Nation-Building, 1888–1916." *New Europe College Ştefan Odobleja Program Yearbook 2019–2020* (2021): 209–246.

Sorescu, Andrei. "The Perils (and Promise) of German Colonization: Civilizational Hierarchies and Anxieties in Nineteenth-Century Romania." *East European Politics and Societies: and Culture* 39, no. 3 (2025): 717–740. https://doi.org/10.1177/08883254251355736.

Sorescu, Andrei-Dan. "Inventing a Prosthetic Bourgeoisie: Romania and the Aromanians, 1848–1906." *Nationalities Papers* 50, no. 4 (2022): 661–683. https://doi.org/10.1017/nps.2021.11.

Sorescu, Andrei-Dan. "National History as a History of Compacts: *Jus Publicum Europaeum* and Suzerainty in Romania in the mid-Nineteenth Century." *East Central Europe* 45, no. 1 (2018): 63–93. https://doi.org/10.1163/18763308-04501004.

Sorescu, Andrei-Dan. "'The Same Causes Occasioning the Same Effects': The 'Jewish Question,' the 'Chinese Question' and the Global Precedents of Exclusion in Late Nineteenth Century Central Europe." *Journal of Global History* 19, no. 2 (2024): 281–300. https://doi.org/10.1017/S174002282300030X.

Sorescu, Andrei-Dan, and Raul Cârstocea. "The (Great) Numbers Game: Demographic Anxieties and Quotas in Nineteenth- and Twentieth-Century Romania and the Global Antisemitic Imaginary." In *Quotas: The "Jewish Question" and Higher Education in Central Europe, 1880–1945*, edited by Michael Miller and Judith Szapor, 67–109. New York: Berghahn, 2024.

Taki, Viktor. *Russia on the Danube: Empire, Elites, and Reform in Moldavia and Wallachia, 1812–1834*. Budapest: Central European University Press, 2021.

Thomas, Martin, and Andrew Thompson. "Empire and Globalisation: From 'High Imperialism' to Decolonisation." *The International History Review* 36, no. 1 (2014): 142–170. https://doi.org/10.1080/07075332.2013.828643.

Tipei, Alex R. *Unintended Nations: France's Empire of Civilization, Southeast Europe, and the Post-Napoleonic World*. Montreal: McGill-Queen's University Press, 2025.

Voința Națională. "Lupta cea mare." December 18, 1894, 1.

Vulturescu, Gr. *Legea minelor: Respuns D-lui G. G. Disescu. Publicat în ziarul* Voinţa Naţională *din 1895 cu un Apendice coprindend parte din lucrarea lui D. P. Marţian. Coloniştii Germani şi România*. Bucharest: Tip. Gutenberg, Joseph Göbl, 1895.

Yoon, Jong-Pil. "Establishing Expansion as a Legal Right: An Analysis of French Colonial Discourse Surrounding Protectorate Treaties." *History of European Ideas* 46, no. 6 (2020): 811–826. https://doi.org/10.1080/01916599.2020.1722725.

Younis, Musab. *On the Scale of the World: The Formation of Black Anticolonial Thought*. Oakland: University of California Press, 2022.

About the Author

Andrei Sorescu is a cultural and intellectual historian of nineteenth-century Europe, with a focus on Romania in a transnational context. His current research interests include the role of self-comparison in nation-building, the historical semantics of the "colonial," the transnational and global dimensions of anti-Semitism, and the critical study of temporalities.

Part II

(Transportation) Infrastructure, Corruption Debates, and Colonization

5. Accusations of Tyranny: Mobilizing the Ottoman Imperial Past in Service of West European Informal Empire in the Early Greek State

Alex R. Tipei

Abstract: Leaders of the early Greek state angled to modernize their country via infrastructure projects. The men governing this war-torn, agrarian state, however, found themselves short on capital. They consequently sought out Western European benefactors. Competing with one another for foreign support, they claimed that their political rivals misused funds or abandoned initiatives paid for with international monies. These men often characterized themselves as conduits of modernity and their competitors as tyrants—a term coded as unmodern (un-European) and Ottoman (or Muslim). This chapter explores how, in these contests, Greek politicians used the concept of tyranny to assimilate their rivals with the Ottoman past during an era best understood as one of transition from formal (Ottoman) to informal (Western European) empire.

Keywords: Greece, France, Corruption, Infrastructure development, Modernization, Ottoman legacy

Politicians in the early Greek state were eager to distance themselves and their country from the recent Ottoman past. One strategy they adopted was an intense program of modernization that stressed administrative, transportation, education, and public health-related infrastructure projects. They intended, among other things, for these programs to tangibly and symbolically render Greece more like its Western European patrons. The

Marton, Silvia, and Andrei-Dan Sorescu, eds. *Infrastructural Development, Corruption, Xenophobia, and Colonization in Central and Southeastern Europe*. Amsterdam: Amsterdam University Press/Central European University Press, 2026.
 https://doi.org/10.5117/9789048577279-6

men governing this new, war-torn, agrarian state, however, found themselves short on capital. To carry out their ambitious agenda, they presented themselves to potential benefactors in places like France, England, and Switzerland as the eager subjects of informal empire, willing to accept unequal economic, diplomatic, and discursive relationships in return for quick cash and political backing. They also competed among one another for this foreign support, claiming that their political rivals misused funds or abandoned initiatives that had received financial backing from abroad. In the letters, articles, and pamphlets that they wrote for Anglophone and Francophone correspondents and readers, Greek leaders often characterized themselves as conduits of modernity and civilization and their local political and economic competitors as tyrants—a term coded as unmodern (or un-European) as well as Ottoman (or Muslim).

In this chapter, I explore how Greek politicians used the concept of tyranny to assimilate their rivals with the Ottoman past. This was one of several discursive strategies—others of which I have discussed elsewhere—that they used to win the support of Western European patrons.[1] Their aims were largely political; they sought to forge alliances with benefactors and discredit their rivals, rather than necessarily accomplishing projects or enacting policy on the ground. Depicting their adversaries as tyrants functioned as a shorthand to denote all that was corrupt, self-interested, and unmodern. It built on a preexisting rhetorical repertoire that permitted commentators to equate the Ottoman Empire with ideas of bad government and social decay elsewhere—for instance, in Spain.[2] Here, I begin by surveying recent literature on the concept of tyranny to locate a mutually shared definition of the term that Grecophone leaders and their Francophone interlocutors deployed. I then outline the semantic weight that the idea of the tyrant had in the immediate post-independence context. Finally, I outline some of the ways public figures working in or concerned with the Greek state pegged accusations of tyranny to the mismanagement of infrastructure projects to alternatively retain or gain the backing of Western European benefactors. I argue that by linking tyranny to the legacy of the Ottoman Empire, they marshaled the concept to ensure the continued financial and political assistance of foreign allies during a period that can best be understood as

1 Alex R. Tipei, *Unintended Nations: France's Empire of Civilization, Southeast Europe, and the Post-Napoleonic World* (Montreal: McGill-Queen's University Press, 2025), 129–139, 159–189.

2 Juan Luis Simal and Darina Martykánová, "Ferdinand and the Sultan: The Metaphor of the Turk and the Crisis of the Spanish Monarchy in the Early Nineteenth Century," *Contributions to the History of Concepts* 10, 1 (2015): 1–26.

one of transition from formal (Ottoman) to informal (Western European) empire.

My (admittedly restrained) case study focuses on accusations of tyranny in the Francophone media during the Greek War of Independence (1821–1828) through the aftermath of the 1831 assassination of Ioannis Kapodistrias (1776–1831), an independent Greece's first governor. I aim to construct a conceptual framework and semantic repertoire that will permit further exploration of how these events were depicted in other linguistic contexts—such as the Grecophone and Anglophone media of the time. This schematic will also allow for analysis of debates shaped by similar dynamics in and beyond the early Greek state in the Ottoman/post-Ottoman Southeast European space. It helps to flesh out a story about how diplomats, businessmen, and public figures from Western Europe imposed hegemonic norms on local leaders in the increasingly autonomous and independent territories in the Balkans via soft power. This, in turn, underwrote asymmetric political and economic relationships and discursively reinforced cultural and biopolitical hierarchies. In a word, these exchanges were a key tool in creating informal colonies or groupings and regions dependent on and beholden to their benefactors in Western Europe over the course of the nineteenth century.[3]

Tyranny as a Shorthand

A range of scholars have developed a renewed interest in the concept of tyranny, notably over the course of the last decade. Many have crafted

3 On informal empire, soft power, and the concept of coloniality generally see John Gallagher and Ronald Robinson, "The Imperialism of Free Trade," *Economic History Review* 6, no. 1 (1953): 1–15; Niall Ferguson, "Hegemony or Empire?," *Foreign Affairs* 82, no. 5 (2003): 154–161; Joseph S. Nye Jr., *Soft Power: The Means to Success in World Politics* (New York: Perseus Books, 2004); Aníbal Quijano, "Coloniality of Power and Eurocentrism in Latin America," *International Sociology* 15, no. 2 (2000): 215–232. For discussions of informal empire in the (former) Ottoman lands and Southeastern Europe see, among others, Tipei, *Unintended Nations*; Alexandre Massé, *Un empire informel en Méditerranée: Les consuls de France en Grèce et dans l'Empire ottoman: images, ingérences, colonisation (1815–1856)* (Paris: Classiques Garnier, 2019); Richard S. Horowitz, "International Law and State Transformation in China, Siam, and the Ottoman Empire during the Nineteenth Century," *Journal of World History* 15, no. 4 (2004): 445–486; Daniel Chirot, *Social Change in a Peripheral Society: The Creation of a Balkan Colony* (New York: Academic Press, 1976); Ştefania Costache, "Westernization as Tool of Inter-Imperial Rivalry: Local Government in Wallachia between Ottoman Control and Russian Protection (1829–1848)," *New Europe College Yearbook* (2011–2012): 55–92.

semantic histories of the term. Though the precise origins of the word remain fuzzy, in classical antiquity a tyrant was a ruler who held illegitimate power. An ancient tyrant might be good or bad. A benevolent tyrant renounced power after shepherding their people through a period of crisis. A tyrant who refused to give up their position, ruled uniquely in their self-interest, or reduced their inner circle to accomplices and cronies, however, made themselves vulnerable to justifiable tyrannicide.[4] In the Byzantine Empire and larger Orthodox Christian world, political discourses associated the concept of the tyrant with an individual who sought to usurp legitimate authority or, alternatively, with a ruler perceived as "foreign." Later writers in the post-Byzantine space equated the tyrant with poor governance or possibly an unsuccessful bid for the throne. They embedded notions of divine punishment in their deployment of the term, a sentiment that grew more acute during the era of Ottoman rule.[5]

Usages and understandings of tyrant and tyranny also mutated in Western Europe over time. Political philosophers such as Machiavelli (1469–1527) and Hobbes (1588–1679) identified unchallengeable rule as a crucial characteristic of good governance. From their point of view, a prince's tyranny was irrelevant. For Hobbes, tyrannophobia was a negative position, whereas Machiavelli asserted that observers could only recognize tyranny retrospectively, making the category essentially useless. Montesquieu (1689–1755), as several historians and political scientists suggest, conflated tyranny and despotism. His political philosophy increasingly connected both concepts with Orientalist ideas about non-European cultures, including the Ottoman state. Other philosophers and political actors followed suit, leading Melvin Richter to contend that the distinction between tyranny and despotism had all but totally eroded during the French Revolution.[6]

After 1815, political thinkers, especially in France, debated notions of tyranny, despotism, and dictatorship anew as they sought to hash out the

4 For recent work on the concept of tyranny, see Andrew Fiala, *Tyranny from Plato to Trump: Fools, Sycophants, and Citizens* (Lanham: Rowman and Littlefield, 2022); Walter R. Newell, *Tyranny and Revolution: Rousseau to Heidegger* (Cambridge: Cambridge University Press, 2020); Aoife O'Donoghue, *On Tyranny and the Global Legal Order* (Cambridge: Cambridge University Press, 2021).

5 Nikos Panou and Hester Schadee, eds., *Evil Lords: Theories and Representations of Tyranny from Antiquity to the Renaissance* (Oxford: Oxford University Press, 2018); Moisés Prieto, ed., *Dictatorship in the Nineteenth Century: Conceptualisations, Experiences, Transfers* (Abingdon: Routledge, 2022).

6 O'Donoghue, *On Tyranny*, 41; Panou and Schadee, *Evil Lords*, 6; Melvin Richter, "A Family of Political Concepts: Tyranny, Despotism, Bonapartism, Caesarism, Dictatorship, 1750–1917," *European Journal of Political Theory* 4, no. 3 (2005): 231.

legacies of the French Revolution and Napoleonic Empire. Some asserted that Napoleon had reigned as a tyrant; others—Benjamin Constant (1767–1830), for instance, per Richter—disapproved of applying the term tyrant to the defeated emperor. Weighed down by its classical baggage, the label of tyranny obfuscated the novelty and modernity of the emperor's regime, in Constant's view. Here, for example, we can think of Napoleon's massive overhaul of the French bureaucratic infrastructure and other modernizing endeavors meant both to encourage efficiency and to solidify state control. Following Richter, a new "family of political concepts" formed in this discursive space. The coinage of Bonapartism connected ideas traditionally rooted in the notion of tyranny to a distinctly modern and, importantly, French form of the state and its administration. Its near-synonym, Caesarism, universalized the constellation of ideas bound up in Bonapartism, rendering it applicable to regimes beyond France.[7]

Few writers, if any, applied terms like Bonapartism or Caesarism to the Sublime Ottoman Porte during the Greek War of Independence. Instead, they relied on the older notions of tyranny and despotism to criticize the sultan and his state. In a speech urging French intervention in the conflict—first delivered in the French chamber and later revised for publication—Constant spoke repeatedly of the Ottomans as "barbarians," "cruel," "violent," and "despotic." He postulated that the empire had been "fashioned over the last several centuries by the necessary and pernicious union of despotism and anarchy [...]." He contrasted "its stationary genius" to the "progress of [Western European] civilization [...]."[8] Dominique Dufour de Pradt (1759–1837), a French statesman and diplomat, wrote numerous pamphlets that approved of the Greek cause. He told readers that "the [Ottoman] pashas exercise [their] tyranny far from the eyes of their masters, living off of extortions made possible by the eternal absence of order [...]."[9] Similarly, he described the Grecophone Porte-appointed princes in the Danubian Principalities—often referred to as Phanariots—as "valets of the

7 Richter, "A Family," 235–240.

8 Benjamin Constant, *Appel aux Nations Chrétiennes en faveur des Grecs* (Paris: Treuttel et Würtz, 1825), 6, 7, 9, 11–12, 13, 15.

9 Dominique Dufour de Pradt, *L'Europe par rapport à la Grèce et à la reformation de la Turquie* (Paris: Béchet ainé, 1826), 131. On the Phanariots, see Christine Philliou, *Biography of an Empire: Governing Ottomans in an Age of Revolution* (Berkeley: University of California Press, 2011); Bogdan Murgescu, "'Phanariots' and 'Pământeni': Religion and Ethnicity in Shaping Identities in the Romanian Principalities and the Ottoman Empire," in *Ethnicity and Religion in Central and Eastern Europe*, ed. Maria Crăciun and Ovidiu Ghitta (Cluj: Cluj University Press, 1995); Andrei Pippidi, "Phanar, Phanariotes, Phanariotism," *Revue des études sud-est européennes* 13, no. 2 (1975), among others.

tyrants of Constantinople" who "accumulated extortions as small tyrants occupied with assuring the protection of other tyrants [in Istanbul] who measured patronage in terms of the gold they produce […]."[10] Importantly, de Pradt suggested Islam and the Ottoman system were antithetical to modern consumer culture and commercial relations; therefore, the Porte would always remain disinterested in economic development and, we might deduce, in the kind of investment in infrastructure that underwrites it.[11] In both Constant and de Pradt's renderings, the Ottoman system bore the hallmarks of classical tyranny—bad governance (or anarchy), an illegitimate hold on power, and a state administration that sustained itself through the self-interest of individual actors or, in their words, "petty tyrants." Neither de Pradt nor Constant considered the empire "civilized" or ascribed the modernizing impulse embedded in novel labels like Bonapartism and Caesarism to the Porte.

De Pradt and Constant, like other contemporaries concerned with Greek independence, used these accusations to Orientalize and even racialize the Ottoman Porte as well as Muslims more generally. Connecting what they viewed as the Porte's inherent tyranny and despotism to the Ottomans' "Asian origins," they drew a line between political and human characteristics, on the one hand, and climate and geography, on the other. Describing the Ottomans as impervious to civilization, they essentialized their supposed tyranny. Evoking what they portrayed as the empire's stilted development—which they explicitly tied to trade and consumption as well as the modern infrastructure which facilitated capitalist forms of exchange—these French writers attributed such perceived failings in large measure to the Muslim faith of the sultan and many of his subordinates. Public interventions like these amounted to an early articulation of the key tropes of modern Islamophobia. They likewise built up an image of the Ottoman state as incapable of embracing not only political but also economic and technological modernity.[12]

10 De Pradt, *L'Europe par rapport*, 155.

11 Dominique Dufour de Pradt, *De la Grèce dans ses rapports avec l'Europe* (Paris: Bechet, 1822), 65–69, 83.

12 See, for example, de Pradt, *De la Grèce*, 37; *Du système permanent de l'Europe à l'égard de la Russie et des affaires de l'Orient* (Paris: Pichon et Didier, 1828), 146; *Vrai système de l'Europe relativement à l'Amérique et la Grèce* (Paris: Bechet, 1825), 75, 78, 77, 235, 264; *L'Europe et l'Amérique en 1822 et 1823 (deuxième partie)* (Paris: Béchet, 1824), 197; Constant, *Appel aux Nations*; Alex R. Tipei, "Korais's Greece and Napoleon's Empire: The Egyptian Campaign, Race Science, and the Europeanization of an Idea," in *From the Napoleonic Age to the Age of Empire: Empire after the Emperor*, ed. Thomas Dodman and Aurélien Lignereux (London: Palgrave, 2023).

The Grecophone, Orthodox leaders of the insurgency also marshaled the concept of tyranny to legitimize the conflict inside the Ottoman Empire. The first wartime constitution, promulgated at Epidaurus in 1822, referenced the Greeks' supposed antiquity and superior level of civilization, as well as their Christian faith, to frame the conflict as a "natural," or justified, war instead of a revolution. It further rationalized the validity of the revolt by emphasizing the "tyranny of the Ottomans."[13] After the Siege of Tripolitsa (1821), Demetrios Ypsilantis (1793–1832), a former Russian officer, proclaimed he had traveled to the Peloponnese to fight against the "tyranny of the Turks." Ypsilantis, who likewise publicly announced his intention to combat anyone who impeded justice, extended the notion of tyranny beyond religious or ethnic divisions. Depicting both the Muslim and Orthodox notables appointed by the Porte as "tyrants," he exhorted Peloponnesians to identify the best men to take over the administration.[14] His address implied that integration into the Ottoman administrative system might render even a Greek-speaking Orthodox Christian a tyrant. Such individuals were incapable of developing a modern bureaucracy or infrastructure in the new country. Ypsilantis's comments, furthermore, pointed to a potential critique that different factions among the insurgents, and later the early state's cadres, could levy at their political adversaries.

Similarly, Alexandros Mavrokordatos (1791–1865), a Phanariot with close ties to continental diplomatic circles who later became the prime minister of Greece, penned his *Coup d'oeil sur la Turquie* for Friedrich von Gentz (1764–1832), Prince Klemens von Metternich's (1773–1859) right-hand man. A primer on the situation in Ottoman Europe and the potential outcomes of the conflict, the text warned that unlike in antiquity, when one man acted as a tyrant, the Ottoman administration multiplied the existence of tyrants: "by as many monsters as there are pashas, ayans, agas, [and] voivodes, who all also have the right of life or death over the trembling subjects of the chief tyrant [the sultan]." In his view, the creation of an entire hierarchy of tyrants constituted a systemic characteristic of Ottoman rule. He noted the uniqueness of the Ottoman state, which "could not be compared to any of the administrative systems we know today, or that history has transmitted to us." Unlike the ancient model of Rome, there was no virtuous Titus to compensate for the cruelties of Tiberius; nor could he reasonably contrast

13 Anton Freiherrn von Prokesch-Osten, *Geschichte des Abfalls der Griechen vom Türkischen Reiche im Jahre 1821 und der Grünung des Hellenischen Königreiches aus Diplomatischem Standpuncte* (Vienna: Carl Gerold's Sohn, 1867), 2: 258.

14 The proclamation appears in von Prokesch-Osten, *Geschichte des Abfalls der Griechen*, 2: 247.

the Porte's despotism to that of Russia, since "Christian morales softened [...] mores and called [Russians] to the path of civilization."[15] Relying on the concept of tyranny to make his case, Mavrokordatos insisted that the Ottomans, due to their Muslim faith and non-European culture, represented a total deviation from the European continental order and stood outside of a linear understanding of history, in which he and other advocates of the insurgency located Greece and the Greeks.

In short, deployment of concepts like tyranny by both Grecophone figures and their Western European advocates during the war helped establish a mutually shared definition of the term, making it a useful touchstone in debates meant to win over foreign patrons during the following period.[16] In this dialogue, tyranny denoted bad governance and illegitimate authority. It also carried an Orientalizing connotation and transmitted a sense of the unmodern. A tyrannical regime, unlike one associated with the newer notions of Bonapartism or Caesarism, lacked the capacity to follow modern economic, political, or administrative practices. Its leader would not oversee massive overhauls of infrastructure, institutions, and state organization as Napoleon had but depended on arbitrary and arcane practices and only engaged in reforms that immediately served their own interests. Importantly, in the context of the early Greek state, foreign patrons could hardly trust a tyrant or their administration to properly employ the financial support that they might receive.

Tyranny and the Kapodistrian Regime

In 1828, the Greek National Assembly elected Kapodistrias governor of a newly independent Greece. A former Russian diplomat, the Ionian-born Kapodistrias commanded a thick network of contacts among the continent's political, cultural, and economic elite. Many observers inside and outside of Greece hailed his appointment as a cause of celebration and hope for the new state's future, especially as he was supremely well positioned to ramp up investment and interest in Greece across Europe.[17] Kapodistrias would

15 Alexandros Mavrokordatos in von Prokesch-Osten, *Geschichte des Abfalls der Griechen*, 1: 5–6.

16 Alex R. Tipei, "Audience Matters: 'Civilization-Speak,' Educational Discourses, and Balkan Nationalism, 1800–1840," *European History Quarterly* 48, no. 4 (2018): 658–685.

17 On Kapodistrias, see C. M. Woodhouse, *Capodistria: The Founder of Greek Independence* (London: Oxford University Press, 1973). In Greek, a comprehensive recent treatment of Kapodistrias's life is Christos Loukos, *Iōannēs Kapodistrias: mia apopeira historikēs viographias*

serve a seven-year term while Greece's protecting powers hammered out a deal regarding its formal status and final borders and eventually searched for a king to name to its throne.

Kapodistrias also had a great deal of assistance from the protecting states and their representatives. The French army, for example, remained on the ground. In addition to performing military functions, French forces helped conduct land surveys, built communication infrastructure, and trained Greek recruits for a modern military.[18] Philhellene volunteers—including Louis-André Gosse (1791–1873), a Paris-trained medical doctor from Geneva, and Henri Dutrône (1796–1866), a lawyer with close ties to Philhellenic organizations in France—worked in the administration. They took on projects like the expansion of the quarantine and healthcare systems, a network of primary schools, and other modernizing initiatives. They often controlled money raised in places like London and Paris via loans and philanthropic donations. This gave their views enormous weight among Greek officials and stakeholders back home in France and Switzerland.[19] Privately, Kapodistrias recognized how this support rendered Greece an informally colonized space. Yet in both public pronouncements and personal correspondence, Kapodistrias also noted how sorely the new state needed these patrons.[20] After all, the agrarian, war-torn country required much investment and manpower to build educational, medical, military, and transportation infrastructure.

Many Greek public figures who had fought in the war and participated in early constitutional and national assemblies bristled under Kapodistrias's leadership. As the governor attempted to centralize the administration, modernize the state's institutions, and develop infrastructure, his adversaries

(Athens: MHET, 2023). See also Ioannis Kapodistrias, [Jean Capodistrias], *Aperçu de ma carrière publique depuis 1789 jusqu'à 1822* (Paris: La Vouivre, 1999). For a brief overview of Hellenophones in the service of the Tsar, see Thomas Gallant, *The Edinburgh History of the Greeks, 1768 to 1913: The Long Nineteenth Century* (Edinburgh: Edinburgh University Press, 2015), 19–22; for a more extensive discussion, see Stella Ghervas, *Réinventer la tradition: Alexandre Stourdza et l'Europe de la Sainte-Alliance* (Paris: Honoré Champion, 2008).

18 Thanos Veremis, "Kapodistrias and the French: The Formation of a Regular Greek Army." *Conference on War and Society in East Central Europe During the Eighteenth and Nineteenth Century* (New York: The Center for European Studies, Graduate School and University Center, City University of New York, 1979); Massé, *Un empire informel*; Jean-Pierre Pellion, *La Grèce et les Capodistrias pendant l'occupation française de 1828 à 1834 par le général de division Pellion: L'avenir du royaume grec est dans sa marine marchande et son commerce* (Paris: J. Dumaine, 1855).

19 See Tipei, *Unintended Nations*, 160–169.

20 Pavlos Petridis, *O Iōánnīs Kapodístrias, 1776–1831: O koryfaíros Ellīnōas Eyrōraíos* (Athens: Govostis, 1992), 123; von Prokesch-Osten, *Geschichte des Abfalls*, 4: 139–141.

contested his policies and authority through public debate, political machinations, and martial violence. To critique his policies and actions, they routinely depicted him as a tyrant. For example, Kapodistrias's political rivals charged him with attempting to create what they termed a government of "family," appointing his brothers, Augustinos (1778–1857) and Viaros (1774–1842), to key posts and stacking state bureaucracies with transplants from his native Ionian Islands—surrounding himself, as a tyrant might, with acolytes and cronies. They especially took issue with the creation of novel administrative positions and the replacement of older appointees with new ones, whom Kapodistrias and his inner circle considered loyal to the state. They accused Kapodistrias of trying to play powerful local notables against one another—for example, provoking tensions between the Mavromichalis and Murzinos clans in the Mani region of the Peloponnese.[21]

Opposition leaders likewise ticketed Kapodistrias with seeking to concentrate power in his own hands, giving him the freedom to govern as a tyrant. They alleged that he used an assembly at Argos in 1828 to consolidate his standing. Indeed, more than twenty years later, General Jean Pierre Pellion (1793–1864), a French officer who served in Greece and was highly critical of Kapodistrias, reiterated this accusation. Pellion told readers in his memoir of the period that Kapodistrias packed the assembly with his supporters. He claimed the congress served to "fool Europe" and gave Kapodistrias a pretext to replace it with a commission—an unscrupulous move without parallel in Pellion's estimation. Though the assembly considered several propositions, some meant to bolster relations with foreign benefactors—the distribution of national lands to philhellenes and a monument to the French Army, for example—it also invested Kapodistrias with "dictatorial" powers, per Pellion. While the French general admitted these measures may have been necessary in the post-war context, he nonetheless asserted Kapodistrias really intended to enhance his own position. When Kapodistrias delayed calling a new national assembly a few years later, critics accused him of governing in the same arbitrary manner as the Ottomans.[22]

Additionally, as Greece's protecting powers considered candidates for the country's throne, Prince Leopold of Saxe-Coburg-Saalfeld, the future king of Belgium, emerged as a top contender. Kapodistrias engaged Leopold in a

21 Some observers additionally argued Kapodistrias had not actually dismantled the old administrative system but simply improved it. See *Lettres et documents officiels relatifs aux derniers événements de la Grèce, qui ont précédé et suivi la mort du Comte Capodistrias: Publié par plusieurs membres de l'ancien comité grec de Paris* (Paris: Firman Didot, 1831), 51.
22 Pellion, *La Grèce*, chapter 4.

lengthy correspondence. In his missives, the governor detailed the political obstacles facing the Greek state as well as the dire postwar economic context and immense need for foreign investment. Contemporary observers and latter-day historians have hotly debated the governor's intentions and the impact of his letters. Those who view Kapodistrias's legacy positively posit that the governor angled to ensure the prince had a full grasp of Greece's situation and the work that lay ahead. Critics of the Ionian statesman suggested that Kapodistrias aspired to the crown himself. Regardless of Kapodistrias's "true" intentions, which historians can only speculate about, his political rivals, like subsequent commentators and scholars, pointed to the controversy as proof of the governor's tyrannical tendencies.[23]

Kapodistrias and his allies, like their political rivals, strove to cultivate clients and appoint their own loyalists to key positions. They engaged both local and European publics to do so, playing to interested parties that might hold Greek leaders and their state accountable. One way we can understand this very public debate is less in terms of the (very real) political infighting and violence taking place on the ground and more in terms of the strategic rhetorical positioning these actors pursued in the foreign press. Greek notables, especially autochthones—or those born within the borders of the new state—had been accustomed to working within the Ottoman framework. As Ali Yaycioglu points out, such officials tended to see themselves as servicers, rather than servants of the state, who could routinely negotiate and renegotiate their relationship and obligations with a central authority. A codified, centralized administration represented a profound rupture with long-standing regional practices.[24] In this environment, Greek leaders had traditionally expended a great deal of energy building and solidifying client networks that would remain loyal to them. During the War of Independence,

23 Among others, see George Finlay, *History of the Greek Revolution* (London: Zeno, 1971); Pellion, *La Grèce*; Woodhouse, *Capodistria*; Christos Loukos, *Iōannēs Kapodistrias mia apopeira istorikis biographias* (Athens: MIET, 2022); Spyridonos D. Loukatou, *Ho Iōannēs Kapodistrias kai hē Heptanēsos Politeia*. (Athens: National and Kapodistrian University of Athens, 1959); Helen (Eleni) E. Koukkou, *Ho Kapodistrias kai hē paideia, 1803–1822. A', Hē Philomousos Hetaireia tēs Viennēs* (Athens: University of Athens, 1958); William P. Kadis, *John Capodistrias and the Modern Greek State* (Madison: State Historical Society of Wisconsin for the Department of History, University of Wisconsin, 1963); Douglas Dakin, "The Formation of the Greek State: Political Developments until 1923," in *Greece in Transition: Essays in the History of Modern Greece, 1821–1974*, ed. John T. Koumoulides (London: Zeno, 1977), 21–63.

24 Ali Yaycioglu, *Partners of the Empire: The Crisis of the Ottoman Order in the Age of Revolutions* (Stanford: Stanford University Press, 2016), 67, 114, 89–92, 97–99; See also Katherine Fleming, *The Muslim Bonaparte: Diplomacy and Orientalism in Ali Pasha's Greece* (Princeton: Princeton University Press, 1999).

for example, local primates periodically changed sides in the conflict when political and economic inducements suited their needs, taking the soldiers under their command with them. Western European observers, moreover, expressed shock when loan money arrived in the territory and, instead of using it to shore up central institutions, Greek notables employed the funds to reward individuals in their own retinues.[25] Thus, while the two factions adopted similar (traditional) practices, they raced to denounce one another, accusing rivals not simply of corruption but also of perpetuating Ottoman customs and styles of governance in the new Greek state—which, to a large extent, was exactly what they were doing. Through these denunciations, the parties looked to assimilate each other's political maneuvers with the Ottoman past. The label of tyrant let them accomplish this discursive task.

Kapodistrias and his collaborators likewise employed the concept of tyranny to characterize resistance to their administration. Addressing members of the national army, Kapodistrias began a speech: "The tyranny which for several centuries has been the scourge of this unfortunate Greece has made way for a fatal anarchy, which has retarded the progress of the fight that the nation has taken on to recover its independence and freedom."[26] Other officials in his administration employed similar rhetoric to explain the political wrangling and violence that plagued post-independence Greece. In a letter to the governor, appointees from the district around Corinth commented on how "long calamities taught [the men of this region] the corrupt and tyrannical principles that guided the conduct of [the notables who ignited intrigues]." The missive connected current conditions to the long-standing use of Ottoman administrative practices in the area as well as the presence of Ottoman troops during the war—phenomena that the letter's authors suggested had caused moral decay among local leaders.[27]

Such letters, addresses, and analyses by Greek actors regularly appeared in both the Grecophone and Francophone press. Western European observers likewise echoed ideas connecting unrest and poverty to an Ottoman legacy of "tyranny." In letters to his mother, Gosse regretted the political machinations of Greek politicians, which he linked to the history of Ottoman rule. The Francophone doctor submitted that these "defects" needed to be combated. Gosse's British collaborator, Leycester Stanhope (1784–1862),

25 Finlay, *History of the Greek Revolution*, 1: 434–437. On the history of loans to the Greek state, see Andreas Andreadis, *Istoría ton Ethnikón Daníon* (Athens: Estia, 1904). On the distinction that developed between gifts and bribes in Britan, see Bruce Buchan and Lisa Hill, *An Intellectual History of Political Corruption* (London: Palgrave, 2014), 111.

26 *Lettres et documents officiels*, 107.

27 *Lettres et documents officiels*, 120.

similarly mentioned in his correspondence how the Greeks remained unchanged after an era of "tyranny." He especially underlined the need for a modern, disciplined army to rein in the martial violence and establish order. He and Gosse worked together to identify trustworthy collaborators who would be unlikely to engage in corruption.[28]

Deploying the concept of tyranny had a dual function in these texts. First, it served a didactic purpose—contextualizing the continued violence in Greece and the massive administrative and economic challenges the state faced. Auguste Lagarde, a Frenchman who had spent three years in Greece mapping the Peloponnese, rhetorically asked in a letter published in Paris: "If your finances are in a wretched state, let the blame fall first on you [Greece], after eight years of war and plunder; then ask some of your leaders: 'who was it that made you rich?'"[29] Gosse, in a dispatch back home, described a lack of patriotism and indifference to public affairs as well as the deployment of non-standard administrative practices among the Greeks. Citing the tyranny of the Ottoman past, he asked: how could they be expected to behave otherwise?[30]

Yet, Gosse, Lagarde, and others also thought the Greeks had an innate ability to become "civilized" and modernize their country. They suggested that the Greeks simply needed guidance from Western European patrons. In this respect, appeals to the lasting effects of Ottoman "tyranny" also helped them construct an image of Greece as a heterotopia.[31] This was a second function of tyranny in these cross-continental debates. As the Christian descendants of antiquity, the Greeks merited independence and (re)integration into the European civilizational family. As a society that—due to its Ottoman past and the mismanagement of its current leadership—had not followed a proper line of civilizational development, the Greeks required assistance. Positioning Greece as a heterotopia validated the exercise of Western European soft power and attempts at informal colonization of the country. Greek leaders could not be trusted to distribute money given to them, implement technologies shared with them, or administer their own state efficiently. They required the tutelage of their more advanced patrons. Remember, Kapodistrias himself had identified this dynamic early on in

28 Leycester Stanhope, *Lettres du Colonel Stanhope sur la Grèce* (Paris: Corneille, 1825), 179–180.

29 *Lettres et documents officiels*, 262–265.

30 Louis-André Gosse, *Lettres à ma mère pendant un séjour en Grèce, 1826–1830 (édition 1920)* (Paris: Édition Atar, 1920), 23.

31 Artemis Leontis, *Topographies of Hellenism: Mapping the Homeland* (Ithaca: Cornell University Press, 1995), 43–44; Michel Foucault, "Des espaces autres," *Architecture/Mouvement/Continuité* 5 (1967): 46–49.

his term as governor.[32] He had also recognized the acute need for Western European political and economic support. How keenly the Greek political class as a whole understood this became abundantly clear in the contest for power on the ground and backing from abroad that erupted after the governor's assassination.

The Assassination of a Tyrant?

Petros Mavromichalis (1773–1848) was among Kapodistrias's most powerful rivals. The head of a Maniot family who had fought in the conflict with the Porte and participated in the wartime administration, he served in the senate during the governor's administration. In response to Kapodistrias's centralization program, Mavromichalis's family and their allies led an uprising in 1830 against the appointed governor of his region, Laconia. Various factions had engaged in civil war during the conflict for Greek independence, and continual revolt plagued the early state. These were events that Kapodistrias and his allies attributed to the legacies of the Ottoman past when they discussed the culture of "tyranny" that still permeated the region. Seeking to put an end to the turmoil, Kapodistrias had Mavromichalis arrested. The governor refused to negotiate with the Maniote chieftain as a one-time Ottoman appointee might have reasonably expected him to do. In response, Mavromichalis's relatives orchestrated Kapodistrias's assassination in October 1831, shooting and stabbing the governor on his way to an early morning Church service in the Greek capital of Nafplio.

The governor's murder reverberated across the new Greek state as well as the rest of Europe. Following the scandal, both pro- and anti-Kapodistrian public figures in Greece angled to position themselves in relation to the fallen statesman's legacy. Much of this back-and-forth played out in the Western European press, as Greek public figures looked to curry the favor of patrons abroad whose backing might lend them political legitimacy in Greece and ensure the continued flow of foreign financial assistance. Both sides in this strategic confrontation grounded their arguments in an evaluation of how "modern" Kapodistrias was and how successful his reform agenda had been. They often focused on the development (or lack of development) of transportation, administrative, and bureaucratic infrastructure, engaging in what Malte Furhmann has previously described in this volume

32 Cited in Petridis, *Iōánnīs Kapodístrias*, 123; Tipei, *Unintended Nations*, 159.

as "infrastructural shaming" and "infrastructural boasting."[33] Use of the concept of tyranny in these debates, like that of corruption, provided anti-Kapodistrians with a frame of reference to critique the governor that their foreign audience might easily grasp.

Opposition leaders and their allies in Western Europe portrayed Kapodistrias and his administration as corrupt and driven by nepotism and vice. Dutrône ridiculed the idea that the governor's administration could be considered enlightened for the Francophone public. Writing in the voice of autochthonous opposition Greeks (notably those born within the borders of the new state, such as members of the Mavromichalis family), he exclaimed: "'How is it possible!' the natives say; 'These are examples of those venerated populations, examples undoubtedly chosen among the best, because they gave them to us as models! [...] Our barbarianism is worth more than their civilization."[34] Opposition leaders and their foreign allies justified their objections to Kapodistrias and his policies, as well as their depiction of his comportment, by portraying him as an uncivilized barbarian and describing his administration as tyrannical. Dutrône drew additional comparisons between Kapodistrias's administration and the Venetian Council of Ten. He scoffed that the slain governor would have seamlessly fit in with this inquisitorial oligarchy.[35] These arguments attempted to undercut the perception of Kapodistrias as a modernizing force in Greece, likening him instead to a retrograde form of governance. Citing efforts to elaborate an educational infrastructure, Dutrône submitted that Kapodistrias and his collaborators had talked a great deal about bringing the country into line with Western European standards, practices, and institutions but had regularly abandoned their efforts to do so, once they had secured the political and financial backing of foreign patrons, letting plans and projects fall by the wayside.[36] Similarly, Pellion recalled how Kapodistrias neglected the advice of benefactors representing more modern societies. He taxed the governor with accepting "the money of France, but not its influence"[37] and with squandering the resources the French army and philhellenes

33 Malte Furhmann, "On Hegemony and Connectivity: A Short History of Infrastructure-Political Struggles and their Discourses" in this volume.

34 Henri Dutrône, *Extraits de la correspondance de M. Dutrône avec M. le Président Capodistria* (Paris: E. Cassin, 1831), 19.

35 Dutrône, *Extraits de la correspondance*, 40. The Council of Ten was a governing body in Venice from 1310 to 1797. They held significant power over state security and, in 1539, introduced state inquisitors, to whom Dutrône referred.

36 Dutrône, *Extraits de la correspondance*, 40.

37 Pellion, *La Grèce*, 149.

had furnished Greece. The implication was not that Kapodistrias did not understand what his administration needed to do to modernize the new state, but that he and his entourage failed to enact measures out of their own self-interest—a hallmark of corrupt and tyrannical behavior.[38]

Kapodistrias's allies countered these allegations in the French press. Stamatis Voulgaris (1774–1842), one of Greece's first urban planners and a close associate of Kapodistrias, reversed the charges. In a volume he published in Paris, Voulgaris argued that only the primates who had wielded their power "tyrannically" in Greece had lost their positions during the governor's restructuring of the administration. He stressed Kapodistrias had attempted to rein in these notables for the good of the country.[39] Voulgaris additionally underscored Kapodistrias's lack of self-interest—a quality that the governor's rivals routinely ascribed to tyrants. Similarly, Jean-Gabrielle Eynard (1775–1863), a noted Swiss philhellene who helped found the National Bank of Greece, told readers across Europe that Kapodistrias had not been a tyrant but rather "the enemy of small tyrants." He, too, highlighted Kapodistrias's disinterested dedication to the new country, referencing his refusal to accept a salary and his tireless work on behalf of the country. Eynard remarked that many of the Greeks he corresponded with, as well as Hellenophones living in France, agreed with his assessment of Kapodistrias's character as virtuous and his actions as intended purely in pursuit of the country's interests.[40]

More colorfully, Lagarde, the French cartographer, compared "the head of the Greek government who showed himself without a retinue, on foot, dressed like one of the least of Paris's bourgeois; accessible and affable to all" to the local primates "pompously encircled by a band of clients and valets, [who] insolently strutted around with their guns and golden vests [...]." Evoking images of the Ottoman world, Lagarde employed them to Orientalize the local elites. He went on to describe the primates as "[...] men of blood and plunder, men of yesterday and nothingness; others," specifically stressing their alterity. He continued: "Jealous of preserving a hereditary exploitation, exaggerating their rights perhaps because of their sacrifices, a sort of mortgage, in their eyes, on possession of the country, freed from the Turks from whom they had leased it [...]." Lagarde linked the primates to the Ottoman system and then depicted these men as "promising themselves in

38 Buchan and Hill, *Intellectual History*, 108–109.

39 Stamati Voulgaris, *Notice sur le comte Jean Capodistrias, président de la Grèce suivie d'un extrait de sa correspondance*, 2nd ed. (Paris: Delaunay, 1832), 1, 10.

40 *Lettres et documents officiels*, 14, 18, 20–21, 24.

good faith to enjoy [the country] as suzerains, as owners, that is to say as idlers, doing nothing and at the expense of all that produces […]."[41] Lagarde's portrayal marshaled orientalist tropes as well as characteristics such as bad governance, lack of productivity, and violence, which contemporary discourses both ascribed to the Ottoman Porte and consistently connected to the concept of tyranny.

Lagarde and other pro-Kapodistrians highlighted the governor's role as a modernizing force in the recently established country. In a pamphlet penned for Francophone audiences, the Ionian Voulgaris depicted Kapodistrias's modernizing agenda as an attempt "to progress Greek civilization,"[42] while Eynard noted in another text published in Paris that "[Kapodistrias] had been the link between civilized Europe and Greece."[43] All these men supported their claims with reports on the reform of the Greek administration, the expansion of the country's network of roads, and the development of education and healthcare systems. Government documents and letters by Greek officials and Western European philhellenes detailed these achievements.

Lagarde enumerated the institutions and projects initiated by Kapodistrias's administration that had come under attack, defending them against "infrastructural shaming."[44] He argued that the small number of school instructors corresponded to the needs and capacities of the state and projected the establishment of a polytechnic school in short order. He explained that the library, formally closed to the public, could be accessed upon request. He described the orphanage Kapodistrias had personally established on Aegina for children who had lost their parents in the conflict. Lagarde asserted that while construction remained incomplete on a key road between Argos and Napflio, the built section was more than serviceable. The same compendium, which featured several essays by Eynard, also contained reports penned by officials from various ministries, including Andreas Moustoxydis (1785–1860) and I. P. Kokkonis (1796–1864), which surveyed the growth of the school system. A summary of the Fourth National Congress, held in 1829, and written by the president of the assembly, appeared. Documents that described the state's budget, plans to repay loans, and reforms of the military accompanied the text. A letter penned by a group of Greek

41 *Lettres et documents officiels*, 260.

42 Voulgaris, *Notice sur le comte*, especially 1–2, 24–25. For some of Kapodistrias's modernizing endeavors, see Veremis, "Kapodistrias and the French"; Dimitris, Loules, *The Financial and Economic Policies of President Ioannis Capodistrias, 1828–1831* (Ioannina: Ioannina University Press, 1985); Koukkou, *O Kapodistrias*.

43 *Lettres et documents officiels*, 14, 16, 28, 306.

44 Furhmann, "On Hegemony and Connectivity."

leaders, including Theodoros Kolokotronis (1770–1843), noted that some French periodicals refused to publish their missive in the context of this political competition. They discussed the construction of roads and the creation of a postal system, among other modernizing projects. Dozens of other official documents and testimonials by Greek politicians and French philanthropists backed the volume's pro-Kapodistrian argument. Drafters of these texts took great pains to emphasize the modernization projects Kapodistrias undertook for the good of the country and refute claims that he ruled as a tyrant.[45] All of this added up to a qualified form of "infrastructural boasting."[46]

Here again, in this back-and-forth, divergent opinions derived more from a given author's perspective and framing than disagreement over the facts. Some argued, as Lagarde did, that a well-made road, though unfinished, represented progress. Alternatively, a critic might focus on the fact this project had yet to reach completion. Perhaps the number of schools corresponded to the needs of the state, or at least to the limited resources at its disposal. Or, following the thinking of detractors like Dutrône, it could signify a shortfall, since Kapodistrias had promised to educate all Greek children yet had taken Western European money for this endeavor without reaching his stated goal. The real prize in these debates concerned politics, not policy or implementation. Discussions of critical infrastructure, thus, both highlighted the material situation in the new Greek state and had a symbolic function, allowing local leaders to showcase their priorities and "civilizational" abilities. Pro- and anti-Kapodistrians shared the same aims: to position themselves in respect to the governor's legacy, to convince foreign patrons that their camp sought to truly enact a modernization program (which would be beneficial to public interests rather than private concerns), and to shore up the future support of these benefactors. The success of these public interventions had less to do with what was happening on the ground and more to do with who could secure the trust and assistance of powerful and wealthy patrons across the continent.

Finally, the opposition's deployment of the term tyranny had another possible implication. If Kapodistrias had been a tyrant, this suggested that, like the Ottomans, he was incapable of acting as a civilizational actor. His administration, therefore, lacked the ability to modernize Greece and instead impeded its development. His assassination, if conceived of as a tyrannicide, consequently might have served the welfare of Greece. An

45 *Lettres et documents officiels.*
46 Furhmann, "On Hegemony and Connectivity."

unstated subtext, then, of the anti-Kapodistrian use of the concept of tyranny was the legitimation of the governor's removal, even via violent means. The Mavromichalis clan had not acted (uniquely) for their own profit on this reading of the events but also in the interests of the Greek state and its citizens. Though a discursive outlier, as recently as the revolutionary debates over regicide, Saint-Just had proposed tyrannicide as a justification for executing the king.[47]

Anti-Kapodistrians had to tread carefully here. The governor had maintained a vast network of contacts and had commanded the respect of many key figures across the continent up to his death. His allies systematically portrayed him as a martyr to Greek development. They could not simply come out and say Kapodistrias's death benefited Greece. They could, however, associate the purported tyranny of his regime with his successors—notably his brothers and the other Ionians who had populated his administration. Dutrône offered a model for how to proceed. While his pamphlet did not entirely spare Kapodistrias from posthumous attack, he focused his ire on the governor's administration as a whole and especially on Kapodistrias's brothers who remained in power. He suggested that they and the other Ionians had duped Kapodistrias. Placing the lion's share of blame for aberrant, corrupt, and tyrannical behavior on the governor's subalterns, he constructed a case for their removal and for Western European support of the opposition faction.[48]

Tyranny in Retrospect

In the nearly two centuries since Kapodistrias's assassination, historians have continued to debate assessments of his tenure as governor. Their evaluations largely depend on which set(s) of actors they choose to give credence to and what texts they explore as well as their own intellectual and political dispositions. Thinking through the radically different portrayals of Kapodistrias that have appeared in both the historiography and the press in the 1820s and 1830s suggests an alternative approach to this history. Specifically, in some ways, whether Kapodistrias or his political opponents acted as unmodern tyrants or as paragons of civilization and modernity had less importance than their ability to convince allies in France (and across Western Europe) that they did. The concepts of tyranny and tyrant

47 Richter, "A Family," 234.

48 Dutrône, *Extraits de la correspondance.*

supplied these figures with one discursive strategy to Orientalize their rivals as Ottoman-like officials. It functioned as a shorthand for all that was unmodern, corrupt, and chaotic. It undercut faith in the ability of leaders to execute projects that would create new educational, healthcare, military, economic, and physical infrastructure.

By mobilizing terms like tyranny and tyrant in these debates, Greek politicians—and the patrons abroad also—sometimes unintentionally reinforced a perception of Greece as a heterotopia. The long Ottoman legacy, supposedly one of tyranny, meant that Greek administrators remained susceptible to the practices of self-interested bad governance. They required the assistance of Western European benefactors to rectify the situation. This dynamic created ample space for the French (and other foreign) philhellenes and governments to impose their own vision of modernity on the new Greek state and shape it in accordance with their own political and economic interests. Accusations of tyranny, then, helped paper over the transition from formal Ottoman imperialism to informal Western European colonization in Greece.[49]

Bibliography

Andreadis, Andreas. *Istoría ton Ethnikón Daníon*. Athens: Estia, 1904.

Buchan, Bruce, and Lisa Hill. *An Intellectual History of Political Corruption*. London: Palgrave, 2014.

Chirot, Daniel. *Social Change in a Peripheral Society: The Creation of a Balkan Colony*. New York: Academic Press, 1976.

Constant, Benjamin. *Appel aux Nations Chrétiennes en faveur des Grecs*. Paris: Treutel et Würtz, 1825.

Costache, Ștefania. "Westernization as Tool of Inter-Imperial Rivalary: Local Government in Wallachia between Ottoman Control and Russian Protection (1829–1848)," *New Europe College Yearbook* (2011–2012): 55–92.

Dakin, Douglas. "The Formation of the Greek State: Political Developments until 1923." In *Greece in Transition: Essays in the History of Modern Greece, 1821–1974*, edited by John T. Koumoulides, 21–63. London: Zeno, 1977.

De Pradt, Dominique Dufour. *L'Europe et l'Amérique en 1822 et 1823* (*deuxième partie*). Paris: Béchet, 1824.

De Pradt, Dominique Dufour. *L'Europe par rapport à la Grèce et à la reformation de la Turquie*. Paris: Béchet ainé, 1826.

49 See Tipei, *Unintended Nations*; Massé, *Un empire informel* among others.

De Pradt, Dominique Dufour. *De la Grèce dans ses rapports avec l'Europe.* Paris: Bechet, 1822.

De Pradt, Dominique Dufour. *Du système permanent de l'Europe à l'égard de la Russie et des affaires de l'Orient.* Paris: Pichon et Didier, 1828.

De Pradt, Dominique Dufour. *Vrai système de l'Europe relativement à l'Amérique et la Grèce.* Paris: Bechet, 1825.

Dutrône, Henri. *Extraits de la correspondance de M. Dutrône avec M. le Président Capodistria.* Paris: E. Cassin, 1831.

Ferguson, Niall. "Hegemony or Empire?" *Foreign Affairs* 82, no. 5 (2003): 154–161.

Fiala, Andrew. *Tyranny from Plato to Trump: Fools, Sycophants, and Citizens.* Lanham: Rowman and Littlefield, 2022.

Finlay, George. *History of the Greek Revolution and the Reign of King Otho.* 2 vols. London: Zeno, 1971.

Fleming, Katherine. *The Muslim Bonaparte: Diplomacy and Orientalism in Ali Pasha's Greece.* Princeton: Princeton University Press, 1999.

Foucault, Michel. "Des espaces autres." *Architecture/Mouvement/Continuité* 5 (1967): 46–49.

Gallagher, John, and Ronald Robinson. "The Imperialism of Free Trade," *Economic History Review* 6, no. 1 (1953): 1–15.

Gallant, Thomas W. *The Edinburgh History of the Greeks, 1768 to 1913: The Long Nineteenth Century.* Edinburgh: Edinburgh University Press, 2015.

Ghervas, Stella. *Réinventer la tradition: Alexandre Stourdza et l'Europe de la Sainte-Alliance.* Paris: Honoré Champion, 2008.

Gosse, Louis-André, *Lettres à ma mère pendant un séjour en Grèce, 1826–1830 (édition 1920).* Paris: Édition Atar, 1920.

Horowitz, Richard S. "International Law and State Transformation in China, Siam, and the Ottoman Empire during the Nineteenth Century." *Journal of World History* 15, no. 4 (2004): 445–486. https://dx.doi.org/10.1353/jwh.2005.0126.

Kadis, William P. *John Capodistrias and the Modern Greek State.* Madison: State Historical Society of Wisconsin for the Department of History, University of Wisconsin, 1963.

Kapodistrias, Ioannis [Capodistrias, Jean]. *Aperçu de ma carrière publique depuis 1798 jusqu'à 1822.* Paris: La Vouivre, 1999.

Koukkou, Eleni (Helen) E. *O Kapodistrias kai hē paideia, 1803–1822: A', Hē Philomousos Hetaireia tēs Viennēs.* Athens, 1958.

Leontis, Artemis. *Topographies of Hellenism: Mapping the Homeland.* Ithaca: Cornell University Press, 1995.

Lettres et documents officiels relatifs aux derniers événements de la Grèce, qui ont précédé et suivi la mort du Comte Capodistrias: Publié par plusieurs membres de l'ancien comité grec de Paris. Paris: Firmin Didot, 1831.

Loukatou, Spyridonos D. *Ho Iōannēs Kapodistrias kai hē Heptanēsos Politeia*. Athens: National and Kapodistrian University of Athens, 1959.

Loukos, Christos. *Iōannēs Kapodistrias: mia apopeira historikēs viographias*. Athens: MHET, 2023.

Loules, Dimitris. *The Financial and Economic Policies of President Ioannis Capodistrias, 1828–1831*. Ioannina: Ioannina University Press, 1985.

Massé, Alexandre. *Un empire informel en Méditerranée: Les consuls de France en Grèce et dans l'Empire ottoman: images, ingérences, colonisation (1815–1856)*. Paris: Classiques Garnier, 2019.

Murgescu, Bogdan. "'Phanariots' and 'Pământeni': Religion and Ethnicity in Shaping Identities in the Romanian Principalities and the Ottoman Empire." In *Ethnicity and Religion in Central and Eastern Europe*, edited by Maria Crăciun and Ovidiu Ghitta, 196–204. Cluj: Cluj University Press, 1995.

Newell, Walter R. *Tyranny and Revolution: Rousseau to Heidegger*. Cambridge: Cambridge University Press, 2020.

Nye, Joseph S., Jr. *Soft Power: The Means to Success in World Politics*. New York: Perseus Books, 2004.

O'Donoghue, Aoife. *On Tyranny and the Global Legal Order*. Cambridge: Cambridge University Press, 2021.

Panou, Nikos, and Hester Schadee, eds. *Evil Lords: Theories and Representations of Tyranny from Antiquity to the Renaissance*. Oxford: Oxford University Press, 2018.

Pellion, Jean-Pierre. *La Grèce et les Capodistrias pendant l'occupation française de 1828 à 1834 par le général de division Pellion: L'avenir du royaume grec est dans sa marine marchande et son commerce*. Paris: J. Dumaine, 1855.

Philliou, Christine M. *Biography of an Empire: Governing Ottomans in an Age of Revolution*. Berkeley: University of California Press, 2011.

Pippidi, Andrei. "Phanar, Phanariotes, Phanariotism." *Revue des études sud-est européennes* 13, no. 2 (1975): 231–240.

Prieto, Moisés, ed. *Dictatorship in the Nineteenth Century: Conceptualisations, Experiences, Transfers*. Abingdon: Routledge, 2022.

Quijano, Aníbal. "Coloniality of Power and Eurocentrism in Latin America." *International Sociology* 15, no. 2 (2000): 215–232. https://doi.org/10.1177/0268580900015002005.

Richter, Melvin. "A Family of Political Concepts: Tyranny, Despotism, Bonapartism, Caesarism, Dictatorship, 1750–1917." *European Journal of Political Theory* 4, no. 3 (2005): 221–248. https://doi.org/10.1177/1474885105052703.

Stanhope, Leycester. *Lettres du Colonel Stanhope sur la Grèce*. Paris: Corneille, 1825.

Simal, Juan Luis, and Darina Martykánová, "Ferdinand and the Sultan: The Metaphor of the Turk and the Crisis of the Spanish Monarchy in the Early Nineteenth

Century." *Contributions to the History of Concepts* 10, 1 (2015): 1–26. https://doi.org/10.3167/choc.2015.100101.
Tipei, Alex R. "Audience Matters: 'Civilization-Speak,' Educational Discourses, and Balkan Nationalism, 1800–1840." *European History Quarterly* 48, no. 4 (2018): 658–685. https://doi.org/10.1177/0265691418799547.
Tipei, Alex R. "Korais's Greece and Napoleon's Empire: The Egyptian Campaign, Race Science, and the Europeanization of an Idea." In *From the Napoleonic Age to the Age of Empire: Empire after the Emperor*, edited by Thomas Dodman and Aurélien Lignereux, 89–110. London: Palgrave, 2023.
Tipei, Alex R. *Unintended Nations: France's Empire of Civilization, Southeast Europe, and the Post-Napoleonic World.* Montreal: McGill-Queen's University Press, 2025.
Veremis, Thanos. "Kapodistrias and the French: The Formation of a Regular Greek Army." *Conference on War and Society in East Central Europe During the Eighteenth and Nineteenth Century.* New York: The Center for European Studies, Graduate School and University Center, City University of New York, 1979.
von Prokesch-Osten, Anton Freiherrn. *Geschichte des Abfalls der Griechen vom Türkischen Reiche im Jahre 1821 und der Grünung des Hellenischen Königreiches aus Diplomatischem Standpunkte.* Vienna: Carl Gerold's Sohn, 1867.
Voulgaris, Stamati. *Notice sur le comte Jean Capodistrias, président de la Grèce suivie d'un extrait de sa correspondance*, 2nd ed. Paris: Delaunay, 1832.
Woodhouse, C. M. *Capodistria: The Founder of Greek Independence.* Oxford: Oxford University Press, 1973.
Yaycioglu, Ali. *Partners of the Empire: The Crisis of the Ottoman Order in the Age of Revolutions.* Stanford: Stanford University Press, 2016.

About the Author

Alex R. Tipei is a transnational historian and an assistant professor at the Université de Montréal. She is also a team leader on the European Research Council-funded project "Transnational histories of 'corruption' in South-East-Central Europe" (ERC–2022-AdG no. 101098095), hosted by the New Europe College—Institute for Advanced Study, Bucharest. Her latest publication is *Unintended Nations: France's Empire of Civilization, Southeast Europe, and the Post-Napoleonic World* (Montreal: McGill-Queens, 2025).

6. British Investors and the "Kustendjie Harbor Dues Issue"

Constantin Ardeleanu

Abstract: This chapter examines the diplomatic, legal, and financial aspects related to the dispute between a British railway company and the Ottoman government over the company's right to levy harbor dues at Kustendjie (1862–1878). Drawing on various sources, it traces three phases of the dispute: diplomatic objections to a private company collecting dues in an Ottoman port; negotiations replacing private dues with state collection and compensation; and, after 1870, efforts to reopen the port to public use and curb the DBSR's monopoly, culminating in an investigation, stalled arbitration, and the 1878 transfer of Dobrudja to Romania. The case illuminates concessionary governance in an imperial periphery, showing how legal ambiguity, bureaucratic delay, and Great Power oversight shaped political risk and infrastructural sovereignty.

Keywords: Concessionary infrastructure, Railways, Ottoman Empire, Dobrudja, Imperial periphery

"The first train on the first railway in Turkey leaves the first station"

"An event of no small importance" for the Ottoman Empire occurred on October 4, 1860, when "the first train on the first railway in Turkey" left "the first station" in the small town of Kustendjie (now Constanța, Romania).[1] The opening ceremony was overseen by İbrahim Edhem Pasha (1818–1893),

1 "The Danube and Black Sea Railway," *Daily News* (London), October 22, 1860, 3.

Marton, Silvia, and Andrei-Dan Sorescu, eds. *Infrastructural Development, Corruption, Xenophobia, and Colonization in Central and Southeastern Europe*. Amsterdam: Amsterdam University Press/Central European University Press, 2026.
 https://doi.org/10.5117/9789048577279-7

the Ottoman minister of commerce, who delivered the inaugural speech and commended the "Danube and Black Sea Railway and Kustendjie Harbor Company" (hereafter DBSR) for "the success with which their undertaking had thus far been conducted" (see fig. 1). Following a series of official speeches, a salute fired by an Ottoman corvette, and a religious service, the approximately 250 visitors proceeded to board the DBSR train, which conveyed them to Tchernavoda (Boğazköy in Turkish, today Cernavodă in Romania). The guests had lunch in one of the company's grain warehouses at Tchernavoda and then returned to Kustendjie, where the Ottoman "man-of-war in the offing again hailed the returning train with a 19-gun salute, and in the evening lighted up the scene with blue lights and rockets." The inaugural banquet that evening, attended by approximately 120 guests, was meticulously organized. For three hours, "luxury followed luxury, till human appetite—helped on though it was with faultless wines and all the other *incitamenta famis* known to gastronomy—could bear no more."[2] This account, to borrow a phrase from Malte Fuhrmann's contribution to the present volume, was one of "boasting," in which notions of Western civilizational and market expansion dovetailed with a projected image of Ottoman commitment to infrastructural development *qua* modernization and legitimacy.[3]

The DBSR had compelling reasons for organizing such an elaborate celebration. In a period when the navigation of the Danube Mouths was hindered by a plethora of obstacles,[4] the railway across Dobrudja was regarded as a promising and profitable alternative mode of transportation for passengers and freight between Central Europe and the East. As observed by a press correspondent who was present at the inauguration, "compared with the Sulina [the middle channel of the Danube Delta], and taking Constantinople as the base, the Kustendjie route saves about 280 miles of sea-river" navigation, while also avoiding the unsafe, unhealthy, and economically costly swamps of the Danube Delta. The DBSR expected to divert over the railway "a most formidable slice of the trade of the Danube," which had suffered during the period preceding the Crimean War, when the Danube Delta was under the control of imperial Russia.[5]

2 "Opening of the Danube and Black Sea Railway," *Saunders's News-Letter*, October 24, 1860, 1.

3 Malte Fuhrmann, "On Hegemony and Connectivity: A Short History of Infrastructure-Political Struggles and their Discourses" in this volume.

4 See Constantin Ardeleanu, *International Trade and Diplomacy at the Lower Danube: The Sulina Question and the Economic Premises of the Crimean War (1829–1853)* (Brăila: Istros, 2014).

5 "The Danube and Black Sea Railway," 3.

The investors were extremely optimistic regarding the economic potential of their infrastructure. At the DBSR's general meeting in November 1860, the directors presented a generally positive report. Since November 1, regular rail services had been operating between Kustendjie and Tchernavoda. An agreement had already been reached with the Austrian Danube Navigation Company (Donau-Dampfschiffahrts-Gesellschaft) for the interchange of traffic at Tchernavoda. Furthermore, a similar arrangement was being negotiated with the Austrian Lloyd, whose steamships operated regularly between Istanbul and the inland ports of Brăila and Galați (see fig. 1).[6]

At the subsequent general meeting of the DBSR's shareholders, held in May 1861, further positive developments were announced. The "railway works and accessories" were nearing completion, allowing the resident engineers in Dobrudja to redirect their attention and resources to the harbor of Kustendjie, where modern cranes and equipment had been shipped from Great Britain. In the spring of 1861, two express services per week were initiated between Istanbul and Vienna via the Dobrudja railway. Meanwhile, the directors sought to establish the DBSR as a significant competitor in the lucrative Lower Danube grain trade. Modern corn separators operated by steam power were constructed at Tchernavoda, providing cleaning facilities for grain that "are not elsewhere obtained in the East."[7]

In November 1861, the construction of the harbor was still underway, but ships could safely enter the harbor enclosure and load directly from the jetty. Merchants from prominent mercantile houses in Brăila and Galați had started to use the new infrastructure, and the DBSR's management was happy to announce that "the limit of warehouse accommodation had been reached." All indications suggested that the forthcoming trading season would be a busy and profitable one. Convinced by such good news, the shareholders agreed to a further increase of capital, raising the total registered capital to £475,000.[8]

However, just as the substantial investments of the shareholders and the assiduous efforts of the resident engineers in Dobrudja appeared to be nearing fruition, a significant disagreement emerged between the DBSR and the Sublime Porte. The dispute originated in 1862 in relation to the harbor dues payable by vessels utilizing the recently constructed "artificial" harbor at Kustendjie.

6 "The Danube and Black Sea Railway," *Morning Chronicle*, November 16, 1860, 3.

7 "Danube and Black Sea Railway," *Daily News*, June 1, 1861, 6.

8 "Danube and Black Sea Railway and Kustendjie Harbor," *Daily News*, December 2, 1861, 7.

The present chapter will analyze this dispute, which was referred to in British diplomatic documents as the "Kustendjie harbor dues issue."[9] The DBSR found itself enmeshed in a complex web of diplomatic and legal negotiations, which, at times, posed a significant threat to the railway company's prosperity and even its very existence. The dispute, therefore, serves as a pertinent case study for the diplomatic complexities inherent to the construction and administration of infrastructure in a peripheral region of the Ottoman Empire. It exemplifies the myriad tensions between the interests of profit-oriented British investors and the priorities of the Ottoman central and regional authorities. As Silvia Marton and Andrei-Dan Sorescu also note in their introduction, infrastructure is a site of power play and contestation, deeply integrated into social, aesthetic, material, political—but also, I would add, diplomatic—configurations.[10] And, indeed, as techno-diplomatic systems, the railway and harbor constructed in Ottoman Dobrudja were the subject of considerable negotiations in London and Istanbul. At the same time, the dispute did not remain confined to the domain of diplomatic exchanges but also gained significant public exposure. Both collective actors involved—the British investors and the Ottoman authorities—sought to highlight perceived breaches of contract by the other party. Despite mutual accusations and blame-shifting, the dispute never escalated into a full-blown public "scandal," which would not have served the financial interests of the British capitalists; the circle of scrutiny and visibility remained narrow. Nevertheless, the dispute contributed to the cultivation and growth of mutual distrust between the Sublime Porte and Western European investors during a period when the Ottoman government was experiencing challenges in managing its foreign debt. This ultimately resulted in the declaration of sovereign default on loan repayments in 1875 and the establishment of the Ottoman Public Debt Administration in 1881.[11]

9 The broader context of the railway construction in Ottoman Dobrudja is examined in detail in Constantin Ardeleanu, *Un drum de fier prin "Estul Sălbatic": Dobrogea otomană în zorii capitalismului* (Bucharest: Humanitas, 2025). The present chapter develops the main argument advanced in Chapter 6 of the above-cited work.

10 See Silvia Marton and Andrei-Dan Sorescu, "Researching Infrastructural Development, Corruption, and Colonization in Central-South-East Europe: An Introduction" in this volume. For a more general discussion of the topic, see Hannah Appel, Nikhil Anand, and Akhil Gupta, "Introduction: Temporality, Politics, and the Promise of Infrastructure," in *The Promise of Infrastructure*, ed. Nikhil Anand, Akhil Gupta, Hannah Appel (Durham: Duke University Press, 2018), 1–38.

11 Murat Birdal, *The Political Economy of Ottoman Public Debt: Insolvency and European Financial Control in the Late Nineteenth Century* (London: I. B. Tauris, 2010).

The dispute had three main phases, which help showcase the different conceptualizations and understandings of "private" and "public" interests. During the first phase, European ambassadors to Istanbul contested the principle that a "private company" was entitled to collect dues in an Ottoman port. They advanced the argument that such an action constituted a violation of the Capitulations, the agreements that conferred rights and privileges in favor of foreign subjects who resided or traded in the Ottoman dominions. At the same time, the Sublime Porte aimed to leverage the dispute to assert greater autonomy in imposing tolls for infrastructure works that provided transportation facilities to merchants and shipowners. In the second phase, the British capitalists endeavored to negotiate an additional agreement with the Ottoman government. They sought fair compensation for their expenses incurred in the modernization of the Kustendjie harbor, which served a larger public good, while also maintaining control of the harbor infrastructure, which constituted an essential component of their business model. In their claims, they had to rely on the support provided by British diplomats in both Istanbul and London. While the British diplomats did offer all needed assistance, they also acted in accordance with the broader imperial interests of the British government in the Ottoman Empire. During the third phase, the British investors encountered a more assertive stance from the Ottoman bureaucracy, which sought to curtail the British monopoly over the harbor infrastructure of Kustendjie and open the port to the "public"—that is, local merchants in Dobrudja who sought to export grain independently of the DBSR. Nevertheless, the distinction between private and public, corporate and state interests remained highly ambiguous and malleable, obscured by mutual accusations, polemics, and expectations of substantial financial gains.

Finally, this contribution also aims to provide further empirical insight into a topic of interest to economic historians: the political risks associated with foreign investments in the Ottoman Empire in the nineteenth and early twentieth centuries. While some authors, such as V. Necla Geyikdaği and M. Yaşar Geyikdaği, posit that the investment climate in the Ottoman Empire "was more welcoming as compared to neighboring lands such as Russia and the Balkan countries,"[12] this text shows that British investors in Ottoman Dobrudja had numerous grounds for expressing discontent with the functioning of the Ottoman bureaucracy at both the regional and central levels.

12 V. Necla Geyikdagi and M. Yasar Geyikdagi, "Foreign Direct Investment in the Ottoman Empire: Attitudes and Political Risk," *Business History* 53, no. 3 (2011): 397.

A Railway and Harbor Concession in Ottoman Dobrudja (1855–1861)

The initial proposals for the construction of a canal or railway across Dobrudja emerged in the 1830s.[13] They were prompted by the growing need to facilitate the export of grain through the mouths of the Danube, which was being hindered by the physical and political challenges of navigation through the Russian Danube Delta. By the early 1850s, the so-called "Sulina Question" had become a significant point of contention in British-Russian relations. Consequently, during the Crimean War, Western diplomats proposed solutions to facilitate navigation of the Danube. This explains the resolutions adopted by the 1856 Treaty of Paris, which compelled the Russian Empire to cede the Danube Delta and a portion of southern Bessarabia to the Ottoman Empire and its vassal principality of Moldavia. Additionally, the treaty established the European Commission of the Danube, an international organization tasked with removing natural or artificial obstructions to navigation at the Danube Mouths.[14]

The Crimean War also brought a multitude of entrepreneurs to the Black Sea region, in the footsteps of the Western armies. For these individuals, the region's vast agricultural or mineral resources and its underdeveloped infrastructure presented significant business prospects and a wealth of potential for economic gain.[15] The ruined seaport of Kustendjie and the neighboring region of Dobrudja, the scene of extensive military and logistical operations during the war, were regarded as promising arenas for the injection of Western capital and technology. In 1855, a road was constructed between Rasova (south of Tchernavoda) and Kustendjie by French military engineers with the objective of enhancing the logistical support available to the Western armies fighting in Crimea.[16] In the same year, the British investor Thomas Wilson (1788–1867) devised a plan to construct a canal between the Danube and the Black Sea. Wilson's agent in Istanbul secured

13 Mihail Guboğlu, "Osmanlı İmparatorluğu'nda Karadeniz-Tuna Kanalı Projeleri (1836–1876) ve Boğazköy–Köstence Arasında İlk Demiryolu İnşası (1855–1860)," in *Çağını Yakalayan Osmanlı: Osmanlı Devleti'nde Modern Haberleşme ve Ulaştırma Teknikleri*, ed. Ekmeleddin İhsanoğlu and Mustafa Kaçar (İstanbul, 1995), 217–247.

14 For the context, see Constantin Ardeleanu, *The European Commission of the Danube, 1856–1948: An Experiment in International Administration* (Leiden: Brill, 2020).

15 M. Yakup Bektas, "The British Technological Crusade to Post-Crimean Turkey: Electric Telegraphy, Railway, Naval Shipbuilding and Armament Technologies" (PhD diss., University of Kent at Canterbury, 1995).

16 Tudor Mateescu, "Date noi despre contribuția românească la construirea șoselei Rasova–Constanța (1855)," *Arhiva românească*, n.s., 2, no. 1 (1996): 120–125.

a concession from the Sublime Porte in May 1856, thereby obtaining the necessary authorization to proceed with the construction of the canal. Other entrepreneurs were also pursuing similar projects in the region, resulting in a struggle for influence with the Ottoman government.[17] For reasons of simplicity, speed, and profitability, two British engineers who conducted a survey of Dobrudja in 1856—Charles Liddell (1813–1894) and Lewis D. B. Gordon (1815–1876)—advised Wilson that a railway would be a more feasible infrastructure than a canal.[18] The railway was projected to be relatively short (65 km in length), and it was assumed that the construction would be inexpensive owing to the low cost of labor and construction materials in an Ottoman periphery. The estimated cost for the railway was £200,000, with an additional £100,000 allocated for improvements to the harbor of Kustendjie. Influential capitalists in Great Britain—such as Sir Samuel Cunard (1787–1865) and William Philip Price (1817–1891), who served as chairman and vice-chairman of the company's board of directors—established the DBSR in 1857 and, following renewed negotiations with the Ottoman government, secured a new concession, signed in September 1857.[19]

John Trevor Barkley (1826–1882), who was a "civil engineer" by profession, played the leading role in the construction of the railway and harbor in Dobrudja. His previous experience included serving as manager of the coal mines at Karadeniz Ereğli in northern Asia Minor, but Barkley demonstrated considerable skill also in negotiating the intricate web of interests in Istanbul. This enabled him in 1857 to advance the railway project and secure the concession in less than half a year. In accordance with the stipulations of the agreement, as duly observed by historians John H. Jensen and Gerhard Rosegger, the DBSR assumed considerable obligations, "while the Porte

17 For details, see J. H. Jensen and Gerhard Rosegger, "British Railway Builders along the Lower Danube, 1856–1869," *The Slavonic and East European Review* 46, no. 106 (1968): 109–110. Also see Jensen and Rosegger, "Transferring Technology to a Peripheral Economy: The Case of Lower Danube Transport Development, 1856–1928," *Technology and Culture* 19, no. 4 (1978), 675–702, and Boriana Antonova-Goleva, "Concessions and Mirages along the Lower Danube: The Town of Silistria in the Plans of Foreign Railway Promoters during the mid–1850s," *Journal of Balkan and Black Sea Studies* 5 (2020): 19–47.

18 Başbakanlık Osmanlı Arşivi, HR.SFR.3.00029.00010.001 (Wilson to Fuad Paşa, London, August 23, 1856). Another letter was sent to the Ottoman Embassy in London: Başbakanlık Osmanlı Arşivi, HR.SFR.3.00029.00010.003 (Wilson to Musurus, London, August 1856)—with thanks to Boriana Antonova-Goleva for providing me access to several Ottoman documents on the DBSR.

19 The texts of the concessions are published in Thomas Forester, *The Danube and the Black Sea: Memoir on Their Junction by a Railway between Tchernavoda and a Free Port at Kustendjie* (London, 1857), 215–224.

offered little in return." The railway was to be completed within a three-year period at the expense of the DBSR, with no guarantee of a return on the investment. The concession was granted for a period of 99 years. The Porte was to provide the DBSR with free government land, where available, and to assist in the expropriation of private land, but at the investors' own expense. A separate concession related to the harbor, which was to be completed within five years, encompassed all "its dependencies, quays, magazines, [and] entrepôts," in accordance with plans and designs approved by the Ottoman government. One of the very few direct benefits granted to the DBSR was the right to collect tolls for services rendered, including "light, anchorage, and other dues," from ships engaged in business in the harbor of Kustendjie. This was in accordance with a tariff agreed upon by the Sublime Porte.[20]

Despite the rather severe terms of the 1857 railway and harbor concessions, the investors accepted them because they anticipated significant profits, given their potential to redirect a substantial portion of the lucrative grain trade of the Lower Danube. In light of the highly optimistic forecasts of annual net profits, which ranged from 20 to 40 percent, the founders themselves provided the vast majority of the initial capital.[21]

The construction of the railway and harbor was overseen by John Trevor Barkley, along with his three younger brothers—all of them self-trained civil engineers. Despite some complications with the engineering works, logistical arrangements, or the lack of skilled personnel in an Ottoman backwater—problems about which Henry Charles Barkley (1837–1903) wrote a comprehensive account[22]—the railway was completed on schedule and inaugurated in October 1860. As the British vice-consul to Kustendjie, Frank Fremoult Sankey (1828–1879), noted in February 1861, the infrastructure had serious quality issues,[23] but it was ready to generate substantial financial returns for the British capitalists.

20 Forester, *The Danube and the Black Sea*, 215–224.

21 Jensen and Rosegger, "British Railway Builders," 111–112.

22 Henry C. Barkley, *Between the Danube and the Black Sea; or, Five Years in Bulgaria* (London, 1876).

23 The National Archives of the United Kingdom, Foreign Office (FO), Fund 195, File 689, f. 17–23 (Sankey to Horatio Suter, Kustendjie, February 14, 1861).

Attempts to Collect Dues in an "Artificial" Ottoman Harbor (1862–1866)

The DBSR soon announced its intention to introduce a charge for vessels entering the harbor of Kustendjie, with the new policy coming into effect in April 1862.[24] However, when the time came for the payment of the dues, foreign merchants and shipowners, acting on the advice of their respective consuls, refused either to pay the dues or to abide by the DBSR's regulations. The Capitulations—agreements that conferred rights and privileges to foreign subjects who resided or traded in the Ottoman dominions—they claimed, deprived the Sublime Porte of the right to levy dues in Ottoman ports; consequently, the government had no authority to enter into any agreement with the DBSR involving such payments.[25]

The DBSR lodged a formal complaint with the British Embassy and the Foreign Office, which advised them to pursue a direct resolution with the Ottoman government. Mehmed Fuad Pasha (1814–1869), who served as Grand Vizier from 1861 to 1866, expressed his willingness to resolve the case, which, however, proved to be more complicated than the capitalists thought. A lengthy correspondence ensued, during which the issue became obscured within the intricate web of Ottoman bureaucratic procedures. It was not until August 1864, following renewed protests, that Ottoman Foreign Minister Mehmed Emin Aali Pasha (1815–1871) requested the collaboration of the ambassadors of the Great Powers in the collection of harbor dues at Kustendjie. Aali Pasha noted that there had been no safe harbor at Kustendjie before 1862, and the construction of infrastructure was a costly endeavor. "As a natural compensation for their expenses, they [the DBSR] were given the right to collect the taxes that the imperial government was undoubtedly entitled to establish." Some precedents existed for subletting this service, such as the fees collected at Sulina by the European Commission of the Danube or the service of lighthouses established throughout the

24 The National Archives of the United Kingdom, FO 78/3194 (Kustendjie Harbor dues, claim of the Danube and Black Sea Railway and Kustendjie Harbor Co., vol. 1), f. 11–12 (Tariff of dues to be paid at Kustendjie).

25 The Danube & Black Sea Railway and Custendje Harbor Company, *Narrative of the Proceedings and Circumstances Which Produced the Company's Claim on the Imperial Ottoman Government* (Constantinople, 1873), 3–4 (the brochure is included in FO 78/3194, f. 380–396). The same complaint is made in a memorandum sent by the company: FO 78/3194, f. 41–43 (Price to Earl Russell, London, June 15, 1865).

Ottoman Empire—a service managed by a French company.[26] From an Ottoman perspective, the decision to sublet tax collection to the DBSR did not constitute a legal innovation.

The ambassadors were not inclined to accept these arguments without further clarification from the Porte. Some highlighted concerns regarding the allegedly excessive rate of the proposed dues. Henry Bulwer (1801–1872, ambassador from 1858 to 1865) commended the Ottoman administration for its efforts to modernize the infrastructure but asserted that the tariff should have been established by a joint commission. Additionally, he deemed it unsuitable for the taxes to be "levied by a company. It is the government itself that should appoint agents for this purpose, and it is only with the Ottoman authorities that the consular authorities can deal with the issues involved."[27]

By the end of 1864, diplomats in Istanbul who were interested in the matter had agreed to the principle of allowing the collection of reasonable dues in return for the extraordinary expenses incurred in improving the harbor infrastructure. However, they objected to accepting that a private company would collect dues "in its own name in an Ottoman port." The solution was that the Sublime Porte should assume responsibility for the collection of those dues and provide fair compensation to the investors.[28]

Following the ambassadors' acceptance of the principle, the DBSR intensified their lobbying activities in both London and Istanbul. In the summer of 1865, they submitted a series of reports to John Russell (1792–1878, foreign secretary from 1859 to 1865). In June 1865, for instance, a delegation met with him to request diplomatic intervention, citing it as "the only means of procuring redress."[29] The DBSR claimed that their investments had a beneficial impact on the economic development of Ottoman Dobrudja; in reference to the Tatar migration of the early 1860s, the capitalists asserted that their settlement in Dobrudja was a direct consequence of the British investment in local infrastructure.[30] The influx of an "industrious and

26 *Narrative of the Proceedings*, 3–4. The brochure is included in FO 78/3194, f. 7–9 (Aali Pasha to Henry Bulwer, Constantinople, August 31, 1864).

27 *Narrative of the Proceedings*, 3–4. The brochure is included in FO 78/3194, f. 5–6 (Projet de réponse à Aali Pasha, drafted by Bulwer, undated).

28 *Narrative of the Proceedings*, 3–4. The brochure is included in FO 78/3194, f. 1–4 (Stuart to Russell, Constantinople, December 13, 1864).

29 *Narrative of the Proceedings*, 3–4. The brochure is included in FO 78/3194, f. 35 (C. W. Eddy to Russell, London, June 15, 1865); f. 45 (FO to Eddy, London, June 16, 1865).

30 Nilghiun Ismail and Serdar Göktaş, "Rolul căii ferate Köstence–Boğazköy (Constanța–Cernavodă), Moștenire otomană, în transportul refugiaților crimeeni și cerchezi," *Revista Română de Sociologie* 32, nos. 5–6 (2021): 343–356.

thriving population" led to a significant increase in grain exports, from approximately 30,000 quarters per year prior to the completion of the railway to 600,000 quarters in 1864. Of this total, nearly 250,000 quarters were grown in the surrounding district. Imports exhibited a similarly marked increase, and in 1864 the port of Kustendjie was visited by 477 ships.[31] This economic flourishing, which was sustained by British capital, generated significant revenue for the Ottoman treasury, amounting to no less than £50,000 annually. As the Ottoman officials were unable to fulfill their obligation to allow the DBSR to collect harbor dues, "owing to the objections of foreign powers," the DBSR proposed that a sum of £7,500 per year be allocated for maintaining the harbor and that Kustendjie should be designated a "free port" with additional customs facilities.[32]

In light of renewed assurances of British diplomatic support, general manager Barkley, who ran the Dobrudja business from the DBSR's quarters in Kustendjie, proceeded to Istanbul with the objective of negotiating with the Ottoman government.[33] The latter was not entirely opposed to the resettlement proposal; however, it sought a more general recognition of its right to levy harbor dues to reimburse expenses "incurred in the improvement and maintenance of other ports of the Empire." As the British ambassador, Richard Lyons (1817–1887, ambassador from 1865 to 1867), reported from Istanbul, while it was "plainly desirable that every encouragement should be given to the Porte to undertake public works with a view to the development and promotion of Commerce," it was not prudent "to abandon by a general declaration" the Great Powers' role of exercising some control over how the Ottoman state charged taxes in its ports. It was deemed preferable that, in each specific instance prior to consenting to the imposition of harbor dues, ambassadors determined that the requisite improvements "are really made and maintained, that the dues exacted are fair, moderate and subject to revision, and that they are levied equally on the ships of all nations."[34]

The revised agreement was endorsed by a unanimous vote at a special meeting of the DBSR's shareholders in January 1866.[35] The Ottoman authorities were still engaged in the process of analyzing the document. Meanwhile, in the spring of 1866, the Sublime Porte resumed its endeavor to collect harbor dues at Kustendjie and once again sought the assistance

31 FO 78/3194, f. 41–43 (Price to Russell, London, June 15, 1865).

32 FO 78/3194, f. 47–48, 49–50 (FO to Bulwer, June 22, 1865).

33 FO 78/3194, f. 37 (Eddy to Henry Layard, London, June 26, 1865); f. 51 (FO to Bulwer, June 22, 1865).

34 FO 78/3194, f. 62–66 (Lyons to FO, Constantinople, December 29, 1865).

35 "Danube and Black Sea Railway," *Daily News*, January 12, 1866, 6.

of foreign embassies.[36] The British embassy awaited the establishment of a "fair and satisfactory arrangement" with the DBSR before directing the vice-consul at Kustendjie to facilitate the disbursement of fees imposed on British vessels.[37] The Ottoman government, however, sought to ascertain the actual collection of dues at Kustendjie before embarking upon negotiations pertaining to the new convention. Ultimately, in November 1866, Barkley apprised Lyons that at that juncture British "assent to the collection of the dues by the Ottoman authorities may be given without prejudice to the interests of the Company." This was based on assurances from Ottoman officials that British consent "will facilitate rather than retard a settlement with the Company."[38]

Following further correspondence, the Ottoman government indicated that "almost all the powers" had agreed to pay, thus authorizing the collection of the dues as of January 1867.[39] Lyons transmitted instructions to Kustendjie to permit the imposition of a levy on British shipping, on the condition that analogous dues would be levied without exception.[40]

For a period of five years, the DBSR was engaged in a dispute that pertained to the ambiguities of the Capitulations and the position of Kustendjie as a new "artificial" port that had been established through hydraulic works financed by a consortium of foreign investors. The 1857 convention granted the company the right to construct and manage the harbor infrastructure, but at that time the British investors failed to fully comprehend the significant legal implications of their granted privilege, which posed a challenge to some of the fundamental principles of the capitulatory regime that the Great Powers were striving to uphold in the Ottoman Empire.

Negotiating a New Harbor Convention (1867–1870)

Until approximately 1866, the DBSR was caught up in an intricate web of diplomatic exchanges, as the Sublime Porte sought recognition from the Great Powers of its own right to levy taxes for infrastructure modernization in its ports. Once an agreement had been reached between the ambassadors

36 FO 78/3194, f. 98 (Aali to Lyons, Constantinople, May 19, 1866).

37 FO 78/3194, f. 105–106 (Lyons to Aali, Constantinople, November 5, 1866).

38 FO 78/3194, f. 91–96 (Lyons to Stanley, Constantinople, November 6, 1866).

39 FO 78/3194, 113–114 (the same, December 4, 1866); f. 115–116 (Aali to Lyons, Constantinople, November 28, 1866); f. 117–118 (Lyons to Aali, Constantinople, November 29, 1866); f. 123 (Aali to Lyons, January 3, 1867).

40 FO 78/3194, f. 128–129 (Lyons to Russell, Constantinople, January 7, 1867).

on the principle that fair tolls could be introduced to offset the costs of hydraulic works, the DBSR hoped for a prompt resolution of their claims. But, soon enough, their case would be regarded as an illustrative example of the Ottoman tactics of procrastination.

In a new note to the Foreign Office in May 1867, the British investors complained that the Ottoman authorities persistently evaded their obligations, offering a multitude of excuses.[41] Despite the ostensible progress made in discussions with "the various committees of the various councils," those committees' reports were "either set aside or stifled," and the matter was allowed to lapse until it was once again revived by the DBSR's importunity. But the same round of proceedings would be repeated over and over again.[42]

The DBSR was more active in both lobbying for their cause and in threatening to financially boycott the Ottoman government. On June 19, 1867, a delegation comprising the DBSR's senior management and principal shareholders was received by Fuad Pasha during Sultan Abdülaziz's (1830–1876, on the throne from 1861 to 1876) visit to London. As reported in the British press, the Ottoman foreign minister received the deputation with great courtesy. After reading the mémoire and asking certain questions, he replied that it was in the interest of the Sultan's government "to encourage English enterprise in Turkey" and undertook to facilitate an expeditious resolution to the DBSR's claims.[43]

Barkley proceeded once again to Istanbul, prepared to engage in negotiations. In November 1867, he sought further assistance from Henry George Elliot (1817–1907), the recently appointed British ambassador to Istanbul (1867–1877). Elliot consulted with Fuad Pasha, who informed the ambassador that the proposal was being translated and prepared for further consideration. The issue remained unresolved during the Islamic holy month of Ramadan (January 1868).[44] By 1868, there was a growing sense of exasperation with the Ottoman government's lack of action. During one of the shareholders' meetings, the chairman of the Board of Directors, William Philip Price (1817–1891), articulated the directors' skepticism regarding the Ottoman government's willingness to address their claims. He argued that the Porte would only act when faced with a crisis, "when the representations of the Foreign Secretary would be followed up by some

41 FO 78/3194, f. 130 (Eddy to Stanley, London, May 13, 1867). f. 132–134 ("Statement of a deputation on behalf of the Danube Railway and Kustendjie Harbor Company").

42 FO 78/3194, f. 140–141 (Barkley to Lyons, Constantinople, June 21, 1867).

43 "The Danube and Black Sea Railway and Kustendjie Harbor," *Morning Post*, July 20, 1867, 6.

44 FO 78/3194, f. 157 (Eddy to Price, London, March 12, 1868).

energetic remonstrance."[45] But waiting for a crisis was hardly a solution for capitalists who invested large amounts of money in the DBSR and had still not received dividends on their shares.

Some more lines of action were thus followed. In April 1868, during a meeting of representatives of several British companies operating railways in the Ottoman Empire (including the lines from Ruse to Varna and Izmir to Aydın), the investors prepared another memorandum to the grand vizier. In this document, the British capitalists demanded that justice be rendered to the DBSR without further delay so that all past claims could be settled and future payments duly provided for. The "prospect of obtaining further capital for constructing new works in the country [Ottoman Empire]," the British investors threatened, "depends entirely upon satisfactory arrangements being made with regard to the works already constructed."[46]

Once again, with no progress having been made, in January 1869 the deputation met George Villiers, fourth Earl of Clarendon (1800–1870), the recently appointed foreign secretary (1868–1870).[47] The investors claimed that the DBSR had expended about £120,000 on the construction of the harbor and had sustained a loss of harbor dues, which from 1862 to 1868 amounted to £95,269. "Meanwhile the Turkish Government benefits to the extent of at least £50,000 per annum by the tithes, taxes & customs dues" levied on the DBSR's properties and on "the large foreign population which has been attracted to the neighborhood by the news engaged in this harbor and formation of what was a barren steppe."[48] The investors' unwavering efforts coincided with a visit to London by Davud Pasha (Garabet Artin Davoudian, 1816–1873), the Ottoman minister of public works, who sought to advance a project for the construction of a new railway line within the Ottoman Empire. The DBSR, once again, threatened to appeal to British investors on the stock exchange in opposition to the denotation of any Ottoman shares.[49]

45 "Danube and Black Sea Railway and Kustendjie Harbor," *Morning Post*, August 14, 1868, 6.

46 FO 78/3194, f. 169 ("Resolutions passed at a Meeting of the Representatives of the Danube and Black Sea Railway Kustendjie Harbor Company, of the Varna Railway Company and of the Smyrna and Aidin Railway Company–held at J. Poultry, London, on April 30, 1868"); f. 174 (note from the FO, June 9, 1868).

47 FO 78/3194, f. 179 (FO to Price, January 21, 1869).

48 FO 78/3194, f. 180–182 (Tipping to Clarendon, London, January 29, 1869).

49 FO 78/3194, f. 177–178 (Price to Clarendon, London, January 20, 1869).

The combined effect of lobbying and threats pushed forward the negotiations between the DBSR and the Porte.[50] Ultimately, the DBSR dispatched a new envoy to Istanbul, Frederick Walpole (1822–1876), a Conservative member of the House of Commons and former naval commander.[51] The particulars of the negotiations remain unclear from the sources consulted; however, they culminated in the conclusion of a new harbor convention on October 10, 1870, signed by Walpole and Davud Pasha. The document comprised 12 articles. The 1857 concession was revoked, and Kustendjie was declared a free port. The Ottoman government assumed complete and absolute control of the harbor, and the DBSR was to receive the sum of £112,761 in two installments, with interest at a rate of 8.5 percent per annum. The Sublime Porte granted the DBSR a free lease of the harbor wharves for the remainder of the concession period, while the DBSR was responsible for operating the lighthouse and maintaining the harbor. The DBSR consented to lengthen the jetty to a total of 700 feet, the cost of which, amounting to £30,000, was to be reimbursed by the Ottoman government in three equal annual installments as the work progressed. A harbor master appointed by the Sublime Porte was responsible for enforcing the port regulations drafted by the Ottoman government. The Ottoman harbor master was assisted by a harbor service manager and his staff, who were employed and remunerated by the DBSR but acted under the authority of the Ottoman appointee.[52]

The convention of 1870 aimed to resolve serious problems in the operation of the harbor of Kustendjie, where the DBSR and local elites had diverging economic interests—a common issue in Ottoman infrastructure development at the time, as Fuhrmann's chapter also notes.[53] Resident engineers had worked hard to build the piers in the early 1860s, but the company invested less money in consolidating them and in dredging the harbor enclosure once it became clear that the DBSR would not be allowed to collect harbor dues. By 1870, the total cost of building and maintaining the harbor amounted to £117,443, and the place was considered a sinkhole for British capital. In 1869, for instance, the harbor brought in revenue of £117, while expenditure for its maintenance and operation was £2,067. "In other words, owing to the default of the Turkish Government not only does

50 FO 78/3194, f. 184 (Eddy to Clarendon, London, February 8, 1869); f. 185 (Eddy to Clarendon, London, March 8, 1869); f. 188–190 (Project of Convention); f. 195–196 (FO to Elliot, London, March 18, 1869).

51 *Narrative of the Proceedings*, 5.

52 *Narrative of the Proceedings*, 26–29; the 1870 Convention is also appended in FO 78/3194, f. 244–247.

53 Fuhrmann, "On Hegemony and Connectivity."

this large amount of the Company's capital remain unproductive, but the revenue of the remainder is taxed to the extent of £1,950 per annum to make good the deficiency in the harbor receipts."[54] The DBSR's main source of income was exporting Danubian grain, transshipped from Romanian or Bulgarian ports along the Lower Danube. This was stored and cleaned in the company's warehouses in Tchernavoda, then transported by rail and dispatched by ships, which loaded the grain using the company's own equipment. The DBSR maintained tight control over the harbor area, which was heavily guarded and considered private property. The situation was severely criticized by local elites and merchants in Kustendjie, who were interested in exporting the grain harvested in the district by the industrious Tatar population settled in Dobrudja.[55] Only a tiny section of the port had remained "public," being open to local merchants, but it was poorly maintained and, critics argued, almost unusable. Meanwhile, the DBSR had also gone to great lengths to force local merchants to use its facilities and services, effectively creating a monopoly on trade in Kustendjie. As the Russian consul in Varna, A. A. Olkhin, reported to his superiors in 1865, managing director Barkley was a veritable tyrant, keeping a tight rein on the local economy. Everyone feared him, everyone suffered from his abuses, but no one dared to denounce him openly for fear of paying dearly "for such insolence." Nor were the vice-consuls, themselves merchants, any braver, convinced of Barkley's "certainty of impunity." According to Olkhin, there had been a few attempts to confront Barkley, but they had all failed, provoking the vengeance of the English manager.[56]

An Ottoman Commissioner's Investigation of the DBSR (1871–1872)

The convention of 1870 was supposed to mark a new beginning for both the British investors and the local traders in Kustendjie. But it soon became clear that the parties involved had a very different understanding of the

54 "The Danube & Black Sea Railway & Kustendjie Harbor, Limited," *Herapath's Railway Journal*, February 26, 1870, 220.

55 For a comparison with local elites' views on railway infrastructure in late nineteenth-century Transylvania, see Gábor Egry, "Steel and Wood: Railway Construction and Rent-Seeking" in this volume.

56 Velko Stanev Tonev, ed., *Izvori za istoriyata na Dobrudzha, 1853–1878: chuzhdestranni dokumenti*, vol. 4 (Sofia: Vekove, 2003), 156–157 (Doc. 24, Olkhin to ambassador Nikolay Ignatiev, Varna, April 1865).

terms of the agreement. In September 1871 the Sublime Porte sent a special envoy, Yusuf (Youssouf) Bey, to Kustendjie to investigate the DBSR's alleged evasion of its concession. Yusuf was tasked with investigating five specific issues: (1) the situation of the harbor works, to ascertain both the quality of the hydraulic works and the legality of the company's business practices there; (2) the allegation that the railway, which obstructed the flow of several streams and thus formed pestilential swamps, was a threat to public health in places such as Medjidieh (now Medgidia); (3) the securing of a quarantine station and customs office on land occupied by the DBSR; (4) the mapping of lands fenced off by the DBSR at various stations, as it appeared that the company occupied more land than it was entitled to; and (5) the adherence of the DBSR to requests to rectify identified issues with the railway line, which posed a potential hazard to the local population.[57]

Edward Harris (1835–1916), who replaced Barkley as managing director at Kustendjie, attempted to resist Yusuf's purportedly abusive actions. A series of intense confrontations ensued between Yusuf's group of Ottoman officials and Harris's team, which was primarily composed of British employees. One such scene, described by Harris, is illustrative: in October 1871, when an Ottoman steamship entered the harbor and evaded all payment to the DBSR, Harris had the ship moor at an inconvenient berth and blocked the entrance to that part of the harbor with a train of wagons. On the following morning, 40 to 50 carts laden with local grain (i.e., harvested in Dobrudja) arrived at the harbor, yet they were unable to reach the ship. The Ottoman harbor master relocated the vessel to an alternative berth to facilitate loading. Yusuf demanded the removal of the train, but when the British harbor service manager refused, a group of Turks proceeded to remove the train wagons themselves. That night, Harris relocated the train wagons, ensuring that they could not be moved from their position. The next morning, in order to circumvent the blockade, Ottoman porters carried the grain from the carts to the ships, over a distance of 70 to 80 yards.[58] In response to British resistance, Yusuf blocked "sundry lines of railway in order to prevent the passage of the Company's wagons to the quays" and requested the DBSR's

57 FO 78/3194, f. 226–232 ("Instructions données à Youssouf Bey, envoyé à Kustendjie avec la mission de faire une enquête sur plusieurs questions relatives au chemin de fer & au port de Kustendjie").

58 FO 78/3194, f. 233–239 (Harris to Edhem Pasha, Kustendjie, October 10, 1871); f. 248–250 ("Extracts from letters of General Manager to the Board referring to visit of Yousouf Bey the Turkish Commissioner," October 17, 1871).

employees "to remove their rails from the quays, and to confine their train operations to the station."[59]

In November, Yusuf dispatched a ship to the wharf without the previously mandatory DBSR pilots. A similar incident occurred subsequently when a disagreement arose between a vessel engaged in loading operations for the Ottoman government and one for the DBSR concerning the priority of berthing. As Harris reported, "I am obliged to use stratagems in order to get our ships alongside before the Turks occupy the berths."[60] Similarly, the DBSR was denied the right to utilize its warehousing facilities within the harbor, as Yusuf placed police forces to guard the aforementioned magazines, who forcibly prevented "one of the tenants from placing grain in them—consequently they were abandoned by the merchants and [as of 1873] have not been let since."[61]

Yusuf elucidated the Ottoman government's new interpretation of the status of Kustendjie. The DBSR considered the payments stipulated in the 1870 convention to be an equitable remuneration for the company's right to collect dues from ships. However, the Ottoman government believed that the DBSR's exclusive right to use the wharves ceased when the Ottoman state purchased the harbor.[62] The removal of the British monopoly on the harbor was, for the merchants of Kustendjie, a step toward economic independence.

Yusuf Bey was equally determined to dispossess the DBSR of the excess land it had occupied in Kustendjie and elsewhere along the line.[63] At the Tchernavoda station, for instance, the DBSR held about 75 hectares—"an amount of land for a single station [that] has not been granted to any company in the world."[64] The DBSR, in turn, asserted that when the land was transferred to the company, "the entire valley for 20 miles inland was a valueless swamp, subject to the periodical inundations of the Danube."[65] Yusuf's investigation resulted in yet another severe blow to the company's financial position. The situation was dire, anyway: ten years after the line opened, ordinary shareholders had not received any dividends. In the new

59 *Narrative of the Proceedings*, 7.

60 FO 78/3194, f. 253 (Eddy to Entfield, London, December 29, 1871).

61 *Narrative of the Proceedings*, 19–20.

62 FO 78/3194, f. 240–242 ("Extract from Report of Mr. I. Bach on conversation held with the Imperial Commissioner Youssouf Bey on behalf of General Manager" and "Translation of letter addressed to Mr. Harris by Youssouf Bey under date October 25, 1871").

63 *Narrative of the Proceedings*, 10–12.

64 *Narrative of the Proceedings*, 21–22 (see also the correspondence in the appendix, at 22–25).

65 *Narrative of the Proceedings*, 2–14.

circumstances, getting a return on their investment became even more problematic. In 1873, the shareholders discussed the possibility of closing the line but decided to keep it open and use it in their negotiations with the Ottoman authorities.[66]

In Search of an Arbitration (1872–1878)

The DBSR once more sought diplomatic assistance in both London and Istanbul. The only viable solution to settle the dispute with the Sublime Porte was to act in accordance with article 17 of the original convention (1857), which provided that all questions should be referred to a mixed arbitration commission.[67]

Persuading the Ottoman authorities to accept this meant a return to the use of both carrots and sticks. In 1873, the DBSR appointed a negotiator to Istanbul, Charles Edward Austin (1819–1893), an engineer (and DBSR shareholder) with prior experience in the Ottoman Empire.[68] Additionally, the DBSR cautioned British investors against allocating capital to the Ottoman Empire, coinciding with the Ottoman request for a £30,000,000 loan.[69] But Austin could not achieve much, advised by the British Embassy to comply with the full bureaucratic procedure "prescribed for the particular case by the Grand Vizier." Austin discussed several times each question pertaining to his petition with both the Ottoman Board of Works and the Council of State. With support from Ambassador Elliot, Austin patiently navigated the petition through all these departments. After a period of approximately 18 months, "I succeeded in obtaining from the Council of State a unanimous opinion that the Company was entitled to arbitration on all questions in dispute." The council recommended the payment of £15,000 for the new harbor works, but Grand Vizier Huseyin Avni Pasha (1820–1876) allegedly rejected the decision and discussed the matter again with various legal advisors, who all concurred with the previous ruling. "He then told them they must give him a different opinion and the question was sent back to the Council of State where it remained in petto till the advent to

66 FO 78/3194, f. 294 (Eddy to Granville, London, February 26, 1873).

67 FO 78/3194, f. 222–224 (Price to Granville, London, November 27, 1871).

68 FO 78/3194, f. 305 (FO to Price, London, June 17, 1873); 317–318 (Eddy to Tenterden, London, July 10, 1873).

69 "The Danube and Black Sea Railway," *Railway News,* August 9, 1873, 6.

the Vizirat of [Ahmed] Essad Pasha [1828–1875]" in late April 1875.[70] As a frustrated William Tipping, chairman of the board of directors, put it at an extraordinary general meeting in 1874, the "notion that time meant money did not exist in Constantinople."[71]

In 1875, the company engaged the services of a new special negotiator in Istanbul, Charles Keating Tuckerman (1827–1896), an American diplomat who had previously served as the United States minister resident ("ambassador") to Athens. Tuckerman was the proponent of a more direct confrontation with the Sublime Porte. For the Ottoman government, the new agent was a perfect excuse to revisit matters that were previously negotiated and agreed upon with Austin. Eventually, arbitration was accepted in July 1875.[72]

New disputes emerged concerning the arbitration procedure. Given the political and military circumstances in the empire during the Great Eastern Crisis of 1875–1878, Ottoman officials were reluctant to take up the matter, while for the British investors it was a long-awaited opportunity to push for a resolution of their claims. In September 1876, Tuckerman petitioned for augmented diplomatic assistance in Istanbul, apprising the Foreign Office that the DBSR would otherwise be compelled to seek redress through the British Parliament. "For 12 months the Turks have evaded them [the DBSR's claims] under frivolous excuse, and there exists no valid reason why the [British] government should not require the [Ottoman] Minister of Public Works to go on with the arbitration."[73] The investors themselves expressed a desire for greater involvement from the ambassador in supporting the DBSR. They noted that it was "only when the ambassador informed the Turkish government that further delay would cause grand offence to the British Government that arbitration was officially agreed." However, despite pressure from the British ambassador, new deadlocks arose on issues such as the number of arbitrators and the timing of the appointment of an umpire.[74]

A compromise was not reached until 1877. In a report sent to London in December 1877, the British ambassador to Istanbul, Henry Layard (1817–1894, ambassador from 1877 to 1880), blamed Tuckerman as much as the Porte for the delay, asserting that if the arbitration had not yet been settled, the

70 FO 78/3194, f. 373–374 (Austin to Derby, London, January 15, 1875); f. 424–427 (DBSR to Derby, July 25, 1875).

71 "Danube and Black Sea," *Herapath's Railway Journal*, December 12, 1874, 1376.

72 FO 78/3194, f. 412–414 (Elliot to Derby, Therapia, July 10, 1875); f. 416 (Elliot to Derby, Therapia, July 16, 1875); f. 418 (Savfet Pasha to Elliot, Constantinople, July 14, 1875).

73 FO 78/3194, f. 447–448 (DBSR to Derby, September 6, 1876).

74 FO 78/3194, f. 449–456 ("Statement" by Tuckerman, September 1876); 457 (FO to Elliot, London, September 18, 1876).

responsibility rested solely with the US diplomat. Furthermore, Tuckerman attempted to persuade the ambassador to consider a new course of action that Layard believed would "cause a very bad impression" with the sultan himself. Layard conceded that "a worse agent than Tuckerman could scarcely have been chosen," given that Tuckerman's anti-Turkish bias was widely recognized in Istanbul.[75]

Layard also mentioned another crucial aspect of the Ottoman Empire's position in that moment: given that in 1877 the railway was under Russian control and might remain so, the Porte was understandably concerned about the potential risk "of having an award against it which may saddle it with heavy pecuniary liabilities."[76]

In May 1878, the arbitration commission was finally constituted, with each of the two parties represented by one commissioner. Hasan Fehmi Effendi (1836–1910) represented the Ottoman state, while John Henry Fawcett (1831–1898), a legal expert who had previously served as British consul general in Istanbul and chief judge of the Supreme Consular Court of the Levant, represented the DBSR. The company produced a detailed document setting out its claims, which amounted to £344,260, while the Ottoman state put forward a claim for compensation of £465,745. The arbitration process was intricate and, given the disparate perspectives on the matters at hand, it was likely to culminate in a stalemate and the necessity for an umpire.[77]

Even before the arbitration commission could convene, the fate of Dobrudja had already been determined. The Russian-Ottoman Treaty of San Stefano of March 1878 stipulated that the Ottoman Empire would cede Dobrudja to Russia in exchange for the majority of war reparations. However, Russia would preserve "the right of exchanging them for the part of Bessarabia detached from her by the Treaty of 1856."[78] In July 1878, the Treaty of Berlin recognized the territorial change (Southern Bessarabia in exchange for Dobrudja), which resulted in Dobrudja becoming an initially unwanted province of newly independent Romania.[79] Furthermore, article 51 stipulated that "with regard to public works and other enterprises of a

75 FO 78/3195, unnumbered (Layard to Derby, Constantinople, December 16, 1877).

76 FO 78/3195, unnumbered (the same, December 16, 1877).

77 FO 78/3195, unnumbered (translation of the document, appended to the message from Layard to Salisbury, Therapia, May 5, 1878).

78 Thomas Erskine Holland, *The European Concert in the Eastern Question: A Collection of Treaties and Other Public Acts* (Oxford: Calrendon Press, 1885), 343–344.

79 On Romania's position on Dobrudja, see Silvia Marton, "Le débat politique en Roumanie: mémoires et perspectives en 1912–1913," in *Alerte en Europe: la guerre dans les Balkans (1912–1913)*, ed. Catherine Durandin and Cécile Folschweiller (Paris: L'Harmattan, 2014), 55–83.

like nature, Roumania shall be substituted for the Sublime Porte as regards its rights and obligations throughout the ceded territory."[80]

In November 1878, Layard reported to the British government that the Sublime Porte had declined to proceed with the arbitration based on article 51. Tuckerman voiced his objection to the decision, prompting Layard to query: "Am I to support him?"[81] Subsequently, intricate debates commenced, ultimately culminating in the transfer of the question to Bucharest. Romanian engineers went to Dobrudja to evaluate the value of the railway and the assets owned by the DBSR. The diplomatic negotiations were arduous, and on occasion the Romanians threatened to construct a parallel railway from Rasova to Kustendjie. The British company initially demanded 25 million francs, equivalent to approximately £1,000,000 for their Dobrudja infrastructure. Following further negotiations, about which I wrote at length in another paper,[82] the Romanian government agreed to pay 16.8 million francs, equivalent to approximately £672,000.[83]

Conclusions

The construction of the railway and harbor in Dobrudja was a highly ambitious undertaking, with the objective of revolutionizing Southeast European transport corridors and reshaping the lucrative grain trade of the Lower Danube. Despite the capitalists' adherence to their established timeline, the DBSR was unable to achieve the projected financial profitability. This was due to several factors, including the completion of competing infrastructure in the Danube Delta—which maintained the Sulina branch of the Danube as the primary transport corridor for Romanian grain—and changes in the agricultural structure of modern Romania following the 1864 land reform. Subsequently, throughout the 1870s, the advent of railway networks in Southeast Europe presented the DBSR with new challenges.

80 "Treaty between Great Britain, Germany, Austria, France, Italy, Russia and Turkey for the Settlement of Affairs in the East: Signed at Berlin, July 13, 1878," *The American Journal of International Law* 2, no. 4 (1908), 419–420.

81 FO 78/3195, unnumbered (Decipher from Sir H. Layard, Therapia, November 6, 1878).

82 Constantin Ardeleanu, "Dobrogea's Iron Road towards Romania: Negotiating the Sale of the British-Built Infrastructure in Romanian Dobrogea (1878–1882)," *Revue des Études Sud-Est Européennes* 63 (2025): 275–294.

83 Dumitru P. Ionescu, "Construirea și răscumpărarea liniei ferate Constanța–Cernavodă," *Anuarul Institutului de Istorie și Arheologie "A. D. Xenopol"* 25, no. 2 (1988): 216–217.

British investors sought to exploit one of the few direct benefits afforded to the DBSR by their 1857 harbor convention: the right to collect "light, anchorage, and other dues" from vessels engaged in commerce at Kustendjie, in accordance with a tariff established by the Ottoman government. However, it proved impossible to collect dues that violated the rights and privileges of the maritime powers, at least according to the European ambassadors' interpretation of the Capitulations. The DBSR found itself in a difficult position, caught between the Sublime Porte and the foreign ambassadors. In order to be allowed to collect the promised dues, the British capitalists sought to highlight the development of Dobrudja's economy and the growth of the province's public revenues. In their reports, the investors underlined the role of their infrastructure in promoting the economic growth of an otherwise underdeveloped Ottoman region and asserted their right to a share of the revenues.

The negotiations between the DBSR's representatives and the Ottoman government were lengthy and intricate. The Sublime Porte employed a range of strategies to defer payment, while the private company sought redress in both London and Istanbul. As time passed, the shareholders' frustration intensified. But the DBSR's management was aware that the optimal moment to act would be during a political or military crisis in Istanbul. This would necessitate the use of all available diplomatic resources to resolve their dispute with the Ottoman government.

Upon the conclusion of negotiations in 1870, the Porte adopted a more assertive posture in its dealings with the DBSR, which was accused of violating the terms of its concessions and of placing the interests of its own shareholders above those of the local merchants of Dobrudja. Yusuf's mission to Kustendjie significantly complicated the company's operations. Once again, the British investors initiated a series of complaints directed toward the relevant authorities in London and Istanbul. According to the managing director of the DBSR in Dobrudja, the British investors failed to secure sufficient influence within the diplomatic corps to effectively address the myriad challenges they faced. British diplomats in Istanbul supported the investors but guided DBSR negotiators through the complex maze of Ottoman bureaucracy rather than advising them to take corrupt shortcuts that could have backfired on the investors' long-term interests.

The culmination of these events came in 1875, yet the Ottoman government remained unyielding in its refusal to reach an agreement. Consequently, in 1878, the territorial provisions of the treaties of San Stefano and Berlin effectively rendered an agreement with the Ottoman Empire unnecessary, and the eventual outcome was mutually acceptable to both parties. Unlike what Tipping said in 1874, the notion that time meant money did exist in Istanbul, and Ottoman procrastination proved, in this case at least, profitable to the empire's treasury.

Bibliography

Antonova-Goleva, Boriana. "Concessions and Mirages along the Lower Danube: The Town of Silistria in the Plans of Foreign Railway Promoters during the mid–1850s." *Journal of Balkan and Black Sea Studies* 5 (2020): 19–47.

Appel, Hannah, Nikhil Anand, and Akhil Gupta, "Introduction: Temporality, Politics, and the Promise of Infrastructure." In *The Promise of Infrastructure*, edited by Nikhil Anand, Akhil Gupta, and Hannah Appel, 1–38. Durham: Duke University Press, 2018.

Ardeleanu, Constantin. "Dobrogea's Iron Road towards Romania: Negotiating the Sale of the British-Built Infrastructure in Romanian Dobrogea (1878–1882)." *Revue des Études Sud-Est Européennes* 63 (2025): 275–294

Ardeleanu, Constantin. *Un drum de fier prin "Estul Sălbatic": Dobrogea otomană în zorii capitalismului.* Bucharest: Humanitas, 2025.

Ardeleanu, Constantin. *The European Commission of the Danube, 1856–1948: An Experiment in International Administration*. Leiden: Brill, 2020.

Ardeleanu, Constantin. *International Trade and Diplomacy at the Lower Danube: The Sulina Question and the Economic Premises of the Crimean War (1829–1853)*. Brăila: Istros, 2014.

Barkley, Henry C. *Between the Danube and the Black Sea; or, Five Years in Bulgaria.* London: John Murray, 1876.

Başbakanlık Osmanlı Arşivi, HR.SFR.3.00029.00010.001 and HR.SFR.3.00029.00010.003

Bektas, M. Yakup. "The British Technological Crusade to Post-Crimean Turkey: Electric Telegraphy, Railway, Naval Shipbuilding and Armament Technologies." PhD diss., The University of Kent at Canterbury, 1995.

Birdal, Murat. *The Political Economy of Ottoman Public Debt: Insolvency and European Financial Control in the Late Nineteenth Century*. London: I. B. Tauris, 2010.

Daily News. "The Danube and Black Sea Railway." October 22, 1860, 3.

Daily News. "Danube and Black Sea Railway." June 1, 1861, 6.

Daily News. "Danube and Black Sea Railway and Kustendjie Harbor." December 2, 1861, 7.

Daily News. "Danube and Black Sea Railway." January 12, 1866, 6.

The Danube & Black Sea Railway and Custendje Harbor Company. *Narrative of the Proceedings and Circumstances Which Produced the Company's Claim on the Imperial Ottoman Government.* Constantinople: Office of the Levant Herald, 1873.

Forester, Thomas. *The Danube and the Black Sea: Memoir on Their Junction by a Railway between Tchernavoda and a Free Port at Kustendjie.* London: Edward Stanford, 1857.

Geyikdagi, V. Necla, and M. Yasar Geyikdagi. "Foreign Direct Investment in the Ottoman Empire: Attitudes and Political Risk." *Business History* 53, no. 3 (2011): 375–400. https://doi.org/10.1080/00076791.2011.565514.

Guboğlu, Mihail. "Osmanlı İmparatorluğu'nda Karadeniz-Tuna Kanalı Projeleri (1836–1876) ve Boğazköy–Köstence Arasında İlk Demiryolu İnşası (1855–1860)." In *Çağını Yakalayan Osmanlı: Osmanlı Devleti'nde Modern Haberleşme ve Ulaştırma Teknikleri,* edited by Ekmeleddin İhsanoğlu and Mustafa Kaçar, 217–247. İstanbul: IRCICA, 1995.

Herapath's Railway Journal. "The Danube & Black Sea Railway & Kustendjie Harbor, Limited." February 26, 1870, 220.

Herapath's Railway Journal. "Danube and Black Sea." December 12, 1874, 1376.

Holland, Thomas Erskine. *The European Concert in the Eastern Question: A Collection of Treaties and Other Public Acts.* Oxford: Clarendon Press, 1885.

Ionescu, Dumitru P. "Construirea și răscumpărarea liniei ferate Constanța—Cernavodă." *Anuarul Institutului de Istorie și Arheologie "A. D. Xenopol"* 25, no. 2 (1988): 205–217.

Ismail, Nilghiun, and Serdar Göktaş. "Rolul căii ferate Köstence–Boğazköy (Constanța–Cernavodă), Moștenire otomană, în transportul refugiaților crimeeni și cerchezi." *Revista Română de Sociologie* 32, nos. 5–6 (2021): 343–356.

Jensen, J. H., and Gerhard Rosegger. "British Railway Builders along the Lower Danube, 1856–1869." *The Slavonic and East European Review* 46, no. 106 (1968): 105–128.

Jensen, J. H., and Gerhard Rosegger. "Transferring Technology to a Peripheral Economy: The Case of Lower Danube Transport Development, 1856–1928." *Technology and Culture* 19, no. 4 (1978): 675–702.

Marton, Silvia. "Le débat politique en Roumanie: mémoires et perspectives en 1912–1913." In *Alerte en Europe: la guerre dans les Balkans (1912–1913)*, edited by Catherine Durandin and Cécile Folschweiller, 55–83. Paris: L'Harmattan, 2014.

Mateescu, Tudor. "Date noi despre contribuția românească la construirea șoselei Rasova–Constanța (1855)." *Arhiva românească,* n.s., 2, no. 1 (1996): 120–125.

Morning Chronicle. "The Danube and Black Sea Railway." November 16, 1860, 3.

Morning Post. "The Danube and Black Sea Railway and Kustendjie Harbor." July 20, 1867, 6.

Morning Post. "Danube and Black Sea Railway and Kustendjie Harbor." August 14, 1868, 6.

The National Archives of the United Kingdom, Foreign Office, files 78/3194, 78/3195 and 195/689.

Railway News. "The Danube and Black Sea Railway." August 9, 1873, 6.

Saunders's News-Letter. "Opening of the Danube and Black Sea Railway." October 24, 1860, 1.

"Treaty between Great Britain, Germany, Austria, France, Italy, Russia and Turkey for the Settlement of Affairs in the East: Signed at Berlin, July 13, 1878." *The American Journal of International Law* 2, no. 4 (1908): 401–424.

Tonev, Velko Stanev, ed. *Izvori za istoriyata na Dobrudzha, 1853–1878: chuzhdestranni dokumenti*. Vol. 4. Sofia: Vekove, 2003.

About the Author

Constantin Ardeleanu is a senior researcher at the Institute for South-East European Studies of the Romanian Academy and at New Europe College—Institute for Advanced Study, Bucharest, where his specialty is the modern history of Danubian Europe and the Black Sea region. His most recent monograph is *Steamboat Modernity: Travel, Transport, and Social Transformation on the Lower Danube, 1830–1860* (Budapest: CEU Press, 2024).

7. Steel and Wood: Railway Construction and Rent-Seeking

Gábor Egry

Abstract: This chapter compares late nineteenth-century railway projects by Transylvanian Saxon elites with railway concessions in the Ottoman Empire from the perspective of local actors and their agency. I argue that the logic of these peripheral projects was similar: local elites acquired concessions for railways they did not intend to build on their own. Rather, such actors invited investors from the center or from the West, whom they could only attract with the promise of government guarantees for a safe return on capital. Thus, it was governments that paid for these investments and transferred capital to the peripheries, instead of investors exploiting those areas. The chapter questions colonial capitalism's analytical value for these cases by framing them through rent-seeking.

Keywords: Colonial capitalism, Railway concessions, Peripheral elites, Capital transfer, Transylvania, Ottoman Empire

Introduction

Railways are generally hailed in the historiography of modern economic development as essential elements of infrastructure and preconditions for rapid economic transformation and growth. They provide access and transportation facilities for raw materials, enabling trade at a mass scale across vast landmasses, reducing unit costs and prices, and helping to integrate isolated smaller markets into a larger whole. Contemporaries of their rapid expansion in Europe and North America understood railways in these terms as well, as reflected in debates on how, where, and why they

Marton, Silvia, and Andrei-Dan Sorescu, eds. *Infrastructural Development, Corruption, Xenophobia, and Colonization in Central and Southeastern Europe*. Amsterdam: Amsterdam University Press/Central European University Press, 2026.
 https://doi.org/10.5117/9789048577279-8

should be extended, with the argument often being that, without them, economic backwardness and resulting social ills could not be alleviated.[1] As Malte Fuhrmann has shown in a previous chapter, this was a matter of grounding legitimacy in the provision of public goods and of framing economic development as equivalent to general happiness.[2] At the same time, railways were understood as strategic arteries—not only in economic terms but also as a means of control over territory and power—generating cohesion at home, enabling expansion abroad, and integrating newly annexed and distant territories into the state. Within colonial contexts, such as British India, they arguably played an essential role in the creation of colonial state space, which in turn reconfigured social and economic relations, dislocated existing patterns, enhanced control over colonized societies, and entrenched an extremely exploitative colonial capitalism.[3]

Yet it is also precisely because of this constellation—viewing railways as more than mere business ventures—that their history is marred by the collusion of business and politics, scandals involving bankruptcies and bailouts, and railway companies overstepping their legal bounds with political backing. Moreover, this is a story of contested railway concessions and rights. And as this chapter—along with the others in this volume—demonstrates, the history of railways in Southeastern Europe provides a rich set of examples for these phenomena. Since railway construction began before Balkan nation-states seceded from the Ottoman Empire, it inevitably became part of the story of how European entrepreneurs and states vied for influence.[4] Consequently, the most common interpretation of these endeavors emphasizes the semi-colonial aspects of late Ottoman history, including how railway concessions—the dominant form of Ottoman

1 For an example from Transylvania, see Csaba Gidó on the extension of the railway network to the Székelyföld. Csaba Gidó, "A székely rabonbán és a székely gőzös: Ugron Gábor és a Héjjasfalva-Székelyudvarhely helyi érdekű vasút," *Magyar Gazdaságtörténeti Évkönvy* (2017–2018): 267–282.

2 Malte Fuhrmann, "On Hegemony and Connectivity: A Short History of Infrastructure-Political Struggles and their Discourses" in this volume.

3 Manu Goswami, *Producing India: From Colonial Economy to National Space* (Chicago: University of Chicago Press, 2004).

4 Peter Hertner, "The Balkan Railways, International Capital and Banking from the End of the 19th Century until the Outbreak of the First World War," in *Finance and Modernization: A Transnational and Transcontinental Perspective for the Nineteenth and Twentieth Centuries*, ed. Gerald Feldman and Peter Herner (Abingdon: Routledge, 2016), 125–154; Jacques Thobie, *Intérêts et impérialisme français dans l'Empire Ottoman (1895–1914)* (Paris: Publications de la Sorbonne, Imprimerie Nationale, 1977); Malte Fuhrmann, *Der Traum vom deutschen Orient: Zwei deutsche Kolonien im Osmanischen Reich (1851–1918)* (Frankfurt: Campus Verlag, 2006), 56–79.

railway construction until the Hejaz Railway[5]—were tools of an exploitative economic and political relationship stemming specifically from this semi-colonial context.[6]

Additionally, such railway incentives are easily situated within the context of corruption, including colonial corruption.[7] Most importantly, this concept entails both the practice of corruption—as part of a broader colonial economy and exploitation through asymmetric concessions—and a discursive grasp on what constitutes corruption.[8] The former has also recently become of interest within a growing literature on the "new history of capitalism." The latter is the result of the "Othering" of colonies as essentially backward, positing corruption both as a symptom of local backwardness and as a force with which colonizers must contend—as opposed to the supposed eradication of corruption in the metropole.[9]

In this chapter, I intend to complicate this rather straightforward view by placing railway construction by Austrian and Hungarian capitalists in the late Ottoman Empire side by side with railway projects in late nineteenth-century Transylvania—a distant and somewhat backward region of dualist Hungary—between 1890 and 1914. In doing so, I hope to reveal an important commonality between these ventures: the agency of local political and business elites in initiating and profiting from railway construction. By presenting various schemes of rent-seeking in which locals used informal tools to obtain leverage over seemingly more powerful partners (both business and political alike), I will argue that the simultaneous prevalence of these practices in a metropolitan region and in a space qualified as

5 Mostafa Minawi, *The Ottoman Scramble for Africa: Empire and Diplomacy in the Sahara and the Hijaz* (Stanford: Stanford University Press 2016), 99–115; Murat Özyüksel, *The Hijaz Railway and Ottoman Empire: Modernity, Industrialization and Ottoman Decline* (London: I. B. Tauris, 2014).

6 Thobie, *Intérêts et impérialisme.*

7 Ronald Kroeze, Pol Dalmau, and Frédéric Monier, "Introduction: Corruption, Empire and Colonialism in the Modern Era: Towards a Global Perspective," in *Corruption, Empire and Colonialism in the Modern Era: A Global Perspective*, ed. Ronald Kroeze, Pol Dalmau, and Frédéric Monier (London: Palgrave Macmillan, 2021), 1–21; Ronald Kroeze, "Colonial Normativity? Corruption in the Dutch–Indonesian Relationship in the Nineteenth and Early Twentieth Centuries," in Kroeze, Dalmau, and Monier, *Corruption, Empire and Colonialism*, 173–208.

8 Kroeze, "Colonial Normativity," 173–174.

9 Kroeze *et al.*, Introduction 3–5. On the New History of Capitalism, see the symposium in Andrew Popp, ed., special issue, *Enterprise & Society* 25, no. 1 (2024). Jonathan Saha develops the latter view into the argument that local officials' involvement in corruption was an essential element and quality of state formation and the colonial state in Burma, further complicating the notion of backwardness and coloniality. See Jonathan Saha, *Law, Disorder and the Colonial State: Corruption in Burma c. 1900* (London: Palgrave Macmillan, 2013).

semi-colonial challenges the notion that exploitation and national political complicity are products of colonial capitalism alone. Instead, a wide array of factors crucially intersected in these specific cases: the strategic importance of railways, the consensus around their benefits to society, the appeal of becoming a "modern" state, and the significance of informal political relations.[10] To make this argument, I will analyze two railway construction projects—one larger and one smaller—from each of the two empires, focusing on their inception, the rationale of their initiators, the ways they hoped to profit from them, and how they managed business and politics to achieve their goals.

It is important to note that my analysis does not aim to compare specific regional contexts or modalities of railway construction nor the specific policies of countries (or regions) in this regard. Rather, I intend to compare processes occurring within asymmetric relationships that both contemporaries and historians perceive as center-periphery in nature. This type of relationship constitutes the common denominator—the *tertium comparationis*—of the cases I have selected, with the important observation that center and periphery are situated differently. While I could have chosen any two examples, I selected Hungary and the late Ottoman Empire because of the availability of sources and my familiarity with certain actors. Notably, some of the actors—major Austro-Hungarian capitalists—overlap in both cases.

In the case of Transylvania, the center and periphery both comprised elements of the same political unit—the Kingdom of Hungary—whereas in the case of the Ottoman Empire, the center formed a distinct unit: the "Western core," which included Austria-Hungary as a component. While this might seem unconventional or even methodologically problematic at first glance, I contend—and hope to demonstrate—that examining the agency of peripheral actors in these different contexts is exactly what can help deconstruct the category of "colonial." Peripherality in both cases presumes an asymmetry of resources and power relative to both national governments and business elites, while coloniality is associated with the exploitation and appropriation of the resources of the periphery. Thus, the similarities or differences can be directly related to the alleged coloniality of one of the two contexts.

In doing so, I will not treat the cases in equal detail. Rather, I will use the more extensively documented Transylvanian Saxon railway schemes to

10 This is analogous to Kroeze's argument about the complexities of the colonial context and how they foster practices easily depicted as corrupt. Kroeze, "Colonial Normativity?" 188.

discern the underlying logic of political, financial, and informal mechanisms and patterns, and condense it into a model. Using this model, I will then analyze railway schemes in the Ottoman Empire to determine whether and to what extent the defining elements of the Transylvanian Saxon pattern are discernible—including traces and evidence found in archival materials. From this point onward, I will compare the cases and draw my conclusions.

My primary aim is thus a critical reconsideration of the "colonial" as a category for such infrastructure projects, tentatively arguing that as a blanket concept—in the absence of a careful analysis of specific cases—it risks losing its analytical purchase. I seek to demonstrate that it is not always structural or power asymmetries between metropole and colony, or center and periphery, that define the terms of interaction. Nor is it a given that resources flow from the periphery to the center as a result of dubious practices, such as rent-seeking through informal relations or the conscious appropriation of public goods. Corruption is, therefore, of secondary importance to the argument at hand. The practical reason for this is simple: none of the cases in this chapter were discussed publicly as corrupt endeavors, despite a significant pushback around the turn of the twentieth century against the concession model in general.[11] Therefore, it would be methodologically questionable to assess how they *might* have been treated and understood in their contemporary public arena. As such, only more normative and retrospective definitions of corruption would apply to the cases in question. Nevertheless, it is important to note that the railway projects themselves were discussed in public, mostly in terms of progress, development, and regional backwardness—though the "public" in question was largely a narrow one comprising stakeholders drawn from various elite strata. In this way, these projects remained potentially open to accusations of "corruption" drawn from the broader discursive landscape of the period. That they were not exposed, indirectly betrays how arbitrary this discourse remained, despite its claims to normative universality.[12]

11 See, for example, Dr. Kepes Gyula, "A Magyar Helyiérdekű Vasutak Rt. tömeges koncessziói," *Független Magyarság*, January 28, 1904. 11–12; "A viczinális vasutak," *Fővárosi Lapok*, March 8, 1896. 16.

12 Gábor Egry, "Regionális fejlesztés nemzeti fejlesztés: Egy fejlesztéspolitikai és nemzetépítési model a 19. századi Erdélyben," in *Nemzetiség—felelősség: Írások Gáll Ernő emlékére*, ed. György Földes and Zsolt Gálfalvi (Budapest: Napvilág, 2005), 27–54.

Railways in Transylvania and Eastern Hungary

The era of railways in Hungary began in the 1840s, with the line connecting Pest with Vác (1846), a horse-driven line from Pozsony/Bratislava to Nagyszombat/Trnava (1846), and the Vienna–Pozsony line (1848) heralding the dawn of this new age. But despite the incipient state of railway transportation, the politics of railways became a major theme in the debates of the Hungarian *Vormärz* (*reformkor*). Both liberal reformers and their opponents considered railways crucial for transforming a country they saw as mired in backwardness and, alongside maritime shipping, a way of opening it to the world. Thus, one of the first major railway plans—promoted by none other than Lajos Kossuth (1802–1894), the unrivaled leader of the liberals in 1848–1849—envisioned a line between the Great Hungarian Plain and the port of Fiume.[13] Simultaneously, Transylvanian politics, then separate, had its own discussions about railway lines; most notably, newspaper director Gheorghe Barițiu (1812–1893) argued for a line from Arad to his town of Brașov (Kronstadt/Brassó)[14] as opposed to one from Timișoara (Temesvár/Temeschwar).[15]

The construction of railway infrastructure accelerated after the defeat of the Hungarians in the War of Liberation. This occurred under the auspices of the centralizing neo-absolutist regime and with the then-customary method of concessions—although there was a four-year interlude of a state monopoly on railway building. The railway network extended rapidly toward Vienna, to the northeast via Kassa/Košice, and to the east. In this last direction, two major lines competed for completion, offering very different geographic visions of how to link the easternmost parts of Hungary with its center. The Keleti Vasút (Eastern Railways) worked on a line through Nagyvárad/Oradea to Kolozsvár/Klausenburg/Cluj and Brașov/Brassó, opening up the less accessible and mainly agrarian areas along its route. The alternative line was built toward Arad with connections to Temesvár/Temeschwar/Timișoara and later Orsova/Orschowa/Orșova (1879), while

13 András Gergely, *Egy gazdaságpolitikai alternatíva a reformkorban: a fiumei vasút* (Budapest: Akadémiai Kiadó, 1982); István Deák, *The Lawful Revolution: Lajos Kossuth and the "Magyars" in 1848–1849* (New York: Columbia University Press, 1979).

14 Toponyms are first introduced in a multilingual format, then referred to subsequently by their current name. Railway lines and historical counties are referred to by their original name.

15 Olimpiu Boitoș, *Lupta lui G. Barițiu, în 1847, pentru introducerea căilor ferate în Transilvania* (Cluj/Brașov: Academia de Înalte de Studii Comericale și Industriale "Regele Mihai I," Institutul de Cercetări Economice și Sociale, 1945).

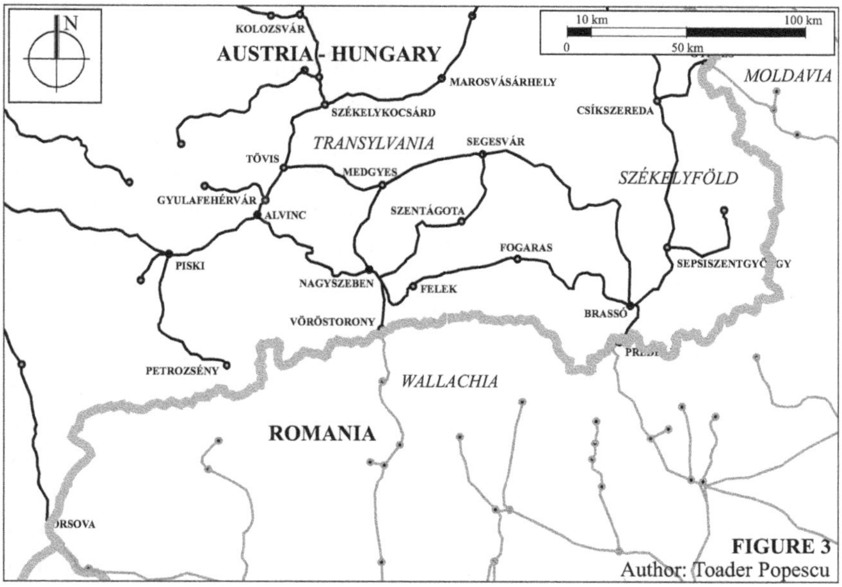

Figure 3. Railway lines in southern Transylvania before World War I

the main branch ran along the rivers Maros and Nagy-Küküllő to reach Brașov/Brassó as well (see fig. 3).[16]

The Compromise of 1867 and the reinstatement of a Hungarian government within what had now become the "dualist" imperial polity of Austria-Hungary did not initially affect railway development; the concession method, combined with generous financial profit guarantees for railway companies, continued to dominate. However, the Keleti Vasút soon ran into financial troubles, and the Budapest government had to bail it out—not least to salvage the connection with Brassó, which the line was scheduled to reach in 1873 to provide the first link with Romania through Transylvania, and, indirectly, with the Ottoman Empire. Following the acquisition of several bankrupt railway companies, a state railway (Magyar Államvasutak, MÁV) was established and expanded into a major transportation company, employing tens of thousands by 1900.

Following a conscious strategy developed by the minister of trade and communications, Gábor Baross (1848–1892), the state took over major lines and introduced a uniform railway tariff for the transport of persons and goods. Around 1890, the MÁV's network reached almost all the major cities

16 Csaba Gidó, "Az erdélyi fővasúthálózat kiépítése," *Pro Minoritate* 25, 3 (2015): 95–113.

of the country and provided transportation along the important routes of commerce in the Kingdom of Hungary and toward other countries. However, this development was uneven, leaving several significant localities—such as the town of Sibiu in southern Transylvania—accessible only through circuitous routes on secondary lines, or even completely neglecting large areas, like the easternmost counties of the Székelyföld.[17] The extension of railway infrastructure to these areas, as well as the building of better lines to substitute for existing connections, was a major concern for both local elites and government politicians, who still conceived of such efforts as a major instrument of development.[18]

Thus, alongside the MÁV's major lines and some of the remaining concessionary railways—such as the Kassa-Oderberg Railway—a second tier of "local interest" lines had developed with state encouragement in the form of subsequent legislation that set out general operating conditions for such companies, as well as the terms for state support.[19] Very often, these lines were promoted by local elites who tried to mobilize financial and political support both locally and nationally and who actively participated in their operation as local railway companies (HÉV). Such lines were often narrow-gauge with limited transportation capacity. Additionally, industrial enterprises, mostly in the forestry sector, joined in railway construction, opening up difficult geographic areas in the mountains. However, these lines also served local needs, with the acquisition of land often conditioned on offering locals privileged access to the lines.[20]

One further issue still plagued debates over railway lines toward the end of the nineteenth century. Although Austria-Hungary had three rail connections with Romania—in Bukovina to the north, at Orșova to the

17 Gidó, "Az erdélyi fővasúthálózat kiépítése." Székelyföld is the easternmost area of Transylvania, inhabited by a predominantly Hungarian-speaking population, while the regions between the Hungarian Great Plains and Székelyföld are ethnically mixed, often with majority Romanian and Saxon inhabitants.

18 Petra Balaton, "The Role of the Hungarian Government in the Development of Peripheries in the Austro-Hungarian Monarchy Focusing the Policy of the Ministry of Agriculture (1897–1914)," in *National Identity and Modernity 1870–1945: Latin America, Southern Europe, East Central Europe*, ed. Viktória Semsey, Petra Balaton, Csaba Horváth, and José Antonio Sánchez Román (Paris: L'Harmattan, 2019), 241–254.

19 Eszter Szabó, "A csallóközi vasút építéstörténete a kezdetektől 1918-ig," *Fórum Társadalomtudományi Szemle*, 6 4 (2004): 98–122, 100–102.

20 Martin Bemmann, "Wood-based Businesses and the Economic Development of Europe in the Nineteenth and Twentieth Centuries: An Introduction," *Jahrbuch für Wirtschaftsgeschichte* 65, no. 2 (2024): 279–305; Csongor Jánosi, "A Hungarian Count's Business in Romania: The Strange Survival of Tișița (1907–40)," *European Review of History/Revue européenne d'histoire* 31, no. 5 (2024): 707–738.

west, and at Predeal in the southern Carpathians—the prospect of building additional lines vexed politicians and businessmen alike. Given the terrain, there were few options for new lines: the Jiu Valley (Zsil-völgy), the Olt River Valley at Vöröstorony/Rotenturm/Turnu Roșu, or Gyimes/Ghimeș, on the border with the historical region of Moldavia. Yet when it came to deciding which one to prioritize, economic, military-strategic, and political considerations clashed, and none was built until the final years of the nineteenth century. A treaty with Romania in 1891 ultimately settled on two new connecting lines: one through Turnu Roșu and one through Ghimeș[21] (see fig. 3).

Finally, although the MÁV dominated railway transportation and owned the most important lines, secondary tracks were still built through a variation of the concession method. A group of individuals would acquire a railway concession for a specific line and turn to construction companies and banks for capital and know-how, often yielding a majority of the company's shares to them. The initial group's retention of founders' and priority shares, however, ensured a secure income for these railway capitalists, with permanent and stable revenue guaranteed through a contract with the MÁV, which essentially took over the operation and paid an annual fee to the owners. Thus, while the MÁV held only 36 percent of the 23,000 km of lines in 1918, it operated almost 84 percent.

By way of comparison, although the railway system of the Ottoman Empire was much less developed than that of the Kingdom of Hungary, the basic features of its construction were analogous to the early phases of the Hungarian experience described above. The era of Ottoman railway construction coincided with the reforms spurred by the Crimean War and the empire's opening to international trade. These new conditions enticed British entrepreneurs (the first railway concession was awarded in 1856) and, later, their French and German counterparts. One aspect of railway construction—both in Anatolia and in the Balkans—was the connection of the empire's most significant ports to the economic hinterland, exemplified by lines running from Smyrna/İzmir to Kasaba and Aydın. Especially after the opening of the Suez Canal, British concerns shifted toward a direct connection with the Persian Gulf via Basra.[22] Nevertheless, the Ottoman

21 Csaba Gidó, "Az erdélyi fővasúthálózat kiépítése," 106.

22 Nevın Coşar and Sevtap Demırcı, "Incorporation into the World Economy: From Railways to Highways (1850–1950)," *Middle Eastern Studies* 45, no. 1 (2009): 20–21; Besnik Emini, "The Efforts of Ottoman Empire to Build Railways in The Balkans," *Journal of Balkan Studies* 2, no. 1 (2022): 51–62.

railway network in its Asian provinces (not only in Anatolia but in Syria and Iraq as well) remained fragmented up until the end of World War I. It lacked major connecting lines, and even the (in)famous Bagdadbahn was still only halfway to completion. Not surprisingly, one of the first reforms of the Young Turk Revolution in 1908 was a public works plan that prioritized railway building, but there was insufficient time to achieve its goals prior to the outbreak of the World War.[23]

Developments in the Balkans were markedly different. While—as we will see—attempts to link important port cities with their economic hinterland were not absent, the strategic development of major connecting lines was the most significant part of the railway construction efforts in this region. This was especially notable in the form of a line linking Constantinople to Europe. What was to become the Orientbahn was first devised in the 1860s. The Belgian baron Maurice de Hirsch obtained the concession in October 1869, and the famous Orient Express ran its first service in August 1882. However, after the Congress of Berlin, the political fragmentation of the Balkans complicated the politics of railway building. National governments understandably had different strategic priorities, while Western entrepreneurs and their own governments struggled to retain their rights and a coherent railway network.[24]

The Ottoman government handed out concessions for specific lines, offering financial guarantees for a fixed annual profit and often property and exploitation rights along the lines as well.[25] Most of the builders were foreign companies—British, French, and Belgian[26]—but Austrian and German firms soon joined in their pursuit of profit. The most important line in the Balkans, the Orientbahn, was initially a Belgian enterprise, but it soon developed into a transnational one, with the growing influence of German owners who—after the gradual independence of the Balkan states—ran into problems with their governments and desperately wanted to divest.[27] In Anatolia, French and British concessionaires built lines around Smyrna and the Sea of Marmara and set their sights on the Ottoman Arab provinces as well. Germans followed suit with the infamous Bagdadbahn,

23 Coşar and Demırcı "Incorporation into the World Economy," 22.

24 Emini, "The Efforts of Ottoman Empire," 57–59.

25 Boriana Antonova-Goleva, "Concessions and Mirages along the Lower Danube: The Town of Silistra in the Plans of Foreign Railway Promoters during the mid–1850s," *Journal of Balkan and Black Sea Studies* 3, no. 5 (2020): 19–47; Thobie, *Intérêts et impérialisme.*

26 Jan Anckaer, *Small Power Diplomacy: Belgium and the Ottoman Empire during the Reign of Lepold I (1831–1865)* (Istanbul: The Isis Press, 2013); Thobie, *Intérêts et impérialisme.*

27 Hertner, "The Balkan Railways."

which was, once more, a transnational enterprise with significant Austro-Hungarian participation. Consequently, many foreign experts in engineering and finance relocated to the Ottoman Empire, often becoming part of its bureaucracy.[28]

What the Ottomans failed to create was a uniform or at least state-regulated system of fares and freight rates or a dominant, state-owned railway company like the MÁV.[29] The Hijaz Railway to Mecca and Medina was the first attempt to do so—curiously, as the private property of the sultan.[30] However, local initiative was not absent from these efforts, even from very early on, as the case of the Ruschuk (Ruse)/Silistra–Varna line between 1856 and 1863 shows (see fig. 1). Intriguingly—as we will see—its story also resembles one of the most important features of later Hungarian developments: the manifest rent-seeking of local figures.

Rent-seeking *á la Hongroise*? Transylvanian Saxon Railway Projects around the Turn of the Twentieth Century

The Transylvanian Saxons—a German-speaking and predominantly Lutheran group—were one of the three privileged feudal nations (*nationes*) of the Principality of Transylvania. Highly educated and urbanized, the Saxon community, less than 250,000 strong, lived mostly in the southern areas of the province: from the cities of Szászváros/Broos/Orăștie through Nagyszeben/Hermannstadt/Sibiu and Medgyes/Mediasch/Mediaș to Brassó/Kronstadt/Brașov, with a smaller settlement area in the north around Beszterce/Bistritz/Bistrița and Szászrégen/Sächsisch-Reen/Reghin. Through the reorganization of the administration after 1848, they gradually lost their feudal autonomy and their own Saxon administrative units.[31] Still, as a relatively wealthy community, they enjoyed the benefits of the electoral census and retained their dominance in local politics. Moreover, their small but not insignificant contingent of MPs—elected by constituencies with exceptionally low numbers of registered voters—usually supported

28 Darina Martykánová and Meltem Kocaman, "A Land of Opportunities: Foreign Engineers in the Ottoman Empire," in *Philosophy of Globalization*, ed. Concha Roldán, Daniel Brauer, and Johannes Rohbeck (Berlin: De Gruyter 2018), 237–251.

29 Coşar and Demırcı, "Incorporation into the World Economy," 23.

30 Murat Özyüksel, *The Hijaz Railway.*

31 Jonathan Kwan, "Transylvanian Saxon Politics and Imperial Germany, 1871–1876," *The Historical Journal* 61, no. 4 (2018): 991–1015.

the government despite occasional conflicts over Hungarian nationalizing policies.[32]

Even though the Saxons were generally considered more conservative than the mainstream Hungarian liberal governments, Saxon leaders were genuinely concerned with the development and modernization of their community. They usually looked toward Germany for inspiration[33] and pursued a long-term institution-building strategy, from schools to banks and rural cooperatives. These mostly provided the community with its own resources, to which were added those drawn from abroad through the Saxons' highly developed banking network.[34] This project also embraced technical development in cities and better transport and communication. Around the turn of the twentieth century, Saxon regions—and Saxon towns in particular—were among the more developed areas of Hungary. However, railway construction was beyond the financial and professional capacity of the Saxon elites. While the first transit line, the Eastern Railways, connected Brașov with Budapest and the Balkans, and an alternative line reached Mediaș and Segesvár/Schäßburg/Sighișoara, other important areas in southern Transylvania around the traditional Saxon capital of Sibiu were left without major railway connections.

The Alvinc–Nagyszeben–Vöröstorony Railway

This infrastructural absence was all the more significant given that Sibiu and Brașov lay on the historical trade route along the Olt River to Wallachia, and local elites hoped to profit from reviving this connection. Railway plans were drafted quite early, but none came to fruition, as it was clear that their success would depend on alignment with the broader railway construction strategies of both Austria-Hungary and Romania.[35] Thus, Saxon elites attempted to persuade policymakers by arguing that the line was important for strategic and international commercial reasons.

The plan was promoted by a group of very prominent Saxons, among them the leaders of the Saxon political party and major financial institutions,

32 Gábor Egry, "Ein anderer Ausgleich: Einiges zur Vorgeschichte des Sachsentages 1890," *Zeitschrift für Siebenbürgische Landeskunde* 29, no. 1 (2006): 51–61.

33 Kwan, Transylvanian Saxon.

34 Gábor Egry, *Nemzeti védgát vagy szolíd haszonszerzés? Az erdélyi szászok pénzintézeti rendszere és szerepe a nemzeti mozgalomban (1835–1914)* (Csíkszereda: Pro Print, 2008).

35 Arhivele Naționale Secția Județeană Sibiu (ANSJ SB), Colecție Familia Bedeus 315. Letter from Karl Wolff, July 19, 1890.

most notably Karl Wolff (1849–1929). Wolff was the party chairman, director general of the Hermannstädter Allgemeine Sparkasse—the largest financial institution in the province—and editor in chief of the *Siebenbürgisch-Deutsches Tageblatt*, the largest Saxon newspaper.[36] The deputy lord lieutenant (*alispán, Vizegespann*) of Nagyszeben County, Gustav Thalmann (1832–1911), also joined the group, which already included Aurel Brote (1842–1897), a prominent Romanian intellectual. Thalmann would become the government representative, or lord lieutenant (*főispán, Obergespann*), in 1892. While the lord lieutenant of Szeben County, András Bethlen—later the minister of agriculture—was not part of the concessionaire group, he actively lobbied on their behalf. The initiators settled on a track from Alvinc/Vințu de Jos through Nagyszeben to Vöröstorony, which also connected the line with Felek/Feleac; through that station on the eastern side of the Olt, it linked with Fogaras/Fogarasch/Făgăraș, where local notables led by the lord lieutenant Mihály Horváth also planned to build a connecting HÉV line (see fig. 3).

The first step—the concession for a shorter part of the line between Nagyszeben and Felek in the form of a HÉV—preceded the treaty with Romania. Thalmann had started planning the railway in 1886, while still deputy lord lieutenant. The group of initiators drafted a construction plan that was sanctioned by the ministry in 1889, coinciding with the plans of the Fogaras group. However, the line seemed to offer profit only if it was not left as a dead end. The Romanian connection was indispensable, and the concessionaire group actively lobbied for it.

Aurel Brote met the Romanian minister of foreign affairs, Alexandru Lahovary, who seemed to be indifferent regarding the connection's exact location. If pressed on this issue, he leaned toward the Jiu Valley for practical reasons.[37] Meanwhile, the Austro-Hungarian army also preferred the Jiu Valley, as did Minister Baross, for whom its coal mines were more important.[38] Thus, the concessionaire group argued that they could build the line by using the resources of the county alone. As Bethlen put it: what was most important was to get the construction permit, as everything else could be settled later.[39] The ministry issued a construction permit in August 1890.[40]

36 Egry, *Nemzeti védgát*; Kwan, "Transylvanian Saxon"; ANSJ SB, Colecție Familia Bedeus 497. 69.

37 ASNJ SB Colecție Familia Bedeus 315. Karl Wolff's letter from July 19, 1890.

38 ASNJ SB Colecție Familia Bedeus 386. 74. András Bethlen to Karl Wolff, July 25, 1890.

39 ASNJ SB Colecție Familia Bedeus 386. 74. András Bethlen to Karl Wolff, July 25, 1890.

40 "Törvényjavaslat, a Nagyszeben-Vöröstoronyi helyi érdekű vasút Nagy-Szebentől Felekig terjedő vonalának megépítésére és üzleti berendezésére szükséges költségek biztosításáról,"

However, the new plans necessitated a recalculation of construction costs, and the revised financing plan is revealing and typical of how risks were transferred to public bodies while the individual profit of the concessionaire group was maximized. The company's stock capital was calculated to cover the estimated construction costs: of the 1.3 million forints (2.6 million kronen), 845,000 forints were allotted for priority shares. The rest was put up by public bodies, among them the county of Nagyszeben (from its public fund for county roads), the city of Nagyszeben (where the initiators held influential political and administrative positions), and the state (which guaranteed a further 75,000 forints as a fee for postal transportation and subsidies for HÉV lines). Originally, the remainder was to be covered by priority bonds redeemable over 50 years, which local financial institutions promised to underwrite. To cover the bank guarantees, Nagyszeben County extended its guarantee for the annual down payment on these priority bonds from the levy of a new 9.25 percent county surtax.[41] If this was the result of negotiations, it came about in a very peculiar way: the group of concessionaires—among them Wolff and Thalmann—were also the leaders of the Saxon banks, dominating both the public and private institutions involved in the railway company.

However, any surtax higher than five percent required government and parliamentary approval, and the ministry feared that by issuing bonds instead of priority shares, the company would use up its potential credit. Thus, even though it promised to build the remaining line to the Romanian border at its own cost, it seemed unlikely that it could attract the necessary capital. As such, the ministry requested the issuance of shares instead of bonds. The solution again was to turn to a public body, and it was again the county which promised to take over the priority shares. Moreover, as the county did not have this amount of capital, it took a loan—from the banks that originally promised to take the guaranteed bonds—and levied the surtax for the annual down payment on this loan.[42]

The line was quickly built and opened in 1892, together with the Felek–Fogaras line (see fig. 3). The next step was the extension of the line to Turnu Roșu and the building of a new connection to Alvinc. As the treaty with Romania obliged Hungary to build the connection by 1897, the government faced significant pressure. For the concessionaires, however, it was all the more important, as the existing line was operated by the company,

Vasúti Közlöny 22, no. 68 (1891): 594.

41 "Törvényjavaslat, a nagyszeben-vöröstoronyi helyi érdekű vasút," 595.

42 "Törvényjavaslat, a nagyszeben-vöröstoronyi helyi érdekű vasút," 595.

and—even though the founders had transferred all risks to the public bodies and banks they managed—potential operational losses loomed over the endeavor. Indeed, with the construction of the Nagyszeben–Felek line, they had likely exhausted the potential capital reserves of the county, the city, and the banks. All priority shares were now held in escrow by the Saxon Nationsuniversität—another public body dominated by the Saxon Volkspartei and thus under the influence of the concessionaire group.[43] Meanwhile, the concessionaires undertook the obligation to build the Romanian connection, likely as an informal condition set by Minister Baross in order to support the initiative and disburse the requested state subsidies.

The solution to these problems was twofold: attracting a new investor-financier and eliminating operational risks. However, the effort was complicated by the fact that, while the extension to Alvinc was important for the Saxons, the government would have been content with the line between Felek and the border alone. Indeed, the ministry seriously considered building this section from the state budget, leaving the Nagyszeben–Felek HÉV out in the cold with a rump line. Thus, the concessionaires simultaneously tried to block the state-led construction, obtain a subsidy from the budget (about two million kronen), transfer the operation of the whole line to the state railways for a fee, and attract a new investor who would cover the remaining construction capital.[44]

This charge was led by none other than Karl Wolff. Wolff was the leader of the Saxon party (Sächsische Volkspartei), and a formal political pact between the government and the Saxons had been in place since 1890. According to this agreement, the Volkspartei retained its independence, and—except for the county of Brassó, where they shared the available seats between the parties—the government would not run against its candidates in either county or parliamentary elections. In exchange, Saxon MPs—their number ranging between 10 and 15—supported the government. As usual, such arrangements implied better access to ministries and high-ranking politicians, although a positive reception of Saxon ideas and proposals was not guaranteed.[45]

This was the case with the Alvinc–Nagyszeben line, too (see fig. 3). Wolff worked on the drafting plans, often participated in the preliminary tracking

43 Ferenc Vas (?) to Leó Lánczy, w/o date, w/o place, Magyar Nemzeti Levéltár Országos Levéltára (MNL OL) PMKB Z 40 1526. t.

44 ANSJ SB Colecție Familia Bedeus. Karl Wolff's letter to Oscar Meltzl, February 1, 1895.

45 For more on this pact, including that Francis Joseph himself helping to facilitate it, see Egry, "Ein anderer Ausgleich."

expeditions in person, and lobbied politicians. In his lobbying efforts, Wolff was assisted by government representative Thalmann and by Bedeus's brother-in-law, statistician and MP Oscar Meltzl (1843–1905).[46]

Key to all these plans was making the MÁV take over the operation. In this case, the owners of the railway could request an annual fee from the MÁV, and the burden of management was lifted from their shoulders. With the promise of an annual fee, they could entice other financial backers and investors, who would provide capital for the construction, as the fees also guaranteed a return on their investment. Baross, who had been a tough negotiation partner in 1890, died in 1892, which probably helped Wolff. Still, Wolff needed all of his connections at the highest levels. He lobbied Prime Minister Sándor Wekerle (1848–1921), who promised the two-million-forint subsidy.[47] However, Wekerle soon left office, and his successor, Dezső Bánffy (1843–1911), had not left good memories with the Saxons from his tenure as lord lieutenant in Beszterce County.

The ministry's plan to build the line between Felek and the border, leaving the Alvinc–Nagyszeben section to the concessionaires, was close to completion in April 1895, when Wolff had an audience with Bánffy. Wolff argued that if they were to be left to build the line as their own effort, they would lose all their credibility with the Saxon public and electorate. He predicted serious consequences for the government, not least the loss of these seats to the Saxon opposition that was still held at bay by the pact of 1890. Bánffy, according to Wolff's missives, was receptive and promised state support.[48] Finally, the government promised a subsidy and the takeover of operations through the MÁV.

Even before securing this agreement, the concessionaires approached the country's largest banks to feel out the chances of making them finance the whole endeavor. Already in 1893, Wolff met with Leó Lánczy (1852–1921), the leader of the second-largest bank in Hungary, the Pesti Hungarian Commercial Bank (Pesti Magyar Kereskedelmi Bank—PMKB), and Zsigmond Kornfeld (1852–1909), the head of the largest one, the Hungarian General Bank (Magyar Általános Hitelbank—MÁH). Kornfeld was quite skeptical, as he did not believe that state support would be granted. Lánczy, whose bank started notable investment into the HÉV sector in cooperation with German partners in 1892, when they established Magyar Helyi Érdekű

46 Egry, *Nemzeti védgát*; see also Karl Wolff's letter to Oscar Meltzl, ANSJ SB, Fond Colecţie Bedeus.

47 ANSJ SB Colecţie Familia Bedeus 367. Karl Wolff's letter to Oscar Meltzl February 1, 1890.

48 ANSJ SB Colecţie Familia Bedeus 367. Karl Wolff's letter to Oscar Meltzl April 19, 1895.

Vasutak Rt.,[49] engaged in negotiations. He estimated the construction costs at about 6.5 million forints and the threshold of rentability at about 4,000 forints per kilometer in revenue—a highly unlikely sum.[50]

Thus, Lánczy came up with a financing plan that was intended to satisfy the concessionaires' desire for stable income, minimize the PMKB's risks and involvement, and secure a stable profit on a loan they gave to the construction company. According to this plan, the MÁV would have paid an annual 350,000-forint rent and the state treasury an additional 100,000-forint transit fee for international transportation along the line. Lánczy also involved himself in the lobbying efforts and made the ministry accept a higher-than-initially-estimated construction cost but wanted to transfer the concession to an affiliate of the PMKB, the Magyar Helyiérdekű Vasút Rt.[51] The concessionaire group, however, did not wish to relinquish control of the company and requested that at least one-third of the board of directors be comprised of the delegates of Szeben County, who also wanted to retain shares in the new company.[52]

These complicated negotiations and transactions resulted in an arrangement that made the state a co-owner of the lines of a new company, Alvinc–Vöröstoronyi HÉV Rt. The state offered half of the calculated construction costs. The MÁV took over the operation, but the company retained all revenues from local traffic. This enabled the public bodies that provided the capital for the first phase of construction to raise new capital for the building of the extension lines,[53] once again leaving the concessionaires free from any obligation.

But how did the concessionaire group profit from the arrangements? The key was separating the construction capital from the stock capital in the form of priority shares. The company issued priority shares for the amount of the construction capital and transferred them to the construction company as offset for its work. The construction company then sold the shares to PMKB gradually as it proceeded with the work, always just enough to cover

49 Károly Lamotte, ed., *A Pesti Magyar Kereskedelmi Bank százéves története 1841–1941* (Budapest: Pesti Magyar Kereskedelmi Bank, 1941), 102–105.

50 Adolf Albrich, "Dr. Carl Wolff und unser Eisenbahnnetz," in *Dr. Carl Wolff als Direktor der Hermannstädter allgemeinen Sparkassa 1885—1910. Festschrift aus Anlass der 25. Jahreswende seiner Erwählung*, ed. Directionsrat der Hermannstädter allgemeinen Sparkassa (Hermannstadt: Josef Drotleff, 1910), 126–130.

51 MNL OL Z 40 1526. t. Naygszeben-Vöröstoronyi HÁV Rt. annual report from 1894.

52 Ferenc Vas (?) to Leó Lánczy, w/o date, w/o place, Magyar Nemzeti Levéltár Országos Levéltára (MNL OL) PMKB Z 40 1526. t.

53 "Egy új nemzetközi vasútvonal," *Pesti Napló*, no. 71. March 12, 1896.

the construction costs of the upcoming section. PMKB paid in advance for all tranches—thus providing the necessary cash flow—but charged 5 percent interest on the advance. Thus, the PMKB not only profited from the interest; it gradually bought up all the priority shares of the company, while the concessionaires retained their ordinary shares, which still brought in revenues.[54] As the PMKB did not have to bother with the operation of the company, in the end both the concessionaires and the bank made only a financial investment.

The line was inaugurated in 1897, but it was not the end of the story for the concessionaires and the Budapest bank. Several other lines in this region of Transylvania—including the Felek–Fogaras and the Fogaras–Brassó connections—were subsequently built following the same template, and at one point the PMKB decided to consolidate these scattered companies, which were only channels for money to shareholders and boards of directors. It is notable that instead of bringing these companies into the PMKB's own investment holding company, the Magyar Helyi Érdekű Vasutak Rt., another company was established through the successive merger of local companies.[55] The statutes of the newly created Erdély-Délvidéki HÉV Rt. seemingly set out strict obligations for the operation of its railway lines—but all were, in fact, operated by the state railways. Its board of directors was, however, quite large for a company whose only work was accounting for its revenues and real estate. In 1911, the board of directors consisted of 35 members and the supervisory board of 10 members, and in 1914, these numbers were 38 and 10, respectively. This was not cheap: these members usually received remuneration as a percentage of the profit, and it was also customary to further provide them with a small number of shares.

However, the composition of the board explains why things turned out this way. Alongside important business figures, it featured mostly public figures. Prominent Saxons (including Wolff) were joined by the leaders of the Romanian community in Făgăraș: Octavian Vasu, Lucian Borcia, Nicolae Șerban, Andrei Micu, Ioan Lupaș, and Ioan Sencea.[56] This involvement of locally significant figures in HÉV companies repeated a general pattern from all over Hungary. The additional twist of the Erdély-Délvidéki HÉV was that the Romanian politicians were not just opponents of the Hungarian government: their party, the Romanian National Party, decried its politics

54 MNL OL Z 40 1526 t. Szerződés tervezet, 71–1452 t. Törlesztési terv.

55 Dr. Hegedűs Loránd, *A Pesti Magyar Kereskedelmi Bank keletkezésének és fennállásának története*, vol. 2, *1892–1917* (Budapest: Pesti Magyar Kereskedelmi Bank, 1917), 257–258.

56 *Nagy Magyar Compass* 40/2. 1912/1913, 964–965, *Magyar Pénzügyi Compass* 1914–1915, 341.

of national oppression and promised to fight against it by every peaceful means. Yet, in this case, the Romanian politicians all received significant income from the Hungarian state by virtue of their political position. They palpably colluded with the government, which nevertheless did not prevent most of them from gaining the credentials of ardent Romanian nationalists, propelling them further in their political careers after 1918. (Vasu, for example, would become the first Romanian prefect of Făgăraş County in 1918, and Borcia an MP in the Romanian Parliament.) That such arrangements might have included favors to the government from the Romanians is clear from the career of Nicolae Şerban, depicted as a renegade by historian and Romanian nationalist leader Nicolae Iorga (1871–1940), who visited the Făgăraş area during the electoral campaign of 1905.[57]

The Nagyszeben–Szentágota–Segesvár Railway

On paper, the second scheme under consideration in the present study was easy to implement, and the breadth of its application seems to confirm this impression. However, if we contrast the Alvinc–Vöröstorony HÉV with the line connecting Sibiu with Szentágota/Agnetheln/Agnita and further with Sighişoara, the comparison makes clear how much the success of such initiatives—both in terms of construction and their rent-seeking aspects—hinged upon informal connections and the mobilization of government support (see fig. 3).

The fate of the Nagyszeben–Felek line has demonstrated how limited the resources of local and county communities were when it came to making capital-intensive investments in infrastructure. While the initiators could mobilize county and municipal resources and even use relatively large financial institutions to boost the credentials of these enterprises, they were well aware of the fact that, without outside investments, regional development plans and the rent-seeking they made possible were hardly attainable.

The origins of the Nagyszeben–Szentágota line also lie in the 1890s, when the Alvinc–Vöröstorony and the Felek–Fogaras lines were successfully built. The area that the proposed line was to serve was quite secluded, lying to the north of the Olt and south of the Nagy-Küküllő. It had no important urban center—the marketplace of Agnita was small and served only as

57 Nicolae Iorga, *Neamul Românesc în Ardeal şi Ţară Ungurească*, vol. 1 (Bucharest: Minerva, 1906), 67, 85, 91.

the administrative center of the district—and no important line of communication traversed the region. Nevertheless, the first initiative for a railway line was floated in 1894, in parallel with the efforts surrounding the Alvinc–Vöröstorony line. The preliminary group of concessionaires (who were awarded the title of preparing a feasibility plan) held its first assembly on July 10, 1894.[58] While a number of important Saxon figures were involved in the simultaneous railway construction efforts, its *spiritus movens* was Gustav Bedeus (1861–1939).

Bedeus held a baronet's title, and his father, Josef Bedeus Jr., was the director general of the Hermannstädter Bodencreditanstalt, the second-largest Saxon bank. Gustav, however, did not attain a high rank in the administration—at the time of his railway-building efforts, he was district chief (*főszolgabíró*). His efforts to build the line were certainly rooted in his experiences of the region's social and economic conditions. But these efforts also show how little leverage and influence he had even within the Saxon community. Compared with the plans and modus operandi of the Alvinc–Vöröstorony concessionaires, the most striking feature was the pettiness of Bedeus's financial maneuvers. To prepare the feasibility plan, the group needed support, but Gustav Bedeus could only rely on the rather meager sum of 500 forints that the Bodencreditanstalt lent him. He tried to use it as leverage to mobilize more money from the local credit cooperative in Agnita and as collateral for a loan of 3,000 forints, while he offered as escrow a mortgage bond of his own.[59] The potential donations from local businesses (a tannery, a salami factory, and an alcohol factory) were of the same magnitude, and when they failed to materialize, he turned to the banks in Sibiu, asking them to commit to an annual donation of 500 forints from their profit.[60] He hoped to use the regular income as collateral for a larger loan, discounted at the rate of 5–6 percent, making the potential amount of the loan a few tens of thousands of forints.[61]

Equally desperate was his plan for financing the construction of the line. From the estimated 870,000 forints, he hoped that municipalities would underwrite about 25 percent. The remaining amount he hoped to collect from the county's road repair fund, from postal transportation fees, and so on. Still, he did not even get close to the target amount, as community

58 ANSJ SB Colecţie Familia Bedeus 496, 18. Finanzplan.
59 ANSJ SB Colecţie Familai Bedeus 496, 16, 17.
60 ANSJ SB Colecţie Familia Bedeus 496, 69.
61 ANSJ SB Colecţie Familia Bedeus 497, 117.

resources were already strained by other endeavors.[62] Thus, the railway construction did not progress at all, although the ministry was ready to issue the construction permit as early as 1895.

Bedeus's next push came in 1899, only this time supported by more important figures, such as Meltzl, who actively engaged in the negotiations with the ministry. Now, the plan was complemented by a proposed merger of the Nagyszeben–Szentágota line with the line connecting Sighișoara and Agnita, thereby linking the Olt and the Nagy-Küküllő valleys. The government conditioned state subsidies on a common endeavor.[63] But even with the planned merger, attracting the necessary capital was almost impossible, despite Bedeus's new, even more desperate ideas for fundraising.[64]

Under such conditions, it is hardly surprising that neither the ministry nor the prospective construction entrepreneurs saw the company as potentially profitable.[65] The solution was, yet again, the takeover of the operation by the MÁV, as a calculation from a prospective construction entrepreneur demonstrated. He presumed that the county would deliver on its financial promises through various schemes. But even under this condition, a reduction of construction costs was necessary. Finally, the revenues were to be split between MÁV and the Nagyszeben–Szentágota HÉV at a ratio of 55–45 to make the prospective dividend close to 5 percent.[66]

It took another four years until the ministry issued the permission.[67] Most importantly, Bedeus had found a new and eager partner in the person of Gusztáv Gratz, a politician of the governing party with good connections within the business elites, who actively lobbied the government on his behalf.[68] While the most important request of the concessionaires was satisfied—the MÁV would operate the line—individual profit from its operation was much less significant than it was from the Alvinc–Vöröstorony line. This reflected not just the difference between the passenger and business traffic on the two lines, but also the recurrent attacks on the concession model, to the extent that in 1904 even the PMKB's Magyar Helyi Érdekű Vasutak Rt. came under public scrutiny. It was claimed that this model deliberately inflated construction costs, often to double the real

62 ANSJ SB Colecţie Familia Bedeus 496, 18.

63 ANSJ SB Colecţie Familia Bedeus 496, 92.

64 ANSJ SB Colecţie Familia Bedeus 496, 92, 497, 135, 203.

65 ANSJ SB Colecţie Familia Bedeus 496, 109, 285, 223.

66 ANSJ SB Colecţie Familia Bedeus 496, 109.

67 "A nagyszeben-szentágotai helyiérdekű gőz- és motoros üzemű vasútra vonatkozó engedélyokirat," *Vasúti Közlöny* 40, 66 (1909): 449–452.

68 ANSJ SB Colecţie Familia Bedeus 449, 64, 66, 68, 70.

ones, because they were the ultimate source of profit.[69] Although Bedeus was clearly less well-networked than Wolff's group, the fact that he put his personal property at risk and pursued a goal that long seemed unattainable shows how much his efforts were motivated by true belief in developmental ideas and a perceived duty to help his poor district. In a way, this shows, if indirectly, how much rent-seeking was a "privilege" of the elite. Moreover, it also speaks to how the supposed "criticality" of a would-be line—to borrow a term used by Jens Ivo Engels in an earlier chapter—was constructed by local lobbies as a means for pursuing their local goals, under the guise of an argument for the very need of networking beyond the local, on a broader regional or transregional scale.[70] Public interest, rather than narrow partisan or local interests, remained the watchword, regardless of where the funding came from or went to.

The story of the lines thus highlights an important aspect of the political economy of the construction of the secondary railway lines with the concession method. With all the profit local elites reaped, in the end this amounted to a flow of capital and revenue from the center, the state coffers, to the periphery. The railway network grew significantly, and the local elites—who never attempted a fact-based assessment of its costs and benefits—easily sold it as development, a victory over backwardness. Discursively, they were aligned with the broader pattern of colonial discourses—but who exploited whom, and to what extent?

Colonial Capitalism *á la Turquoise*?

At the outset, railway construction in the Ottoman Empire did not significantly lag behind Hungary, and its main model was comparable. As Malte Fuhrman and Constantin Ardeleanu also explore in the present volume, the developmental logics underpinning these efforts often involved the capital and expertise of foreign entrepreneurs—quite frequently engineers—who conceived plans based on strategic and commercial considerations, negotiated a concession from the Ottoman government, and built the line at their

69 Dr. Kepes Gyula, "A Magyar Helyiérdekű Vasutak Rt."

70 Jens Ivo Engels, "Writing the History of Critical Infrastructures: Some Conceptual Remarks" in this volume. For regional lobbying for rail network building to avoid peripheralization, see Mihai Chiper, "Regional Competition, Economic Interests, and Military Stumbling-Blocks in the Making of the Moldavian Railway Network" in this volume.

own costs.[71] Such concessions usually entailed some profit guarantee and very often the transfer of land along the line, exploitation rights to natural resources, as well as the eventual transfer of the lines to the state after a long period of operation.[72] Unlike in Austria or Hungary, however, the Ottomans never transitioned to the state railway model; even the Hijaz Railway was built as a specific variety: the private property of the sultan. Thus, blatant rent-seeking through the transfer of the operation to the state was not available. But the wide range of guarantees that concessions offered to investors and operators made rent-seeking a plausible business strategy here, too. Moreover, whether their profit came directly from the state or indirectly was a matter of indifference to the local elites, though the guarantees given to concessionaires and the effects of weak competition on transportation prices also had an effect. Still, as concessions were negotiated with the Ottoman government, good connections mattered.

In the first phase of the Ottoman railway construction, between the 1850s and 1860s, one can already find cases similar to the Saxon schemes outlined above. Boriana Antonova-Goleva has devoted several articles to the story of the Ruschuk/Silistra–Varna railway between 1856 and 1863, mostly focusing on the cooperation between one wing of the Ottoman government and the local Bulgarian and Ottoman elites who—she argues—aspired to build this line as an Ottoman investment.[73] But in light of the Transylvanian examples, an alternative interpretation may also be drawn from the fragmentary source material uncovered. The endeavor was not about building a railway line as an investment by local businesses—or not necessarily building it *alone*—but about local elites getting in a position that would enable rent-seeking for the businesses in question.

71 See the chapters Fuhrmann, "On Hegemony and Connectivity," and Constantin Ardeleanu, "British Investors and the 'Kustendjie Harbor Dues Issue'" in this volume.

72 Boriana Antonova-Goleva, "Foreign Technology and Local Agents of Modernization: The Rusçuk and Varna Railway Project as a Case Study for the Ottoman Modernization," in *Transforming Southeast Europe during the Long Nineteenth Century: Persons and Personalities as Agents of Modernization in the Ottoman and Post-Ottoman Space*, ed. Boriana Antonova-Goleva and Ivanila Masheva (Leiden: Brill, 2024), 353–402; Martykánová and Kocaman, "A Land of Opportunities"; Sevtap Demirci and Nevin Coşar, "Modernisation through Railways: Economic and Social Change in the Ottoman Empire in the Nineteenth Century," *Journal of Balkan and Near Eastern Studies* 23, no. 5 (2021), 684–694; Darina Martykánova, "Mastering the Uncontrollable: The Ottomans and the Use of Modern Technologies," *History of Science* 62, no. 4 (2024): 473–487.

73 Antonova-Goleva, "Concessions and Mirages"; Antonova-Goleva, "Foreign Technology and Local Agents"; Boriana Antonova-Goleva, "'Top-Down' or 'Bottom-Up' Modernisation: Local Railway Entrepreneurs in the Ottoman Empire in the Second Half of the 19th Century," *Turkish Historical Review* 14, no. 1 (2023): 26–57.

Several lines connecting the Danube and the Black Sea, including the Ruschuk/Silistra–Varna railway, were initially negotiated by British groups of investors (see fig. 1). Yet a group of local elites, probably with the tacit encouragement of Grand Vizier Mustafa Reşid Pasha, emerged as an alternative and competitor. The Ottoman government clearly preferred this group, whose members, similarly to the Saxons, belonged to a multipositional elite[74] that had started to collect capital among the Bulgarian and Ottoman business elites—including those in Constantinople, among them the prominent Galata banker Jacques Alléon. With the government's help, they successfully obstructed negotiations with the British Danube and Black Sea Railway Company, which promised to build a line through Dobrudja to the port of Kustendjie/Constanța.[75]

The British employed diplomatic pressure and hired an agent, add John Trevor Barkley, to investigate the background of the rival company. Even though Barkley concluded that the Bulgarians lacked the necessary knowledge to establish a railway company and did not have the funds—only promises for a portion of the necessary construction capital, £160,000 out of £3 million—at one point the Bulgarian-Ottoman group even engaged in negotiations with the British. They demanded representation on the board of directors of the Danube and Black Sea Railway Company without investment, while the British offered shares against investment capital. While no agreement was reached, the British lowered their demands in their negotiations with the Ottoman government, and eventually both concessions were approved in September 1857.[76]

The state offered profit guarantees for the Ruschuk–Varna line, but despite the obvious success of the initiators, its story took another turn. Soon, a Belgian company led by the Ottoman consul in Antwerp, Joseph Posno, took over the construction, which finally ended up in British hands with a group led by William Gladstone. However, the local Bulgarian group remained engaged with the project, actively promoting it and collecting capital.[77] This shift to Posno's group and the seemingly paradoxical activism of the Bulgarian-Ottoman group is probably better explained by the Saxon

74 Judit Pál, "The Study of the Hungarian Elites of the 19th and 20th Century," *Historica: Revue pro historii a príbuzné vedy* 5, no. 2 (2014): 221–237.

75 Antonova-Goleva, "'Top-Down' and 'Bottom-Up'"; Antonova-Goleva, "Foreign Technology and Local Agents."

76 Antonova-Goleva, "'Top-Down' and 'Bottom-Up,'" 46–47.

77 Antonova-Goleva, "'Top-Down' and 'Bottom-Up,'" 46–47. On British investors in general, see J. H. Jensen and Gerhard Rosegger, "British Railway Builders along the Lower Danube, 1856–1869," *The Slavonic and East European Review* 46, no. 106 (1968): 105–128.

analogies above. I would argue that this was not a rupture but rather a manifestation of the rent-seeking aspects of the project. The Ottoman government, which had already sided with the Bulgarians, transferred it to someone who seemed to be able to ensure its realization, just as it had happened in Hungary after the concessionaire group received construction permission.

If we go back to the rent-seeking schemes in Transylvania, one important characteristic is that the initiators never considered the construction at their own cost, nor had they ever planned to come up with more than a fraction of the venture capital from their own resources. First, they immediately involved public bodies, even in the accumulation of the first tranche of capital. Second, even when the company was formally established and the concession holders started underwriting ordinary shares, they had to cover only a part of their nominal value to register the company. They counted on the cooperation of serious investors who would join the company and cover the difference in a way that remained advantageous for their dividend revenue. Finally, they never intended to operate the railways or manage the company beyond accounting for its secure revenues.

This was also true for the Bulgarians, as they only had to present the deposit for the concession and 2 percent of the estimated construction capital—about £60,000.[78] And while in one of his reports Barkley remarked that the Bulgarians did not know how to establish a joint-stock company, their partner in Constantinople, Jacques Alléon, certainly did. The way the Bulgarians negotiated with the Danube and Black Sea Railway Company—only expressing interest in membership in its board of directors, with reimbursement and a customary small number of shares guaranteeing secure income—also suggests rent-seeking. So do Posno's attempt and the continued involvement of the Bulgarians. Concessions were an asset to financialize and to transfer to other companies, just as the Erdély-Délvidéki HÉV (which was eminently a payout company for prominent figures) emerged. Unlike the Bulgarian group, Posno had the necessary business connections in Belgium to mobilize capital and expertise, but he had never been involved in railway construction before. He had various investments in Belgium, but he rarely acted as their actual manager. He seems to have been a middleman in this case as well, and his dealings with the Porte fit the rent-seeking model. Thus, in 1860, he negotiated changes to the concession, and one of his demands was that the Ottoman government would eventually take over one-third of the

78 Antonova-Goleva, "Foreign Technology and Local Agents."

shares, which the Ottomans refused.[79] It is highly likely that the Ottomans also used the Bulgarian-Ottoman group for another purpose—namely, that of putting pressure on other concession-seeking groups. This remained a persistent negotiation tactic up to World War I and beyond.

The Colonizers' Burden?

In a sense, all the railway schemes mentioned in this chapter illustrate the different power relations among the actors involved—in particular, the political agency that individuals could still wield for their own profit, at both the national Ottoman and Austro-Hungarian levels and the level of local politics. The Gemlik–Bursa line, on the other hand, provides us with a case study in which a new railway line was to be built in an area where a French company already held a concession and operated. Thus, the new line promised only limited profitability.[80] However, the Austro-Hungarian ambassador in Constantinople, János Pallavicini, considered it important to establish a foothold for Austro-Hungarian capital in Anatolia. While he did not directly want to interfere with French and British interests (and therefore he dismissed ideas of building lines around İzmir/Smyrna), he saw the intended track of the Gemlik–Bursa line as strategically important because it was close to the Sea of Marmara and the Straits. Thus, he constantly tried to pressure Viennese and Budapest capitalists into bidding for the concession.[81]

This was not an easy task, as the Ottoman government was not ready to provide profit guarantees, and in early 1914, the Ottoman minister of finance, Djavid Bey (1875–1926), reached a wide-ranging agreement with French businesses that offered priority for French investors in the area.[82] Moreover, the Austrian banks involved—the Bodencreditanstalt and the Wiener Bankverein—felt that they could only get a concession if they acquired the majority of the shares of the French Mudanya-Bursa Railway (*Chemin de*

79 Anckaer, *Small Power Diplomacy*, 313–314.

80 See the letter from the Orientbahn to Marckus Wickenburg (Wickenburg Márk), head of the section for trade and commerce in the Ministry of Foreign Affairs, October 1, 1914. Haus-, Hof- und Staatsarchiv Vienna (HHStA), Ministerium des Äussern (MdÄ), Türkei, Bahnbau Türkei 33. Fach 19. For the French concession, see Thobie, *Intérêts et impérialisme*, 341–343.

81 HHStA MdÄ Türkei, Bahnbau Türkei 33. Fach 19.

82 HHStA MdÄ Türkei, Bahnbau Türkei 33. Fach 19. Pallavicini to Foreign Minister Berchtold, July 21, 1914.

Fer Moudania Brousse),[83] which turned out to be hopeless. But there were other interests at play. First, the idea of building the railway came from none other than the section head of the Ottoman Ministry of Public Works, Vondra Bey. Vondra, whose given name was Béla Vondra, was a Hungarian engineer with a long career in Ottoman service. From his key position within the Ottoman administration, he had an overview of concession possibilities and the means of acquiring them. Thus, Vondra managed to get a conditional concession for a feasibility study in February 1914. While the government did not offer a profit guarantee, it allegedly ensured that the state would bear the costs of expropriating the property owners whose lands the tracks would cross while transferring the expropriation rights for natural resources located within 25 kilometers from the line.[84] However, the condition was that Vondra should have the financial backing of a serious group of investors. In this regard, he relied on Pallavicini, who mobilized the ministry's efforts to feel out the views of the Viennese and Budapest banks that eventually led to the failed scheme of buying out the Mudanya-Bursa Railway shares.

Still, Vondra's plan was not dismissed entirely. As the war loosened Ottoman-French ties, an Austro-Hungarian investment group, the so-called Orientgruppe—which included banks involved in the previous attempt—resurrected plans regarding the rail line with Vondra's participation. They did so even though their views on its profitability had not changed significantly, and despite their opinion—which Pallavicini also shared—that French and British businesses would regain their position in the Ottoman Empire after the war. However, as the group consisted of the banks that disbursed a 240-million-krone loan to the Ottoman government in 1916—a significant capital commitment underpinning the Ottoman war effort—it tried to find investment possibilities to secure a return on that amount as well.[85] This circumspect approach and the defeat of the Central Powers brought an abrupt end to similar efforts in October 1918.

The Orientgruppe faced rent-seeking efforts in a very different context, albeit also in northern Anatolia. The Ottoman Ministry of Public Works drew

83 HHStA MdÄ Türkei, Bahnbau Türkei 33. Fach 19. Baron Mittag's telegram to Musulin, June 9, 1914.

84 HHStA MdÄ Türkei, Bahnbau Türkei 33. Fach 19. Pallavicini to Musulin, February 21, 1914.

85 Gábor Egry, "When Imperialists Joined the Nationalists against the West: Post-imperial Business Networks and the Creation of National Economies in the Habsburg Post-imperial Economic Space," in *The War for Anatolia and the Remaking of International Order: Greece, Turkey and the End of the First World War*, ed. Georgios Giannakopoulos, Joseph A. Maiolo, and Gonda Van Steen (London: Bloomsbury, 2025), 203–228.

their attention to the envisaged narrow-gauge railway line from Inebolu on the Black Sea coast to Kastamonu. This line ran perpendicular to the coast through rugged terrain, with deep mountains to cross. All the expert opinions discounted the possibility of operating it profitably, as the area was sparsely inhabited, underdeveloped, and mainly agrarian.[86] The only exception was one large section of inland forests, which, although not yet exploited industrially, was in the sights of one of Austria-Hungary's preeminent forestry entrepreneurs,[87] Baron Hermann Groedel (1856–1930). Thus, it was Groedel's eminent interest to make someone else build a railway line that would transport timber to the Black Sea. Even though the Orientgruppe experts considered the traffic potentially generated by Groedel's enterprise insufficient for a profitable operation, they acted under pressure. The Ottoman ministry clearly favored this line, and on one occasion the region's MP even participated in a meeting with the ministry's railway director to make clear how important it would be politically.[88] The Orientgruppe, however, was able to delay implementation, not least because Groedel himself faced difficulties in securing the concession. Although Pallavicini fully supported him in this endeavor, the concession was only awarded in the fall of 1918. Groedel expressed his gratitude to the ambassador in a letter in September—only to face uncertainty after the armistice.

In the case of the Bursa–Gemlik railway, the rent-seeking aspect of the project was obvious, but its target was much less straightforward than in the case of the Transylvanian lines or the Ruschuk–Varna railway. Vondra's scheme ultimately would have drawn profit from the investors—the Viennese banks and their partner, the Orientbahn—while the latter's shares were in possession of the banks. Thus, it was not enough to mobilize Vondra's contacts in the Ottoman government and administration, as he needed someone to pressure the potential investors. It was ultimately Pallavicini who reframed the endeavor in political terms in order to shift calculations away from purely economic ones. As the plan did not come to fruition, it is hard to make any guesses about who would have had to bear the burden. However, if the Orientbahn's story is any indication, it would have probably been the Austrian and Hungarian governments. Less than a year before Vondra secured the concession, a group of Viennese and Hungarian banks

86 HHStA MdÄ Türkei, Bahnbau Türkei 33. Fach 19 nr 48.

87 Bemmann, "Wood-based Businesses."

88 Magyar Nemzeti Levéltár Országos Levéltára (MNL OL) PMKB Z41 515. cs. 3484 f. 52–53, Abschfrift March 3, 1917. 3484d f. f. 367–376., f. 407; Historische Archiv der Bank Austria, Wiener Bankverein Verwaltungstartsprotokolle, February 12, 1918.

bought the majority of the shares of the Orientbahn from its German owners in a complex diplomatic and financial transaction. The banks requested guarantees from the respective governments for their investment, including an option to sell the shares to them.[89] Thus, at the end of the day, the banks were safe from losses, and it is plausible to conclude that they might have also been able to wrangle similar guarantees for the Gimlik–Bursa line. They had much less chance to do the same with the Inebolu–Kastamonu line, however. Even though Pallavicini actively lobbied the Ottoman government on behalf of Groedel, he was more restrained in doing so with the railway line. In this case, it is more probable that the banks either would have had to reject the plan as a whole or enter a partnership with Groedel, who was, however, in a strong position. In any case, rent-seeking was as much directed at the Austro-Hungarian investors and government as it was at the Ottoman government.

Final Remarks: Rent-Seeking or Colonial Business?

The last point mentioned above is already an indication that simplified concepts of colonial capitalism—resting on a power differential between a colonial power and the Ottoman Empire, with the former relentlessly exploiting the latter's resources by occasionally co-opting local elites—cannot easily capture the nature of these business relationships. Transylvania is far from being a region traditionally subsumed under the label of "semi-colonial" or "colonial," although some more recent literature has attempted to apply the concept to its history.[90] Whatever the economic position of the province as a whole, however, its Saxon regions were much more developed. But more importantly, the rent-seeking schemes of the Transylvanian Saxons (and, for that matter, of other regional elites) are anything but the exploitation of the region's resources for the benefit of the center based on existing power asymmetries. Rather, it was this regional elite that, on terms advantageous to them, diverted resources from the center for investment. And to achieve this, they used informal contacts in and political leverage over the government. Moreover, this form of rent-seeking was ubiquitous in Hungary—and not

89 Historische Archiv der Bank Austria.

90 Manuela Boatcă and Anca Pârvulescu, *Creolizing the Modern: Transylvania across Empires* (Ithaca: Cornell University Press, 2022).

merely a peripheral phenomenon—to the extent that one might see it as a structural feature of the political economy of the country.[91]

None of the Ottoman examples is likewise simply a case of exploitation. Without knowing the details of how the Ruschuk–Varna railway operated, it is impossible to assess how much it facilitated the investment of capital on advantageous terms. But one thing is clear: local Ottoman elites—even if much less integrated into the modern economy than their Saxon peers in Transylvania were—could achieve the same results for themselves, albeit only with the assistance of the government and Constantinople business circles. Moreover, their involvement—while most probably rent-seeking from the outset—still enabled the Ottoman government to secure better conditions for itself vis-à-vis foreign investors.

The latter is even more visible in the case of the two Austro-Hungarian plans for railway investment in the Ottoman Empire. The most important characteristic of the investors in those cases was reluctance to commit capital—as they were aware of the real value of the Ottoman government's promises and their political implications. Again, if only temporarily due to the war, they had much less leverage over the Ottoman government than what classic concepts of colonialism would imply: the investors' best hope was to secure guarantees from their own governments to cover potential losses, having few means to fend off rent-seekers.

While a limited set of comparisons is certainly not enough to draw broad theoretical conclusions, the cases of the Transylvanian and Ottoman railways indicate a more complicated situation, in which individual actors and their intentions are key to understanding the eventual unfolding of investments and the distribution of profit. Rent-seeking seems to be a better explanatory factor than colonialism. Thus, it is tempting to think that the striking analogies between railway construction in Transylvania and the Ottoman Empire eventually hint at the limitations of the explanatory potential of the concept of colonialism and colonial capitalism. To put it rather bluntly: if the core is colonial, what remains of colonialism?

Railway construction is not the only phenomenon that might lead to such a revisionist conclusion. In one article,[92] Özgür Türesay expressed doubts

91 On the salience of HÉV construction, see Sándor Csaba Horváth, "Alternatív periodizáció a magyar vasúttörténetben," in *XV. Apáczai-napok Nemzetközi Tudományos Konferencia, 2011: A gazdasági és társadalmi átalakulás perspektívái Magyarországon: Tanulmánykötet*, ed. Ildikó Lőrincz (Győr: NyME Apáczai Csere János Kar, 2012), 29–42, 33–35.

92 Özgür Türesay, "The Ottoman Empire Seen through the Lens of Postcolonial Studies: A Recent Historiographical Turn," *Revue d'histoire moderne et contemporaine* 60, no. 2 (2013): 127–145.

about the understanding of the late Ottoman Empire as a "replica of modern empires" within its boundaries, exerting colonial rule in its peripheries and engaging in a civilizational discourse of backwardness towards those areas. While this critique is aimed at the postcolonial interpretation of the empire's internal relations, its substance is close to what our insights into railway construction suggest: sometimes, analytical concepts are deployed based on too narrow specifics and hide rather than reveal the logic of concrete historical developments. Instead of speaking of colonial capitalism, at least in these cases, a more basic feature of economic action and interaction—namely, age-old rent-seeking—explains railway construction much better.

Bibliography

Albrich, Adolf. "Dr. Carl Wolff und unser Eisenbahnnetz." In *Dr. Carl Wolff als Direktor der Hermannstädter allgemeinen Sparkassa 1885—1910: Festschrift aus Anlass der 25. Jahreswende seiner Erwählung*, edited by Directionsrat der Hermannstädter allgemeinen Sparkassa, 113–158. Hermannstadt: Josef Drotleff, 1910.

Anckaer, Jan. *Small Power Diplomacy: Belgium and the Ottoman Empire during the Reign of Lepold I (1831–1865)*. Istanbul: The Isis Press, 2013.

Arhivele Naționale Secția Județeană Sibiu (ANSJ SB), Colecție Familia Bedeus.

Balaton, Petra. "The Role of the Hungarian Government in the Development of Peripheries in the Austro-Hungarian Monarchy Focusing the Policy of the Ministry of Agriculture (1897–1914)." In *National Identity and Modernity 1870–1945: Latin America, Southern Europe, East Central Europe*, edited by Viktória Semsey, Petra Balaton, Csaba Horváth, and José Antonio Sánchez Román, 241–254. Paris: L'Harmattan, 2019.

Boitoș, Olimpiu. *Lupta lui G. Barițiu, în 1847 pentru introducerea căilor ferate în Transilvania.* Cluj/Brașov: Academia de Înalte de Studii Comerciale și Industriale "Regele Mihai I," Institutul de Cercetări Economice și Sociale, 1945.

Antonova-Goleva, Boriana. "Concessions and Mirages along the Lower Danube: The Town of Silistria in the Plans of Foreign Railway Promoters during the mid–1850s." *Journal of Balkan and Black Sea Studies* 3, no. 5 (2020): 19–47.

Antonova-Goleva, Boriana. "Foreign Technology and Local Agents of Modernization: The Rusçuk and Varna Railway Project as a Case Study for the Ottoman Modernization." In *Transforming Southeast Europe during the Long 19th Century: Persons and Personalities as Agents of Modernization in the Ottoman and Post-Ottoman Space*, edited by Boriana Antonova-Goleva and Ivanila Masheva, 353–401. Leiden: Brill, 2024.

Antonova-Goleva, Boriana. "'Top-Down' or 'Bottom-Up' Modernisation: Local Railway Entrepreneurs in the Ottoman Empire in the Second Half of the 19th Century." *Turkish Historical Review* 14, no. 1 (2023): 26–57. https://doi.org/10.1163/18775462-bja10007.

Bemmann, Martin. "Wood-based Businesses and the Economic Development of Europe in the Nineteenth and Twentieth Centuries: An Introduction." *Jahrbuch für Wirtschaftsgeschichte* 65, no. 2 (2024): 279–305. https://doi.org/10.1515/jbwg–2024-0015.

Boatcă, Manuela, and Anca Pârvulescu. *Creolizing the Modern: Transylvania across Empires*. Ithaca: Cornell University Press, 2022.

Coşar, Nevın, and Sevtap Demırcı. "Incorporation into the World Economy: From Railways to Highways (1850–1950)." *Middle Eastern Studies* 45, no. 1 (2009): 19–31. https://doi.org/10.1080/00263200802547636.

Deák, István. *The Lawful Revolution: Lajos Kossuth and the "Magyars" in 1848–1849*. New York: Columbia University Press 1979.

Demirci, Sevtap, and Nevin Coşar. "Modernisation through Railways: Economic and Social Change in the Ottoman Empire in the Nineteenth Century." *Journal of Balkan and Near Eastern Studies* 23, no. 5 (2021): 684–694. https://doi.org/10.1080/19448953.2021.1935079.

Egry, Gábor. "Ein anderer Ausgleich: Einiges zur Vorgeschichte des Sachsentages 1890." *Zeitschrift für Siebenbürgische Landeskunde* 29, no. 1 (2006): 51–61.

Egry, Gábor. *Nemzeti védgát vagy szolíd haszonszerzés? Az erdélyi szászok pénzintézeti rendszere és szerepe a nemzeti mozgalomban (1835–1914)*. Csíkszereda: Pro Print, 2008.

Egry, Gábor. "Regionális fejlesztés nemzeti fejlesztés: Egy fejlesztéspolitikai és nemzetépítési model a 19. századi Erdélyben." In *Nemzetiség—felelősség: Írások Gáll Ernő emlékére*, edited by György Földes and Zsolt Gálfalvi, 27–54. Budapest: Napvilág 2005.

Egry, Gábor. "When Imperialists Joined the Nationalists against the West: Post-imperial Business Networks and the Creation of National Economies in the Habsburg Post-imperial Economic Space." In *The War for Anatolia and the Remaking of International Order: Greece, Turkey and the End of the First World War*, edited by Georgios Giannakopoulos, Joseph A. Maiolo, and Gonda Van Steen, 203–228. London: Bloomsbury, 2025, 203–228.

Emini, Besnik. "The Efforts of Ottoman Empire to Build Railways in The Balkans." *Journal of Balkan Studies* 2, no. 1 (2022): 51–62.

Fuhrmann, Malte. *Der Traum vom deutschen Orient: Zwei deutsche Kolonien im Osmanischen Reich (1851–1918)*. Frankfurt: Campus Verlag, 2006.

Gergely, András. *Egy gazdaságpolitikai alternatíva a reformkorban: a fiumei vasút*. Budapest: Akadémiai Kiadó, 1982.

Gidó, Csaba. "Az erdélyi fővasúthálózat kiépítése." *Pro Minoritate* 25, no. 3 (2015): 95–113.

Gidó, Csaba. "A székely rabonbán és a székely gőzös: Ugron Gábor és a Héjjasfalva-Székelyudvarhely helyi érdekű vasút." *Magyar Gazdaságtörténeti Évkönvy* (2017–2018): 267–282.

Goswami, Manu. *Producing India: From Colonial Economy to National Space.* Chicago: University of Chicago Press, 2004.

Haus-, Hof- und Staatsarchiv Vienna (HHStA), Ministerium des Äussern (MdÄ).

Hertner, Peter. "The Balkan Railways, International Capital and Banking from the End of the 19th Century until the Outbreak of the First World War." In *Finance and Modernization: A Transnational and Transcontinental Perspective for the Nineteenth and Twentieth Centuries*, edited by Gerald Feldman and Peter Herner, 125–154. Abingdon: Routledge 2016.

Historisches Archiv der Bank Austria.

Horváth, Sándor Csaba. "Alternatív periodizáció a magyar vasúttörténetbem." In *XV. Apáczai-napok Nemzetközi Tudományos Konferencia, 2011: A gazdasági és társadalmi átalakulás perspektívái Magyarországon: Tanulmánykötet*, edited by Ildikó Lőrincz, 29–42. Győr: NyME Apáczai Csere János Kar, 2012.

Iorga, Nicolae. *Neamul Românesc în Ardeal și Țară Ungurească.* Vol. 1. Bucharest: Minerva, 1906.

Jánosi, Csongor. "A Hungarian Count's Business in Romania: The Strange Survival of *Tișița* (1907–40)." *European Review of History/Revue européenne d'histoire* 31, no. 5 (2024): 707–738. https://doi.org/10.1080/13507486.2024.2409125.

Jensen, J. H., and Gerhard Rosegger. "British Railway Builders along the Lower Danube, 1856–1869." *The Slavonic and East European Review* 46, no. 106 (1968): 105–128.

Kroeze, Ronald. "Colonial Normativity? Corruption in the Dutch–Indonesian Relationship in the Nineteenth and Early Twentieth Centuries." In *Corruption, Empire and Colonialism in the Modern Era: A Global Perspective*, edited by Ronald Kroeze, Pol Dalmau, and Frédéric Monier, 173–208. London: Palgrave Macmillan, 2021.

Kroeze, Ronald, Pol Dalmau, and Frédéric Monier. "Introduction: Corruption, Empire and Colonialism in the Modern Era: Towards a Global Perspective." In *Corruption, Empire and Colonialism in the Modern Era: A Global Perspective*, edited by Ronald Kroeze, Pol Dalmau, and Frédéric Monier, 1–21. London: Palgrave Macmillan, 2021.

Kwan, Jonathan. "Transylvanian Saxon Politics and Imperial Germany, 1871–1876." *The Historical Journal* 61, no. 4 (2018): 991–1015. https://doi.org/10.1017/S0018246X17000486.

Lamotte, Károly, ed. *A Pesti Magyar Kereskedelmi Bank százéves története 1841–1941.* Budapest: Pesti Magyar Kereskedelmi Bank, 1941.

Loránd, Dr. Hegedűs. *A Pesti Magyar Kereskedelmi Bank keletkezésének és fennállásának története.* Vol. 2, *1892–1917.* Budapest: Pesti Magyar Kereskedelmi Bank, 1917.

Magyar Nemzeti Levéltár Országos Levéltára (MNL OL).

Magyar Pénzügyi Compass 1914–1915.

Martykánova, Darina. "Mastering the Uncontrollable: The Ottomans and the Use of Modern Technologies." *History of Science* 62, no. 4 (2024): 473–487. https://doi.org/10.1177/00732753241298134.

Martykánová, Darina, and Meltem Kocaman. "A Land of Opportunities: Foreign Engineers in the Ottoman Empire." In *Philosophy of Globalization*, edited by Concha Roldán, Daniel Brauer, and Johannes Rohbeck, 237–251. Berlin: De Gruyter, 2018.

Minawi, Mostafa. *The Ottoman Scramble for Africa: Empire and Diplomacy in the Sahara and the Hijaz.* Stanford: Stanford University Press 2016.

Nagy Magyar Compass 40/2. 1912/1913.

Özyüksel, Murat. *The Hijaz Railway and Ottoman Empire: Modernity, Industrialization and Ottoman Decline.* London: I. B. Tauris, 2014.

Pál, Judit. "The Study of the Hungarian Elites of the 19th and 20th Century." *Historica: Revue pro historii a príbuzné vedy* 5, no. 2 (2014): 221–237.

Popp, Andrew, ed. Special issue, *Enterprise & Society* 25, no. 1 (2024)

Saha, Jonathan. *Law, Disorder and the Colonial State: Corruption in Burma c. 1900.* London: Palgrave Macmillan, 2013.

Szabó, Eszter. "A csallóközi vasút építéstörténete a kezdetektől 1918-ig." *Fórum Társadalomtudományi Szemle* 6, no. 4 (2004): 98–122.

Thobie, Jacques. *Intérêts et impérialisme français dans l'Empire ottoman (1895–1914).* Paris: Publications de la Sorbonne, Imprimerie Nationale, 1977.

Türesay, Özgür. "The Ottoman Empire Seen through the Lens of Postcolonial Studies: A Recent Historiographical Turn," *Revue d'histoire moderne et contemporaine* 60, no. 2 (2013): 127–145.

About the Author

Gábor Egry is a historian and director-general of the Institute of Political History, Budapest. His research interests include nationalism, everyday ethnicity, identity politics, and economic history in modern East Central Europe. He has authored and co-edited several volumes and numerous articles published in journals such as *European Review of History*, *Slavic Review*, and *Contemporary European History*.

8. Hoping and Coping: Local Anxieties and Railway Development in Romania

Toader Popescu

Abstract: The development of cities included in the Romanian railway network before World War I was heavily influenced by the presence of this infrastructure and by its associated opportunities and restrictions. This interaction also generated significant anxiety at the local level for citizens and public administrations alike, stemming from the anticipation of the negative outcomes brought about either by the advent of railways or because of their absence. The underlying factors were, on the one hand, the unpredictability of the effects that this new system would have on the established way of life and, on the other hand, the perceived lack of influence that both ordinary citizens and municipalities felt they had on decision-making.

Keywords: Railway network, Urban planning, Power asymmetry, Petitions, Public anxieties

Railway and modern infrastructural development started later in the Balkans than in Central and Western Europe. However, when it finally started, railway expansion evolved quickly. In Romania, the last three decades of the nineteenth century and the years leading up to World War I can be seen as something of a "heroic age" of railway construction, as the network developed at a pace that was never again to be matched afterwards.

The newly built infrastructure encompassed most of the nation's cities, and their subsequent development was heavily influenced by the presence

Marton, Silvia, and Andrei-Dan Sorescu, eds. *Infrastructural Development, Corruption, Xenophobia, and Colonization in Central and Southeastern Europe.* Amsterdam: Amsterdam University Press/Central European University Press, 2026.
 https://doi.org/10.5117/9789048577279-9

of railways and the opportunities and restrictions they brought.[1] However, most cities had neither the vision nor the leverage to deal with this novel and complex interaction. These "rules of engagement" had to be developed empirically, without the benefit of a specific doctrine or the support of practical experience, as railway agents—both private companies and public administrations—had their own priorities and focus. This generated great expectations and significant anxiety at the local level for private citizens and public administrations alike. In the present chapter, I intend to explore the roots thereof and to attempt a taxonomy of these anxieties.

In doing so, I will first briefly discuss the development of railway infrastructure in Romania before World War I, focusing on its territorial expansion and the main agents pursuing it. I will then describe the power hierarchies and planning mechanisms involved in this process, showcasing the relative positions of municipalities and the railway administration. Next, I will explore how this framework contributed to generating the aforementioned local expectations and anxieties. In the case of private citizens and informal local communities, I will follow their reactions when faced with impending railway projects, be it one of avoidance and fear of the unknown or one of acceptance and concern for missing out on unique opportunities. In the case of municipalities and counties, I will set these anxieties—which most often had to do with the fear of being shortchanged by the central administration or of being left behind and losing in comparison with their peers—within the planning discourse and practices of the time.

Railway Infrastructure in Romania before World War I: Agents and Territorial Evolution

It was only after 1860 that a significant interest in developing a railway network in the United Principalities of Moldavia and Wallachia started to arise. As the infrastructure in neighboring countries and territories neared the borders,[2] both the Romanian government and foreign companies

1 On how Moldavia was imagined as peripheral by decision-makers in Bucharest, and its subsequent economic peripheralization, see Mihai Chiper, "Regional Competition, Economic Interests, and Military Stumbling-Blocks in the Making of the Moldavian Railway Network" in this volume.

2 By 1865, Austro-Hungarian railways had already reached Lemberg in the north (and were advancing toward Czernowitz), Arad and Oradea in the west, and several towns in the Banat in the south; the Ottoman Empire had inaugurated its first line in 1860, connecting the Romanian border on the Danube to the Black Sea (Cernavodă–Constanța), and the construction of the

started testing the grounds for the implementation of the first railway investments. Between 1869 and 1879, several lines were built and operated by foreign investors: an Austrian company extended its lines from Galicia and Bukovina into northern Moldavia (the Ofenheim concession); a German company managed a controversial railway concession, building some of the core lines before going bankrupt and generating a huge public and political scandal (the Strousberg concession); a Russian company built some lines of important military interest during the Russo-Turkish War, in 1877, but they were dismantled afterward; and a French company took over and finished a failed British attempt to reach one of the main mountain border crossings with Austria-Hungary.[3] Adding up all of these investments and including the Constanța–Cernavodă line—taken over in 1878 from the Ottomans when Romania acquired the province of Dobrudja—the Romanian network accounted for some 1,380 km by the year 1880 (see fig. 2).

After gaining its independence in 1878, it was both a matter of national pride and political obligation for the Romanian state to fully own, operate, and expand its railway infrastructure. By the end of the following decade, it had, through various means, acquired ownership of all the lines in the country. In the meantime, from 1881 onward, the Ministry of Public Works (MLP), through its Romanian Railways Company (CFR), embarked on an ambitious program of public investment, rapidly expanding the existing network.

After almost five decades of sustained investment in the development of the rail network (1869–1916), the total length of Romanian railways reached almost 3,900 km, with around 430 stations. This corresponds to a ratio of 1 km of line for every 35 square km of land—well within the European average despite the late start. More importantly, the normal-gauge passenger and freight network (excluding narrow-gauge industrial or forest lines) was almost entirely state-owned, with a few minor exceptions, placing Romania in a rather unique position worldwide.[4]

It is noteworthy that during these five decades, the territorial expansion of rail infrastructure was shaped by various agents, each with its own interests

Rushchuk–Varna line was underway; and Russian lines were advancing from Kiev toward Odessa. See also the contributions by Constantin Ardeleanu, Gábor Egry, and Malte Fuhrmann in this volume.

3 Toader Popescu, *Proiectul feroviar românesc (1842–1916)* (Bucharest: Simetria, 2014), 39–54.

4 According to a table published by A. D. Webb, *The New Dictionary of Statistics* (Abingdon: Routledge, 1911), and reproduced by J. S. Foreman-Peck, "Natural Monopoly and Railway Policy in the Nineteenth Century," *Oxford Economic Papers*, New Series 39, no. 4 (1987): 714.

and agenda.[5] During the first stage, up until 1880, the main actors involved in choosing and prioritizing routes were, on one side, the representatives of the Romanian state. These were the MLP, charged with the selection and negotiation of proposals coming from private investors, and Parliament, serving as a political debate area, especially on financial matters. On the other side were the foreign companies bidding for railway concessions. Also, less visible and in the background, the European powers manifested their interests either through upholding the claims of some investors against the Romanian state (e.g., the Strousberg affair, as covered by Silvia Marton in the volume's introductory chapter) or through trying to influence the selection or prioritization of certain routes, as covered by Mihai Chiper in his chapter in this volume. Crossing the Carpathians at the easternmost possible point with easy access to the Danube was of high importance for Austria-Hungary, as was opening new routes toward Constantinople, the Black Sea, and the Middle East for Britain, France, and Germany, or ensuring safe and quick access to the Balkans for Russia. One way of looking at this, then, is to acknowledge that interlinking national (or indeed, imperial) networks partly came down to negotiating the creation of nodal points that simultaneously fit visions of "critical" infrastructure for multiple actors.

During the second stage of rail infrastructure development, between 1880 and 1900, the entire network passed into the public realm, as the MLP (through the CFR) became the sole executive authority for building and operating railways—although it held no monopoly over planning decisions. This was the golden age of "predatory planning," with members of Parliament often acting in a biased manner, actively trying to hijack or overrule the government's railway building programs to the advantage of their constituencies; some major landowners, holding political influence enhanced by the electoral system, were also sometimes able to alter, to their own benefit, the pathway of the proposed lines.[6] At this stage, both Romanian state-controlled railway planning at every level and the "central spine" of the network, including most of the important cities, had already been completed. The main economic priority for the expansion of the infrastructure thus became obtaining balanced coverage of the national territory and accessing areas that could produce important resources—either plain areas for agricultural products or mountainous areas for wood, stone, and

5 Popescu, *Proiectul feroviar românesc*, 81–85.

6 On such political, regional, and local interference in rail line construction, see also Chiper, "Regional Competition, Economic Interests, and Military Stumbling-Blocks" in this volume.

minerals. These priorities were almost always cited during parliamentary debates as the main reasons for public investment in railway lines.

Finally, during the third stage of rail development (from the turn of the century to the ourbreak of World War I), the central government tried to withdraw from most railway projects, focusing on operating the existing network and on long-term planning. Local authorities on the county level and private investors were supposed to take the lead in managing railway expansion but generally failed to do so, which prompted the MLP to ultimately retake the initiative by completing unfinished local projects and by investing in lines of military interest, as the regional political climate became increasingly tense.

Urban and Railway Planning: Power Hierarchies and Mechanisms

As the network expanded, towns became, as they did everywhere, its main articulation points. Being the main population and economic centers, they were unavoidable attractors for the railways, as they generated most of the traffic. Conversely, as city administrations quickly became aware, the early presence of railways generated significant local opportunities. Yet what promised to be mutually beneficial was often problematic, given that the specific interests of municipalities and railway builders or operators did not always coincide. The resulting power relations were generally asymmetrical, and a modus operandi had to be reached on a case-by-case basis. I will now show how urban and railway planning responded to this dynamic.

As the pre-war era drew to a close, most of the country's major cities had already been included in the railway network. Compared to Moldavia, Wallachia benefited from a more rational rail network layout, especially in terms of the connectedness of major towns to the main lines.[7] Thirty-five out of the 37 county seats had a railway connection,[8] as did many of the second-tier urban centers. However, in most cases, these municipalities hardly had any viable means for asserting their local interests when faced with what railway builders and planners identified as priorities. These shortcomings were at once political, financial, and technical.

7 See the detailed analysis in Popescu, *Proiectul feroviar românesc*.

8 Only Tulcea and Silistra were left out, both being geographically eccentric and hard-to-reach towns in North and South Dobrudja (recently annexed in 1878 and 1913, respectively).

The 1899 census revealed that out of the 71 urban communes in the country, 60 had fewer than 20,000 inhabitants.[9] The urban population only accounted for some 17 percent of the national total.[10] Their relatively small size meant that municipalities hardly had any significant influence in shaping policies at a national level. Whereas counties could at least assert their interests through elected MPs,[11] municipalities (especially smaller ones) were devoid of a political voice.

Local financial means were also correspondingly weak, preventing cities from actively undertaking any significant public investment linked to railway building; they had almost no leverage in negotiating with the much wealthier railway builders (either private companies or state agencies). For instance, in the 1911–1912 fiscal year, the total municipal spending of the 71 urban communes in the country amounted to 42.5 million lei (38 million of which was spent at county-seat level).[12] During the same period, the central government spent some 505 million lei, out of which 95 million was disbursed by the Ministry of Public Works alone.[13] Municipal spending thus accounted for only 8.5 percent of total government spending, making the financial imbalance quite obvious. Moreover, special municipal credits had to be approved at a higher administrative level,[14] thus further reducing the financial autonomy of cities.

Finally, from a technical and regulatory perspective, urban planning also had less leverage than railway planning. In fact, urban planning, as defined in modern terms,[15] hardly existed in Romania at the end of the nineteenth century. Before 1900, there were practically no specifically trained specialists, no dedicated academic programs, very few specific legal provisions or

9 Leonida Colescu, *Analiza rezultatelor recensământului general al populației României dela 1899* (Bucharest: Institutul Central de Statistică, 1944), 33.

10 Colescu, *Analiza rezultatelor recensământului*, 47. For context, the average rate of the urban population in Europe was about 31 percent, with even the largely rural Austro-Hungarian Empire exhibiting 26 percent.

11 Administrative counties coincided with electoral constituencies for both chambers of Parliament.

12 Ministerul Industriei și Comerțului, *Anuarul statistic al României* (Bucharest: Imprimeria Statului, 1912), 437.

13 *Anuarul statistic al României*, 413.

14 According to the administrative law of 1894, all special and extraordinary communal credits had to be sanctioned by the Ministry of Internal Affairs and, in the case of smaller towns, by the county delegation. See "Legea pentru organizarea comunelor urbane" (July 31, 1894), art. 117.

15 An appropriate definition states that modern urbanism (or urban planning) is a discipline distinguishing itself from previous urban arts through its reflexive and critical character and through its scientific pretense. See Françoise Choay, *L'urbanisme, utopies et réalités: Une anthologie* (Paris: Seuil, 1965), 8.

instruments, no urban planning doctrine, and no professional recognition. This is not to say that practical accomplishments did not exist—modern boulevards and public spaces, parks, allotments, and other such amenities had certainly begun to leave their mark on urban Romania. However, they were not so much the result of the deliberate application of some theoretical model as merely an empirical "trial-and-error" reasoning: urban projects, certainly, but hardly instances of a vision of modern urban planning as such.

On the other hand, railway planning was already a well-established discipline in other parts of Europe by the mid-nineteenth century and was also gaining ground in Romania. Higher education for infrastructure engineers at the National School of Bridges and Roads (established in 1864) included courses in railway planning since at least 1870[16]—so possibly even before the country had its first line in operation—and local railway treatises were part of a broader literature that was also quite rich by 1880. However, these works exhibited little preoccupation with issues related to the interaction between railways and cities, as they were mostly focused on the technical and economic problems posed by railway building.[17]

Taken together, all of the above led to an uneven and unbalanced relation between cities and the almighty railway agents, be they private companies or the public central administration. Municipalities had hardly any means at their disposal for attracting infrastructural development, other than through local members of Parliament trying to hijack public programs—a widespread practice, especially in the 1880s. Once the decision to include a certain city in the national railway network had been taken, local authorities had almost no say in where the stations would be placed.[18] This decision was taken by the railway administration following its own specific criteria, which were mainly economic or technical in nature, such as expropriation costs, land availability, and topographic configuration. Especially in smaller

16 A course in railways is known to have been taught that year by Professor Spiridon Iorceanu, but it is possible it had already been in place for a few years. See Nicolae Şt. Noica, *Şcoala Naţională de Poduri şi Şosele: 125 de ani* (Bucharest: Editura Vremea, 2010), 122.

17 *Drumurile de fer în România* (Bucharest: Impr. Naţională Stephan Rassidescu, 1862); Paul Iacovenco, *Câte-va cuvinte asupra căiloru ferate în România* (Bucharest: Typographia Stephan Rassidescu, 1864); Eugene Henri Gavand, *La Question des Chemins de fer en Roumanie, pour les provinces en deça du Milcov* (Bucharest: Imprimerie Nationale, 1866); Constantin I. Brătianu, *Missiunea căilor ferrate române : Studii politice, economice, sociale şi tecnice* (Paris: Gustave Menard, 1870); Alois Zwiedinek, *Reţeaua căilor ferate române ca basa pentru comerciul general al Europei* (Bucharest: Typographia Curţii, Proprietar F. Göbl, 1879); L. Calmar, *Geografia căilor ferate* (Bucharest: Ed. Căilor Ferate, 1879).

18 On centralization in railway planning and construction in Romania, see Chiper, "Regional Competition, Economic Interests, and Military Stumbling-Blocks" in this volume.

towns, this generally led to choosing peripheral sites near or even beyond town limits. Municipalities were thus relegated to a reactive planning position, having to ensure the integration of these important amenities into the existing urban structure, with railway administrators acting as "absentee landlords"[19] once the infrastructure was in place. In most cases, the main planning challenge that local authorities faced was to link these new "city gates" to the center (through some form of "station boulevard") and to create a coherent and representative public space associated with the station (the "station square").

For several cities, these were the most complex urban operations they had ever had to tackle, and they were of great significance for local communities. Adding to this, the urban areas created by the arrival of the railways were quickly vested with a particular representation in the public mind. This representation was both positive and negative. On one hand, station boulevards and squares were epitomes of modernity, often hosting expensive housing or leisure areas and exhibiting a type of high-end, sometimes monumental, urban architecture. On the other hand, railway areas were still a strange new addition for communities still very much rooted in tradition—small-scale and crafts-based economic activities, local or regional trade, and low-income peripheries. The development of these areas often amounted to a culture shock, as the everyday complexities of railway and travel engendered a combination of respect and fear.[20] This dual representation of railways, set against the unpredictability and lack of precedent in municipalities' responses, proved to be at the root of a wide array of local anxieties.

Citizens and Communities

As a previous chapter in the volume has noted, "anxieties" may be understood for our present purposes not so much in ahistorical diagnostic terms but as projections of possible future outcomes that created tensions in the historical present and prompted polemical reassessment of developmental

19 This term, coined by economist Thorstein Veblen in 1923, refers to people who own or rent out property without being physically present in the region.

20 This phenomenon is well documented in seminal works such as Wolfgang Schivelbusch, *Geschichte der Eisenbahnreise* (Munich: Carl Hanser Verlag, 1977), and Jeffrey Richards and John M. MacKenzie, *The Railway Station: A Social History* (Oxford: Oxford University Press, 1986). For the Romanian case, see Popescu, *Proiectul feroviar românesc*, 93–101.

strategies.[21] It is therefore understandable that, in our case as well, some of the first railway-linked public anxieties preceded their advent and were associated with the effects that they were foreseen to have on established ways of life.

The first line inaugurated in the country, after more than a decade of failed attempts, linked the capital city of Bucharest to the Danubian port of Giurgiu in 1869. The small urban community of Giurgiu (10,557 inhabitants as per the 1859 census)[22] did not present any special interest for the British company building and operating the line, which was understandably more focused on the harbor facilities on the Danube. In order to reach the port, several options were considered, involving either going through the town or around it.

Given the lack of appropriate planning instruments at its disposal and perceiving the railway as an impending threat to local trade, the local community, through its mayor, called directly upon the minister of public works, who had the prerogative of approving the plans submitted by the foreign investor. In two separate petitions,[23] the community insisted on placing the lines and the station well outside the town to avoid any direct influence on local activities. These petitions were couched in a dramatic tone, insisting upon the "calamities" and "disfigurement" threatening Giurgiu and warning about the "complete demise of the whole town" were it to be affected by expropriations.

Another petition, also addressed to the minister of public works but this time sent directly by local citizens, asked for the Bucharest–Giurgiu line not to be built at all because "it would be of no use for Giurgiu or the country, as it would be the only line in Romania and thus would not be able to connect to any others"; and since a road already existed between the two cities, a railway would be a useless luxury, only meant "to transport foreign merchandise and drain the country's finances."[24] This was, to critically deploy a term used by Malte Fuhrmann in an earlier chapter, an odd instance of infrastructure "shaming"[25]: not so much holding authorities responsible

21 Andrei-Dan Sorescu, "The Infrastructural Power of the 'Colonial' as a Concept" in this volume.

22 Colescu, *Analiza rezultatelor recensământului*, 47.

23 Central National Historical Archives of Romania, fund no. 1304 "Ministry of Public Works," file 62 (1865), 1, 168.

24 Central National Historical Archives of Romania, fund no. 1304 "Ministry of Public Works," file 62 (1865), 143.

25 Malte Fuhrmann, "On Hegemony and Connectivity: A Short History of Infrastructure-Political Struggles and their Discourses" in this volume.

for failing to deliver on their promises of development as attacking them for the misuse of public money in promising a development that was perceived as potentially disruptive in and of itself.

On the other hand, other citizens or communities experienced a different kind of anxiety, which became increasingly widespread as time passed and the advantages of being included in the rail network became obvious: the fear of missing out. Understanding the potential opportunities engendered by railways, some 400 citizens of Giurgiu wrote their own petition,[26] also asking for the line not to cross the town and to stop at its limits, but only so that local innkeepers and coachmen could take advantage of the stream of passengers and merchandise heading to the harbor. Here, the issue was neither questioning the usefulness of an infrastructural network nor being excluded from it but rather maintaining and connecting preexisting economic and transportation networks to it as something now critical to local well-being.

This type of anxiety, in fact, predated even the building of the first line. Several years before, with a number of potential investors having submitted their propositions to the government, some local communities experienced a fear of missing out on future opportunities. In 1862, Prince Grigore Bibescu Basarab Brâncoveanu (1827–1886) proposed a line running from Galați to the Austrian border along the Jiu River valley, which meant that the railway would have to cross the Olt River. As the concession was submitted for debate in the General Assembly[27], the citizens of Slatina, a mid-sized town on the eastern side of the Olt, sent a letter to Prince Alexandru Ioan Cuza (1820–1873, r. 1859–1866), accusing Brâncoveanu of acting upon his own personal interests when proposing a crossing at Ipotești, some 10 km south of Slatina, and demanding justice and reparations for their town:[28] "[...] the citizens of Slatina and of the entire county are being robbed of the promised advantage of a railway station by giving preference to the interests of only one individual and building the station on the estate of Prince Brâncoveanu in Ipotești." This, then, was a more straightforward example of "shaming," with explicit corruption allegations in tow.

More than 30 years later, in 1894, the same type of anxiety affected the merchants of the most important commercial area in Bucharest. When the

26 Central National Historical Archives of Romania, fund no. 1304 "Ministry of Public Works," file 62 (1865), 1, 144.

27 The unified parliament established in 1862, in the wake of the unification of the two principalities of Wallachia and Moldavia in 1859.

28 Central National Historical Archives of Romania, fund no. 1304 "Ministry of Public Works," file 125 (1862), 41–42.

site of a new railway station near the Obor market was being negotiated, a rumor had it that the Assan brothers, influential industrialists of the time, were willing to pay a substantial contribution in order to operate the line and have the station moved near one of their factories, away from the main commercial area. The Obor merchants, as well as another important competing industrialist, Andrei Popovici, sent several petitions to Bucharest City Hall, painting a catastrophic picture of what might come to pass:[29] "Woodworkers, grain merchants, industrialists, owners great and small, all ruined! The entire Black District[30] sentenced to death!"

Expropriations were another anxiogenic factor associated with railway development. A decision for expropriation was either directly generated by the territorial needs of the infrastructure (lines, stations, yards, etc.) or by urban transformations following these investments. If the former was generally perceived as unavoidable (since the decision-making level was quite removed from the affected parties), the latter appeared more amenable to negotiation, given that such operations were undertaken by local authorities—at a level closer to that of the ordinary citizen—and tended to unfold at a slower pace. If anxieties associated with the prospect of being expropriated are easily understandable, the opposite kind also existed—the fear of missing out on the opportunity to cash in on a property.

Many examples serve to illustrate the former case. For instance, in Focşani, when the project for the new station boulevard was drawn up in 1881, several potentially affected citizens reacted vehemently. Most of them were affluent merchants from the commercial Armenian quarter, which the proposed thoroughfare was slated to traverse; they challenged the public benefit of the expropriations, accusing some members of the city council of having hidden personal interests in the matter, charging the planners with being incompetent, and asking for the entire project to be scrapped.[31] "The projected boulevard is of no general interest, as it only serves the estates of some members of the city council, who were very eager to vote for it," and "the propositions of the architect are more than arbitrary, proving his complete ignorance or bad faith." The notion of public versus partisan interest, even if not articulated in political terms, was thus part and parcel of this discourse of blame and denunciation.

29 Bucharest Municipal Historical Archives, "Bucharest City Hall—Technical Services" fund, file 1062 (1894), 16, 19–28, 30.

30 One of the five administrative divisions of Bucharest, covering the entire northeastern part of the city.

31 Vrancea County Historical Archives, "Focşani City Hall" fund, file 43 (1881), 7–13, as indicated in Ramona Miron, "Bulevardul Gării din Focşani: Contribuţii," *Cronica Vrancei* XIII (2012): 15–16.

On the other hand, let us consider Ploiești, a city well known at the time for its rebellious propensity. In 1871, just as an abortive attempt at establishing a "republic" had brought it into the national limelight,[32] 107 of its citizens—from priests to fiddlers—asked for the direction of a station avenue under consideration by the municipality and under negotiation with the Superior Technical Council of the Ministry of Public Works to be altered. As the properties of this diverse group of neighbors were to be partly expropriated, even a slight alteration would have served to increase their value.[33] However, although the local administration accepted and upheld this amendment before the Superior Technical Council, the latter nevertheless approved the initial proposal, which was cheaper and easier to implement.[34] These examples demonstrate how a growing public of self-identified citizen stakeholders felt emboldened to hold decision-makers accountable, not shying away from engaging with them at various levels and accusing them—albeit through petitions that were not themselves necessarily meant to be publicized—of imperfectly serving a public interest that was, in fact, very much private.

Counties and Municipalities

The anxiety of being left behind and of missing out on opportunities was not limited to private citizens. Municipalities were also affected, and this tendency only grew stronger as the rail network expanded and included more cities. Before 1880, when the core lines were built by foreign private companies via concessions on prenegotiated paths decided upon together with the MLP, there was very little that local administrations could do to influence these choices. However, the situation changed after the Romanian state took control of the existing network and became responsible for expanding it through government-run programs.

After a short trial period (1880–1882), when the MLP (through the CFR) established its technical and administrative capacity for building new

32 Silvia Marton, *"Republica de la Ploiești" și începuturile parlamentarismului în România* (Bucharest: Humanitas, 2016).

33 Paul D. Popescu and N. Petrescu-Redi, *Bulevardul Independenței* (Bucharest: Fiat Lux, 1999), 22.

34 Irina Paveleț, "Contribuţii privind istoria Bulevardului Independenţei, Ploieşti," Muzeul Virtual Ploieşti, accessed October 22, 2024, https://muzeuvirtualploiesti.cimec.ro/?page_id=131.

railways,[35] an ambitious program consisting of a package of six new and longer lines[36] was presented for approval to the lower chamber of Parliament. These were secondary lines, branching out from the core network and of roughly equivalent importance. Because the program was entirely state-funded, several MPs—especially those representing counties with little or no access to railways—saw this as an opportunity to amend the law and extend the proposed lines or add new lines of local benefit.[37] Since these new additions were mainly politically motivated, some of them had to follow difficult or improper paths. For instance, the Filiași–Târgu Jiu and Vaslui–Iași lines included segments of exaggerated slope, which led to operating problems and even contributed to a very serious accident some decades later.[38]

In 1885, the MLP brought another proposal before Parliament. This time, the package was smaller, consisting of just four shorter lines, mainly aiming to serve areas of some economic importance.[39] The MPs' tactic was repeated, as they proposed no less than thirteen other lines, most of them of little or no economic significance. For instance, one MP upheld his proposal for a Roșiori–Zimnicea line by claiming that "it would help introduce the products of national manufacture to the villagers of this historic county and replace current, Bulgarian ones," while another stated that "a healing fountain would flood the entire country" if the spa town of Băile Slănic were to gain a railway connection.[40] Local interest was thus reframed as national interest, in line with an understanding of infrastructure as a conduit for public happiness and prosperity. In spite of government opposition and lack of funding, six of these segments were voted into law,[41] jeopardizing the entire program.

35 The Buzău–Mărășești line and the three so-called "salt lines" (Câmpina–Doftana, Buda–Slănic, and Adjud–Târgu Ocna).

36 Râmnicu Vâlcea–Corabia, Titu–Târgoviște, Bacău–Piatra, Bârlad–Vaslui, Dolhasca–Fălticeni, and Docolina–Huși.

37 Filiași–Târgu Jiu, Golești–Câmpulung, Huși–Prut, and Vaslui–Iași. See *Desbaterile Corpurilor Legiuitoare*, 97–100 (April 18–22, 1882): 1608–1682.

38 The Ciurea accident, on the Vaslui–Iași line, caused some 1,000 victims in 1917. It was the worst railway disaster in European history up to that point (and possibly to this day). The steep slope of the Grajduri–Ciurea segment had been a well-known problem long before the accident; see Mihail Romniceanu, "Calea ferată Vaslui-Iași: Alegerea traseului," *Buletinul Societății Politecnice* IV, no. 4 (July-August 1888): 447–474.

39 Târgu Ocna–Moinești, Pitești–Curtea de Argeș, Târgoviște–Lăculețe, and Piatra–Tarcău.

40 *Desbaterile Corpurilor Legiuitoare*, 80 (March 15, 1885): 1471–1476.

41 Craiova–Calafat, Focșani–Odobești, Târgu Ocna–Băile Slănic, Bucharest–Oltenița, and Dorohoi–Iași.

Some two and a half decades later, after abandoning the "state-funded package" strategy and trying several different approaches,[42] the MLP came back to Parliament with a third set of proposed lines. Because this proposal came after a decade of failures and slow-paced railway expansion,[43] many MPs saw it as a last chance to have their constituencies included in the network and not miss out again. What followed was a competition between counties, with calls for mending "historical injustices" and for getting compensation for past missed opportunities, which prompted one speaker to declare, "I have the impression that all of this is turning into a joke."[44] The debates were finally put to rest by the minister of finance, who announced that if any of the proposed amendments were to be voted into law, the government would withdraw its project entirely. This episode, and all the similar ones before it, goes to show that the lack of established planning mechanisms that would include the voice and interests of local administrations contributed to the generalized impression that railways were just a government handout, only hijacked by local representatives making themselves heard during parliamentary approval.

Another type of "institutional anxiety" that some municipalities experienced was that of being shortchanged by the central authorities during the planning process. This phenomenon was limited to the larger cities (and to Bucharest in particular), as in other cases there was neither major spatial nor chronological overlap between railway planning and urban response. Stations were placed near or beyond city limits, and railway authorities hardly had any significant involvement in the subsequent operations, which were left entirely up to the local authorities, both in terms of planning and of implementation. Since almost no city other than the capital required a second station (at least not before World War I),[45] there generally was no occasion for further conflict.

In the case of Bucharest, the complexity of the network and the constant need for adapting city growth patterns to the shifting priorities of the railway

42 For details, see Popescu, *Proiectul feroviar românesc*, 62–65, 67–72.

43 After 1900, the MLP tried to devolve railway building to the counties—an attempt that led to few accomplishments. The first decade of the twentieth century saw an average of just 29 km of new lines per year, the lowest such average since railway building started in Romania.

44 *Desbaterile Adunării Deputaţilor*, 37 (February 4, 1910): 418–427.

45 Except for Ploieşti (where a station of much lower importance was inaugurated in 1908 for a local line) and Râmnicu Sărat, Piatra Neamţ, and Slobozia (where the functions of the initial stations were taken over by newer ones located elsewhere). Special stations were also built in some Danubian port towns, away from the main station (Turnu Măgurele, Zimnicea, and Giurgiu).

administration generated an unstable situation, with both parties involved following their own objectives. That the infrastructure in question was "critical" was beyond doubt, yet just how it would be assembled into a complex network that could integrate the needs of the city was open to debate. The balance of power between the railway and city administrations shifted over time, sometimes leading to complex negotiations as the municipality faced the constant anxiety of being shortchanged in the process.

The first two stations were inaugurated shortly after one another, in 1869 and 1872, respectively. As they were operated by different companies,[46] they were placed far away from each other, in the optimal places for the directions they were serving. Their locations were chosen by their private operators with the approval of the MLP but without consulting local authorities. The usual technical and economic criteria were prioritized, meaning that peripheral and vacant sites—involving few expropriations and located on flat terrain with little risk of flooding—were preferred.

After 1880, when the Romanian state took possession of all the lines in the country, the MLP took it upon itself to draw up plans for a more efficient, unified railway network for Bucharest through various technical studies drafted to primarily serve its own needs.[47] The idea of a new Central Station thus became a priority for the railway administration, which expropriated the land and, in 1892, organized an architectural competition to select a fitting project.[48] And yet, throughout this process, consultations with the municipality were minimal, on the level of mere formalities.

After 1893, Bucharest City Hall became increasingly assertive in defense of its own interests. This coincided with some important legislation being passed by Parliament—the 1893 law for the financing of public works in Bucharest and the 1894 law for the organization of urban municipalities.[49]

46 The Filaret station, inaugurated in 1869, served the Bucharest–Giurgiu line, operated by a British company administered by John Trevor Barkley and John Staniforth, and was built at the southern limit of the city; the Târgoviște station, inaugurated in 1872, was part of the German Strousberg concession and was located in the northwestern outskirts.

47 Căile Ferate Române, *Memoriu: Necesitatea construirei unei gări de călători, transformarea gărei de mărfuri, sporirea atelierului central și construirea clădirei administrațiunei centrale din București: Studiul proectelor după care se pot efectua aceste lucrări* (Bucharest: Tipografia Curții Regale, F. Göbl Fii, 1890); Mihail Romniceanu, "Raportul către Direcțiunea general a căilor ferate, privitor la fixarea definitivă a locului pe care urmează a se construi gara centrală," *Analele Ministerului de Lucrări Publice* I, (1893): 108–109.

48 "Concursul internațional pentru proiectarea unei gări centrale la București," *Analele Architecturei și ale Artelor cu care se legă* III, nos. 9–10 (1892): 161–166.

49 "Legea pentru crearea unei Casse a lucrărilor orașului București" (May 3, 1893); "Legea pentru organizarea comunelor urbane" (July 31, 1894).

Both these laws gave the local authorities in Bucharest significant leverage and charged them with drafting a general planning document for the city (a "General Systematization Plan"[50]). The municipality was thus able to articulate its own vision as to the place of railways in the urban structure, which did not always coincide with the intentions of the MLP.

A first clash occurred in 1894, when, in an unusual turn of events, the MLP asked for the opinion of City Hall. As a new station was planned for the northeastern part of the city, City Hall was to weigh in on its location and on the path that the new line serving it might take.[51] What ensued was a prolonged and complex negotiation stretching over twelve years, with the municipality not willing to yield to the will of the railway administration or to miss out on the opportunity to increase the commercial attractiveness of the station area—or on the opportunity for the new line also to serve purposes of local importance.[52] In the end, the implemented solution still predominantly reflected the will of the railway administration, but this episode marks a watershed moment. From this point on, the municipality was no longer willing to accept without question the outcomes of railway planning.

As work progressed on the General Systematization Plan—at first with a public competition in 1906 (with no clear winners) and then within the municipal planning department under the coordination of engineer Cincinat Sfințescu—even the project for the Central Station, which had long seemed unavoidable, began to be questioned. Sfințescu upheld an alternative solution involving several scattered, specialized stations and a main line along the Dâmbovița River. A large urban park was proposed for the site previously reserved for the Central Station, as Sfințescu pushed for decisive action by the municipality to persuade the MLP to abandon its flagship project and the entire railway scheme associated with it.[53]

50 "Systematization" is a term often used to describe the practice of urbanism in twentieth-century Romania, underlining its scientific pretense; the "Systematization Plan" ("Systematization Sketch" during socialist times) was the most comprehensive planning instrument throughout this period—see Dana Vais, "Systematization: A Key Term in 20th-Century Romanian Urbanism," *Urban Planning* 7, no. 1 (2022): 207–222, and Toader Popescu and Irina Calotă, "Planurile de sistematizare ale orașelor din România înainte de al Doilea Război Mondial," in *Confluențe arhitecturale* (Bucharest: Editura Ozalid, 2018), 249–276.

51 Bucharest Municipal Historical Archives, "Bucharest City Hall – Technical Services" fund, file 1062 (1894), 1.

52 Popescu, *Proiectul feroviar românesc*, 168–171.

53 Cincinat Sfințescu, *Studiu asupra Planului General de Sistematizare al Capitalei urmat de un Anteproect de lege asupra stabilirei, construirei, desvoltărei și sistematizărei comunelor* (Bucharest: Tip. "Jockey Club," 1919), D.48 – 51, B.42.

Conclusions

The first decades of railway building in Romania had a significant impact on urban and economic development but also generated significant anxiety for several of the involved parties. The general underlying factors were, on the one hand, the unpredictability of the effects that this new and strange system would have on the established way of life and, on the other hand, the perceived lack of voice or influence that both ordinary citizens and local administrations had on the decisions made by the central authorities. In the case of citizens and communities, this perception was amplified by the distance that separated them from national decision-making. In the case of municipalities, it was the power asymmetry between them and the central administration and the subordinate position that urban planning held in relation to railway planning.

From a typological perspective, these anxieties fell into one of two main categories. They were generated by the anticipation of the negative outcomes that might happen either because of the advent of railways or because of their absence. The former was generally caused by a fear of the unknown, while the latter had to do with the apprehension of missing out on opportunities or of being shortchanged in the process. Citizens and administrations coped with these anxieties in different ways, using whatever means at their disposal: writing petitions, forging alliances at lower levels, hijacking decision-making processes, negotiating with railway administrations, or asserting themselves through alternative proposals. In the end, all these anxieties and coping mechanisms were indicative of some of the major shifts that the "golden age" of railways brought at every level of Romanian society.

Bibliography

Brătianu, Constantin I. *Missiunea căilor ferrate române: Studii politice, economice, sociale și tecnice*. Paris: Gustave Menard, 1870.

Bucharest Municipal Historical Archives. "Bucharest City Hall—Technical Services" fund, file 1062 (1894).

Căile Ferate Române. *Memoriu: Necesitatea construirei unei gări de călători, transformarea gărei de mărfuri, sporirea atelierului central și construirea clădirei administrațiunei centrale din București: Studiul proectelor după care se pot efectua aceste lucrări*. Bucharest: Tipografia Curții Regale, F. Göbl Fii, 1890.

Calmar, L. *Geografia căilor ferate*. Bucharest: Ed. Căilor Ferate, 1879.

Central National Historical Archives of Romania. Fund no. 1304 "Ministry of Public Works," files 125 (1862), 62 (1865).

Choay, Françoise. *L'urbanisme, utopies et réalités: Une anthologie.* Paris: Seuil, 1965.

Colescu, Leonida. *Analiza rezultatelor recensământului general al populației României dela 1899*. Bucharest: Institutul Central de Statistică, 1944.

"Concursul internațional pentru proiectarea unei gări centrale la București." *Analele Architecturei și ale Artelor cu care se legă* III, nos. 9–10 (1892): 161–166.

Desbaterile Adunării Deputaților. 37 (February 4, 1910).

Desbaterile Corpurilor Legiuitoare. 97–100 (April 18–22, 1882) and 80 (March 15, 1885).

Drumurile de fer în România. Bucharest: Impr. Națională Stephan Rassidescu, 1862.

Foreman-Peck, J. S. "Natural Monopoly and Railway Policy in the Nineteenth Century." *Oxford Economic Papers,* New Series 39, no. 4 (1987): 699–718.

Gavand, Eugene Henri. *La Question des Chemins de fer en Roumanie, pour les provinces en deça du Milcov.* Bucharest: Imprimerie Nationale, 1866.

Iacovenco, Paul. *Câte-va cuvinte asupra căiloru ferate în România.* Bucharest: Typographia Stephan Rassidescu, 1864.

Ministerul Industriei și Comerțului. *Anuarul statistic al României.* Bucharest: Imprimeria Statului, 1912.

Miron, Ramona. "Bulevardul Gării din Focșani. Contribuții." *Cronica Vrancei* XIII (2012): 9–21.

Noica, Nicolae Șt. *Școala Națională de Poduri și Șosele: 125 de ani.* Bucharest: Editura Vremea, 2010.

Paveleț, Irina. "Contribuţii privind istoria Bulevardului Independenţei, Ploieşti." Muzeul Virtual Ploieşti. Accessed October 22, 2024. https://muzeuvirtualploiesti.cimec.ro/?page_id=131.

Popescu, Paul D., and N. Petrescu-Redi. *Bulevardul Independenței.* Bucharest: Fiat Lux, 1999.

Popescu, Toader. *Proiectul feroviar românesc (1842–1916).* Bucharest: Simetria, 2014.

Popescu, Toader, and Irina Calotă. "Planurile de sistematizare ale orașelor din România înainte de al Doilea Război Mondial." In *Confluențe arhitecturale,* 249–276. Bucharest: Editura Ozalid, 2018.

Richards, Jeffrey, and John M. MacKenzie. *The Railway Station: A Social History.* Oxford: Oxford University Press, 1986.

Romniceanu, Mihail. "Calea ferată Vaslui-Iași: Alegerea traseului." *Buletinul Societății Politecnice* IV, no. 4 (July-August 1888): 447–474.

Romniceanu, Mihail. "Raportul către Direcțiunea general a căilor ferate, privitor la fixarea definitivă a locului pe care urmează a se construi gara centrală." *Analele Ministerului de Lucrări Publice* I (1893): 108–109.

Schivelbusch, Wolfgang. *Geschichte der Eisenbahnreise*. Munich: Carl Hanser Verlag, 1977 [translated as *The Railway Journey: The Industrialization of Time and Space in the 19th Century*. Berkeley: The University of California Press, 1986].
Sfințescu, Cincinat. *Studiu asupra Planului General de Sistematizare al Capitalei urmat de un Anteproect de lege asupra stabilirei, construirei, desvoltărei și sistematizărei comunelor*. Bucharest: Tip. "Jockey Club," 1919.
Vais, Dana. "Systematization: A Key Term in 20th-Century Romanian Urbanism." *Urban Planning* 7, no. 1 (2022): 207–222. https://doi.org/10.17645/up.v7i1.4791.
Vrancea County Historical Archives. "Focșani City Hall" fund, file 43 (1881).
Zwiedinek, Alois. *Rețeaua căilor ferate române ca basa pentru comerciul general al Europei*. Bucharest: Typographia Curții, Proprietar F. Göbl, 1879.

About the Author

Toader Popescu is assistant professor at the "Ion Mincu" University of Architecture and Urbanism in Bucharest. His book *The Romanian Railway Project (1842–1916)* (Bucharest: Simetria, 2014) on the influence railways had in the development and modernization of Romanian cities won the Science Book of the Year award of the Association of Romanian Publishers.

9. Regional Competition, Economic Interests, and Military Stumbling-Blocks in the Making of the Moldavian Railway Network

Mihai Chiper

Abstract: Beginning with the early stages of Romania's railway network expansion in the nineteenth century, Moldavia became a relative "loser" compared to Wallachia, the country's two main provinces. Initially, it was constrained by Austrian economic interests through the Ofenheim concession, which set the stage for path dependency and blocked the formation of a north–south trade route that would have been vital for grain exports. Later, budgetary favoritism toward secondary railway lines further disadvantaged the region, as funds were directed mainly to Wallachia. Finally, centralist military strategies, aimed at limiting potential Russian invasion routes toward Bucharest, hindered Moldavia's transport development. This strategic rationale fostered long-term marginalization and reduced state interest in investing in the region's infrastructure.

Keywords: Railway expansion, Path dependency, Regional marginalization, Regional economy, Moldavia, Wallachia

Introduction

In the nineteenth century, the idea of building a railway network found enthusiastic support among Romanian politicians, who saw it as an essential

---

Marton, Silvia, and Andrei-Dan Sorescu, eds. *Infrastructural Development, Corruption, Xenophobia, and Colonization in Central and Southeastern Europe*. Amsterdam: Amsterdam University Press/Central European University Press, 2026.
 https://doi.org/10.5117/9789048577279-10

means for fostering national territorial unity and economic development.[1] As soon as railways were equated with a vector of progress, they attracted a substantial amount of time and interest from the political world and generated an imaginary that quilted together the impact of speed, progress, and the Romanianization of an emerging transport network as a matter of "national pride."[2] Their effects were perceived as overwhelmingly positive: they were to unify and fortify a nascent state and swiftly bring it in line with developed European nations. In economic terms, railways were ascribed the capacity to raise the overall level of wealth and income, and two kinds of integrative effects were expected: (1) the internal integration of the two formerly divided principalities of Moldavia and Wallachia; and (2) the integration of the Romanian economy among those of Europe's most prosperous nations.

Yet what is missing from this overwhelmingly optimistic picture—as is still the case with European historiography on the topic more broadly—is a closer look at the losers. They were those who—to employ terms used by historical actors themselves—were affected by a series of "inconveniences" that threatened to upset the "economic equilibrium" and lead to depopulation; chief among those were the chosen trajectories of the railways, bypassing and isolating them.[3] Once more, as elsewhere in Europe, the Romanian case overlooked the relationship between, on the one hand, the building and running of railways and, on the other, the differences they generated in terms of regional economic development. Nineteenth-century European economists were scarcely concerned with how railway development might affect regional economies and seemed, for the most part, in agreement that it would stimulate industrial or agricultural growth and facilitate the emergence of unified national markets.[4] Nor, in fact, have economists' and historians' retrospective assessments of the concrete outcomes of railway building and its exponential growth across Europe led to a shared

1 Ion Ghica, *Scrieri economice*, vol. 1 (Bucharest: Editura Asociației generale a economiștilor din România, 1937), 149; *Răspuns adressat de D. P. Mavrogheni la articolul publicat in Românul sub titlul Căile ferate: Partea a V-a* (Bucharest: Typografia Stephan Rassidescu, 1864); D. P. Marțian, *Opere economice*, ed. N. Marcu and Z. Ornea (Bucharest: Editura Ştiinţifică, 1961), 314–317; G. Barițiu, *Memorialu pentru cel dintâi drum de feru in România unită* (Brașov: Cu Tipariulu lui Ioanu Gött, 1862); Victor Slăvescu, *Vechi proiecte de căi ferate in Moldova 1846–1851* (Bucharest, 1942); Const. Botez, Dem. Urmă, Ion Saizu, *Epopeea feroviară românească* (Bucharest: Editura Sport-Turism, 1977), 22–26.

2 "Drumurile de fer in România," *Românul*, March 20–21, 1867, 234.

3 *Monitorul Oficial al României* (hereafter MOf), no. 116, May 24/June 5, 1868, 738.

4 MOf, no. 116, May 24/June 5, 1868, 740. See also Albert Schram, *Railways and the Formation of the Italian State in the Nineteenth Century* (Cambridge: Cambridge University Press, 1997), 94.

theoretical model for assessing the economic impact thereof.[5] What is, however, certain is that this impact was economically significant, shaping urban development and creating new trade routes. Along with a reduction in transaction costs by lowering tariffs, the technological superiority of railways over all other hitherto existing means of transportation brought about major improvements in terms of regularity, safety, and speed.[6]

Infrastructure building serves to highlight the competing and contradictory influences of various stakeholders whenever route selection becomes the issue of the moment. Technological complexity aside, infrastructure also presupposes a legal framework underpinning it, a funding system, decision-making structures, and control mechanisms. As such, infrastructure is an expression of state power, and its routes, in the final instance, are reflective of power relations between various socioregional groups engaged in a competition for both state resources and the ability to shape sociocultural reality itself—through the very decision of including or excluding a given locale, area, or group from infrastructural development.[7]

Relative historiographical disinterest notwithstanding, what is at stake here is how the second half of the nineteenth century witnessed the emergence of developmental inequalities, leading to the emergence of relative losers and net winners. The relative losers were the regions and urban centers that, although traversed by a given railway network, failed to secure the most adequate and efficient connections—oftentimes forced to face restrictive effects on their development. By contrast, net winners benefited, from a spatial point of view, not only from proximity to the networks in question but also from preeminence in establishing dominant economic routes before other centers had the opportunity to secure similar preferential conditions for industry, trade, and services. To this, we might add, winners also had the remodeling of time on their side: the new, "modern" relationship with time brought about by the restructuring and acceleration of socioeconomic rhythms had an impact on human experience and on the division between

5 Schram, *Railways and the Formation of the Italian State*, 152.

6 Schram, *Railways and the Formation of the Italian State*, 152.

7 Paul N. Edwards, "Infrastructure and Modernity: Force, Time, and Social Organization in the History of Sociotechnical Systems," in *Modernity and Technology*, ed. Thomas J. Misa, Philip Brey, and Andrew Feenberg (Cambridge, 2003), 185–225; Jens Ivo Engels, "Writing the History of Critical Infrastructures: Some Conceptual Remarks" in this volume. See also Silvia Marton and Andrei Sorescu, "Introduction: Colonial Anxieties, Corruption Scandals and Xenophobia in Nineteenth-Century Infrastructure Development in Romania," *East European Politics and Societies* 39, no. 3 (2025): 680–690.

a dynamic center touched by "progress" and the sluggishness of losers untouched by the magic of technological modernity.[8]

The obvious tension between winners and losers raises the question of whether railways did indeed bring about what so many had hoped they would: political and cultural integration between regions through increased communication and, no less importantly, the leveling of income disparities across the board. A more detailed discussion of the negative effects of railway building would thus help, at least in part, to nuance the axiom whereby railways were a "unifying" force—as well as to call into question many prevailing hypotheses in neoclassical economics, which posit that income levels converge when cheaper transportation lowers transaction costs.

In the case of Moldavia, we hold that the economic peripheralization stemming from the overall architecture of the Romanian railway network—itself, in essence, a project of socio-spatial ordering[9]—was only a continuation of previous patterns of peripheralization caused by political and administrative centralization in Bucharest, the capital. This reality, however, has often been glossed over in light of prevailing convictions regarding the impact of railway networks on regional economies.[10] As such, data on interregional railway traffic in Romania has never received a detailed analysis and has only been presented in terms of *national* statistics.[11] Moreover, while Romanian historiography has yet to examine railway and business history in an integrated manner, it is nevertheless rather evident that, given the backward nature of Romania's agrarian economy, neither supply nor demand was significant enough. Few opportunities for trade between the north and south of Romania existed, as their economies—mostly centered on agrarian goods—were not complementary but similar in nature. Governments failed to adopt coherent policies, which, in turn, generated uncertainty for foreign investors.[12] The various mixed systems resulting from compromises between proposals made by the state and concession companies led only to limited

8 Edwards, "Infrastructure and Modernity."

9 Engels, "Writing the History of Critical Infrastructures."

10 Schram, *Railways and the Formation of the Italian State*, 79. On peripheral economies and railway construction, see also Constantin Ardeleanu, "British Investors and the 'Kustendjie Harbor Dues Issue'" in this volume.

11 For instance, *Tablou comparativ intre veniturile și cheltuielile căilor ferate pe anii 1877–1890*, in *Buletin statistic general al României* 1 (1892), noiembrie-decembrie (Bucharest: Imprimeria Statului, 1893), 389.

12 On the sequestrating of the Lemberg–Cernăuți–Iași network (1888), see Eduard Sava, *Istoricul căilor ferate român. Perioada dintre anii 1885–1902* (Bucharest: Tipografia G.A. Lăzăreanu, 1904), 37–41.

success and, overall, to an unsatisfactory level of economic performance.[13] Nonetheless, it is worth pointing out that, in spite of recurring allegations of "colonialism" in Moldavian political discourse at the time, the critical situation of local railways was *not* singled out as a symptom thereof. Rather, such accusations were more common when it came to the administration (the "colonization" of Wallachian clerks in Galați harbor, for instance), the judicial branch (the overstaffing of local courts with Wallachian judges), or the officer corps[14] (and its bias toward promoting Wallachians to higher ranks).[15]

Nationwide Gains and Regional Sacrifices

While the watchword of the day was the unification of the national market, as per the tenets of classical liberal economics, the commercial and economic interests of Moldavia and Wallachia were neither overlapping nor convergent in the first decades after the Union of 1859. In spite of administrative centralization, two distinct and non-complementary economies remained: both produced the same agricultural goods and employed similar methods for their processing. Interests were still dictated by the structure of regional grain economies, locked in a competition over the core objective of transporting their output to Danubian harbors at competitive prices and maximizing profits on exports. These profits, traditionally correlated with access to roads, rivers, and ports, were now beginning to be influenced by railway proximity as a disrupting factor. The fierce competition between landowners and estate administrators over connections for the export of goods became apparent as soon as railways were first debated and projected.[16] Those who managed to secure a station close to their estates gained enormous advantages, as was the case with those in the south, better connected to

13 G. Christodorescu, *Economia națională și căile noastre ferate* (Constanța: Tipografia "Ovidiu" H. Vurlis, 1906).

14 For instance, see the famous speech of member of Parliament N. Voinov (1834–1899), *DAD*, no. 11, December 1, 1890, 71–72.

15 On colonial anxieties in nineteenth-century Romania, see Andrei-Dan Sorescu, "The Infrastructural Power of the 'Colonial' as a Concept" in this volume.

16 Ecaterina Negruți-Munteanu, *Dezvoltarea agriculturii în Moldova între anii 1848–1864*, in V. Popovici et al., *Dezvoltarea economiei Moldovei 1848–1864* (Bucharest: Editura Academiei Republicii Socialiste România), 1963, 97–102, 110–116; Apostol Stan, *Agricultura românească in faza finală a clăcășiei 1831–1864* (Bucharest: Editura Academiei Române, 1994), 147–159.

the Wallachian harbor town of Brăila on the Danube.[17] Such high stakes fueled rivalries and constant maneuvering whenever routes and resources for railway building were on the table.

Additionally, regional interests remained disjunct due to geographic and geopolitical factors that persisted throughout the first two decades after 1859. Moldavia, and especially its northernmost part, lay between the railway networks of Habsburg Central Europe and the Russian Black Sea. This was the setting for the privately-run Ofenheim concession and the state-built Iași–Ungheni railway (see fig. 2).[18] On the other hand, in direct competition with Moldavia, Wallachian politicians aimed to ensure that the southern half of Romania would become the main railway corridor between Central Europe and the Orient.[19] Bucharest thus considered that a junction with Austria-Hungary at the westernmost point of Vârciorova–Orșova would offer Wallachia the longest possible stretch of the route to Istanbul (see fig. 2). However, this plan came to naught due to the rapid eastward advancement of the Austro-Hungarian rail system, and Romania was forced to accept a connection in the Carpathians at Predeal rather than at the Danube's point of entry into the country.[20]

Up until 1865, Moldavian politicians such as Mihail Kogălniceanu (1817–1891) sought to find concessionaires for Moldavian railways whenever they held ministerial positions.[21] That these plans also came to naught due to political upheavals serves as a reminder of how infrastructure embodies power and reflects the shifting balance between political blocs. Thus, the Șuțu–Salamanca concession, which was to connect Galați harbor to northern Moldavia via a line along the Trotuș River valley, was suspended with the abdication of Prince Alexandru Ioan Cuza (1820–1873, r. 1859–1866) in 1866 and annulled by the incoming government.[22] While railway development plans initiated by the Moldavians were stalled, those in Wallachia stayed on schedule: 1869 saw the opening of the first line in Romania, connecting

17 Constantin Bușe, *Comerțul exterior prin Galați sub regimul de port franc (1837–1883)* (Bucharest: Editura Academiei Republicii Socialiste România, 1976), 123, 140.

18 Dumitru P. Ionescu, *Construirea liniei ferate Iaşi–Ungheni, AIIXI* 23, no. 1 (1986): 283.

19 See also Ion Ghica, *Scrieri economice*, vol. 3 (Bucharest: Editura Asociației generale a economiștilor din România, 1937), 63–64.

20 Toader Popescu, *Proiectul feroviar românesc (1842–1916)* (Bucharest: Simetria, 2014), 46. On the development of the railway network in Transylvania, see also Gábor Egry, "Steel and Wood: Railway Construction and Rent-Seeking" in this volume.

21 Dumitru P. Ionescu, *Proiecte şi concesiuni feroviare in vremea lui Cuza Vodă,* in AIIX 31 (1994): 239–272. See also Bucur Țincu, "Contribuții la istoria căilor ferate din România: Probleme în perioada 1859–1869," *Studii: Revistă de istorie* 24, no. 5 (1975): 951–962.

22 Botez, Urmă, and Saizu, *Epopeea feroviară*, 39.

Bucharest to the Danubian port town of Giurgiu (the J. T. Barkley–J. Staniforth concession) (see fig. 2).[23]

Aware of the fact that tacit stalling would hamper their chances, the Moldavians attempted to negotiate as part of the constitutional debates of 1866, introducing an amendment whereby the Court of Cassation was to reside in Iași "until railways are completed."[24] While this proved to be an unsuccessful bid, it was nevertheless symptomatic of the deepening distrust among Moldavian political circles, irritated by the Wallachians' unfulfilled promises.[25] This tense moment only prefigured parliamentary debates in 1868 over the construction of a core railway network—the Strousberg–Ofenheim concessions (see fig. 2). Although perceived to be the result of nationwide political effusion and a turning-point in state-building, the debates preceding the vote openly reflected anxieties over the territorial hierarchies that proximity to, or distance from, the railway network threatened to create.[26] These fears were especially palpable in parts of northern and eastern Moldavia that had not been connected to the network or had merely been promised branch lines. And they were deepened by the fact that railways brought with them not just connection but also competition. For Moldavia's great landowners, this was a source of existential stress and brought into public debate the question of how competitive pricing and profitability were tied to access to the network. This concern was visible not only at a local level but also articulated itself on a regional level as the expression of traditional solidarities.

Tensions thus ran high, and in 1868, debates were deeply marked by a transactional mindset. Such was their economic and commercial importance that it trumped the principle of "political unification" and "cementing the union" between Moldavia and Wallachia—ostensibly the leitmotif of a concession spanning the entire country. Significantly, the Chamber of Deputies sacrificed the Buzău–Mărășești line, which would have covered the shortest possible distance between Bucharest and Iași. A "political" line was abandoned in favor of much longer routes converging toward the mouths of

23 Const. C. Mănescu, *Istoricul căilor fierate din România*, vol. 1 (Bucharest: Ed. Direcțiunii Generale a Căilor Fierate Române, 1906), 383–491.

24 ANIC, fond Parlament, d. 351, f. 9.

25 For a fuller account, see Mihai Chiper, "'Valahii ne-au înșelat': Chestiunea compensațiilor Iașului in acțiunea politică moldovenească după unire (1862–1874) (1)," *Archiva Moldaviae* 10 (2018): 155–162.

26 See also Silvia Marton, "Shades of Dependency and the Discourse on 'Corruption.' Railway Concessions in Romania in the Nineteenth Century," *East European Politics and Societies* 39, no. 3 (2025): 691–716.

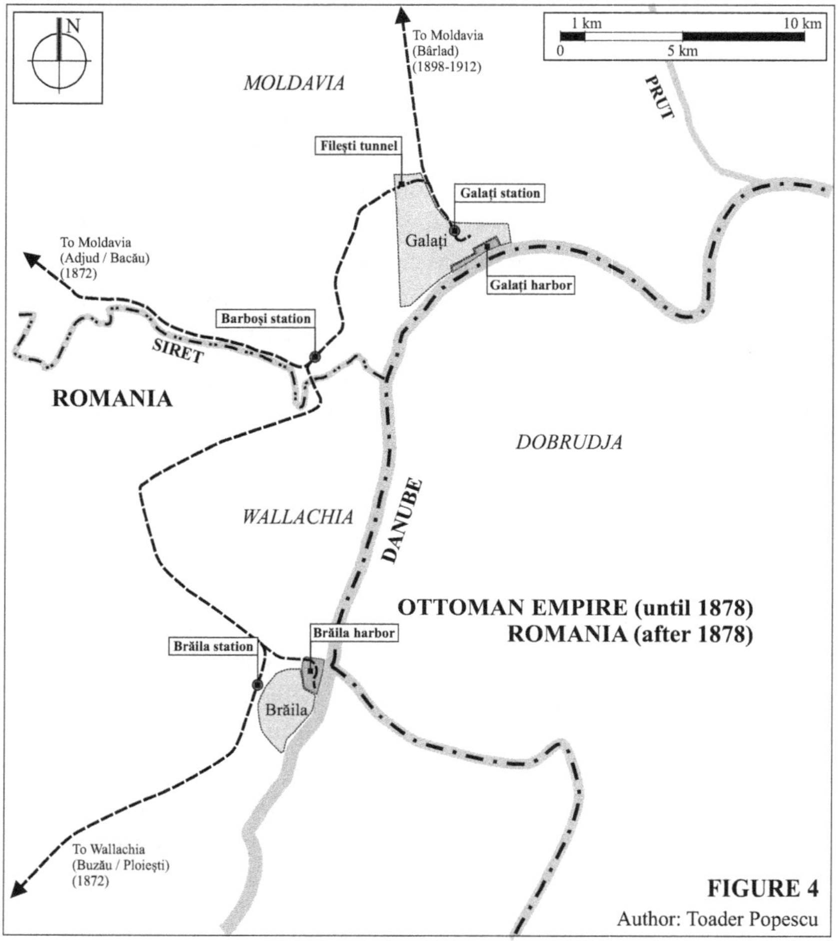

Figure 4. Railway system in the Brăila – Galaţi region in Romanian at the end of the nineteenth century

the Danube—namely, Galaţi for Moldavia and Brăila for Wallachia—clearly revealing the primacy of economic rationales in the project (see fig. 4). Prosperity was ultimately conditional upon the cost of exporting grain through these harbors, which linked the country to major international trade networks. In sum, we may already note how railways could produce very different effects, depending on the scale at which we judge them: positive outcomes at the macro level of the national economy came at the expense of absolute or relative losses in terms of competitiveness—be it economic, urban, or human—at the micro or regional level.

Railways as Existential Stress and "Instruments of Ruin": The Strousberg and Ofenheim Concessions, 1868

The parliamentary debates on the Ofenheim and Strousberg concessions provide us with an entry point for considering how the economic and commercial interests of the two formerly separate principalities differed. The two main topics under negotiation were (1) the length of the networks in Moldavia and Wallachia in proportion to their respective territories and (2) the trajectories of the lines. The newly inaugurated regime of Carol Hohenzollern-Sigmaringen (1839–1914, r. 1866–1914) was eager to offer an example of transparency and good practice, particularly as opaqueness and corruption had been criticized as the root causes of his predecessor Alexandru Ioan Cuza's downfall. As minister of finance, Ion C. Brătianu (1821–1891) explained to the Chamber that the previous regime's experience demonstrated the need for accountability in infrastructure; in his opinion, it was not the concession itself that had spelled Cuza's end but "the revolt of the country, knowing that it had been awarded behind closed doors."[27] Now, in this supposedly more accountable setting, Wallachian and Moldavian MPs accused each other of "not knowing" the hierarchy of cities in the other's region, their economic importance, and their specific interests. The debates grappled with both differentials in political power and a desire for an equitably drawn network. In principle, the government and the commission set up to negotiate with the concessionaires had agreed upon a total of around 900 km for the entire network with Ofenheim and Strousberg (see fig. 2); the question now was how this length of track would be divided between Moldavia and Wallachia without public debt rising to unsustainable levels and without annuities destabilizing the state budget. Restrictive financing conditions, however, created insurmountable dilemmas: choosing one route from among two or three options meant, as we shall now see, creating the conditions for path dependence.

Although in previous years major political figures such as Ion C. Brătianu or Ion Ghica (1816–1897) had proposed that economic studies be undertaken for determining the best possible network layout from a geographic and demographic standpoint, the accusation now leveled was that "what the concessionaires had proposed" carried more weight than the country's economic needs.[28] Yet this was not strictly true. The routes proposed by

27 MOf, no. 120, May 29/June 10, 1868, 770. See also Marton, "Shades of Dependency and the Discourse on 'Corruption.'"

28 MOf, no. 116, May 24/June 5, 1868, 738.

concessionaires had not been drawn up at the last minute but were themselves the result of earlier projects that had been under negotiation with Romanian authorities, even if ultimately unsuccessful.[29] These projects already incorporated some of the demands made by Romanian representatives and left open the political—rather than economic—question of how the network might be drawn, one that MPs sought to answer by attempting to harmonize the economic imaginaries of the two provinces.

The debate was, on the whole, centered on two main lines: (1) Czernowitz–Galați, along the Siret in Moldavia and (2) Turnu-Severin–Bucharest–Brăila–Galați in Wallachia (see fig. 4 and fig. 2).[30] It was estimated that the greater national and international flux—and thus the more important of the two lines—would be in Moldavia, owing to the scheduled opening of the Suez Canal and the role that the line would play in connecting the West to the Orient, channeling Central European traffic to Galați, and ultimately to the Black Sea and the Mediterranean.[31]

The Ofenheim–Strousberg network was to be split between 355 km in Moldavia and 650 km in Wallachia, the latter already benefiting from ongoing works on the 62 km of the Bucharest–Giurgiu Barkley–Staniforth concession inaugurated in 1869 (see fig. 2).[32] While the Chamber of Deputies established the route in broad strokes, the finer, final details were the prerogative of the concessionaires—a fact that would prove problematic for the Romanian authorities. The concessionaires' priority was to find cost-effective solutions, which meant not only shoddy workmanship but routes that disadvantaged major urban centers. Notably, this was the case with Galați harbor, cut off from what had once been the historic Moldavian trade route, now diverted toward Brăila—to the advantage of the neighboring Wallachian port town. Another revealing and problematic example was the Ofenheim section of the concession: acting as a Iași-bound extension of the Czernowitz–Suceava line in Austrian Bukovina, it bypassed the town of Botoșani, the country's fourth-largest town and an important economic center.

The MPs' interest in railway development was readily demonstrated by how well informed they were on the most recent debates and theories: some went beyond merely praising the advantages that infrastructure could bring and warned of the risks of neglecting economically relevant

29 L. Boicu, "Transporturile în Moldova între 1848–1864," in V. Popovici *et al*, *Dezvoltarea economiei Moldovei 1848–1864*, 467–470; Popescu, *Proiectul feroviar românesc*, 20–23, 25–29.

30 MOf, no. 115, May 23/June 4, 1868, 733.

31 MOf, no. 115, May 23/June 4, 1868, 733; no. 119, May 28/June 9, 1868, 762.

32 MOf, no. 118, May 26/June 7, 1868, 753.

communities and areas, turning railways into "instruments of ruin." For instance, Ion Codrescu, a Moldavian MP (Tutova County), quoted French engineer Albert A. Perdonnet when drawing attention to the importance of network planning:

> A railway is a magnet which attracts all the commercial and industrial activity of the country in a very wide circle. While it invigorates the regions it crosses, it impoverishes those it leaves behind, and if it is badly laid out, not only does it waste public money, but it can also cause the most unfortunate disturbance in the distribution of national wealth and, while cause of prosperity for some parts of the country, become an instrument of ruin for many.[33]

Debates had to adjudicate between two competing axioms. One held that a route had to follow the shortest distance between two points, with branch lines connecting cities along the way to the main line. This system, first used in France, was ultimately abandoned after Charles-Joseph Minard[34] discovered that the revenue thus generated by the railways was smaller than if the main line took a longer route through various towns—the second axiom.[35] The Ofenheim concession in northern Moldavia had been planned according to the former of the two paradigms: a short route that was favorable to the concessionaire and to the Austro-Hungarian Empire but unfavorable to the regional economy of Moldavia. Connecting Suceava to Galați by following the valley of the Siret River southward, it bypassed population centers such as Botoșani, Iași, and Bârlad (see fig. 2). The Strousberg takeover of the concession only worsened the situation, as traffic to the harbor of Galați was rerouted to Brăila. By contrast, the Strousberg concession in Wallachia was far more advantageous to the local economy, as major urban centers lay directly on the main line. Had the route been modeled on the same principles as in Moldavia, major towns such as Ploiești and Craiova—or even Bucharest itself—would have been connected only via secondary branch lines.

33 MOf, no. 116, May 24/June 5, 1868, 738. The original quote is taken from Auguste Perdonnet, *Traité élémentaire des chemins de fer*, vol. 1 (Paris: Langlois et Leclercq, 1858) [1855], 77.

34 Charles-Joseph Minard (1781–1870) was one of the best-known French engineers specializing in statistics, graphs, and maps of transit fluxes; see Theodore M. Porter, *Trust in Numbers: The Pursuit of Objectivity in Science and Public Life* (Princeton: Princeton University Press, 1995), 122–123.

35 MOf, no. 116, May 24/June 5, 1868, 738.

Debates made plain the regional origins of proposals for railway routes. Parliament received petitions first from Moldavian counties and then from the south.[36] "A great number of citizens" pleaded with the Chamber that "its approval for [the building of] railways be its utmost concern, so that this general desire be more swiftly accomplished, and that a highly urgent necessity be met."[37] Often, MPs framed their interventions in terms of their good faith—understood as the absence of personal pecuniary motives—using phrases such as "the route does not pass by my estate," "without private interest," not "impelled by localism" (A. Stolojan), or "free from any material interest" (M. Kogălniceanu). Moldavian MPs argued that the Vaslui–Bârlad–Tecuci line would be "thrice as important" as the Bucharest–Craiova line (C. Ursache),[38] while Wallachian MPs sought to turn Oltenia into a transit hub between East and West (Ion C. Brătianu, minister of the interior).[39]

These discussions patriotically reframed railways as "cementing the union" and as "rightful compensation for Moldavia's suffering" (C. Ursache) but were also used by the Moldavians as a veiled threat—for instance, when they declared that they would accept to cast their vote in favor of a "just" proportional allocation of kilometers for the region: "and we shall come and demand that they be built, kilometer by kilometer" (M. Kogălniceanu).[40] Kogălniceanu went so far as to threaten that the Moldavians would boycott any vote prioritizing the Strousberg concession in Wallachia over the Ofenheim concession in Moldavia, as proposed by Dimitrie Brătianu (1818–1892), the minister of public works.[41] Mutual recriminations soon followed, with Moldavian and Wallachian MPs trading accusations of ignorance regarding each other's cities, economies, and regional interests. For instance, Petre P. Carp (1837–1919) argued that Moldavians had a greater stake in railway construction, since the Wallachians already had access to the Danube as a convenient thoroughfare—whereas the "difference that nature, that God himself" had created now had to be rectified.[42] When D. Brătianu proposed replacing the Târgu-Frumos–Iași line with a shorter Roman–Iași line for the sake of saving seventeen kilometers of track, Kogălniceanu retaliated with

36 MOf, no. 113, May 20/June 1, 1868, 720.
37 MOf, no. 113, May 20/June 1, 1868, 720
38 MOf, no. 116, May 24/June 5, 1868, 739.
39 MOf, no. 120, May 29/June 10, 1868, 770.
40 MOf, no. 116, May 24/June 5, 1868, 739.
41 MOf, no. 120, May 29/June 10, 1868, 771.
42 MOf, no. 117, May 25/June 6, 746.

a counter-amendment proposing equivalent cuts in Wallachia.[43] He further argued that the Bucharest–Craiova main line would skew the allocation of railway kilometers in favor of the southern region, despite the fact that Moldavian counties were in greater need: "Moldavia is divided by three parallel river valleys from top to bottom, and a rail route through one of them is of no use to the others."[44] At the vote before the introduction of amendments, Kogălniceanu noted that negotiations had resulted in a 712 versus 290 km advantage for Wallachia —"that part of Romania that also benefits from the Danube."[45]

Discontent was particularly visible among Moldavian MPs regarding the route taken by the Ofenheim concession along the Siret Valley. Their hope was that, following Charles-Joseph Minard's reasoning, it could be rerouted to connect Moldavia's main urban and economic centers. As MP P. Suciu attempted to argue in an unsuccessful bid for amending the law, (1) the Siret Valley already benefited from both a road and a river, (2) the proposed Ofenheim route did not link any major cities, and, (3) it crossed only four counties, compared to the six that his alternative route would have connected—thus serving a larger population and transporting more grain.[46] However, extending the route all the way to Botoșani in the north would have meant that the Moldavian railways would connect to the Austro-Hungarian network at a different point[47] from the one previously agreed upon with Ofenheim.[48] At that time, Ofenheim had already become the concessionaire of the Czernowitz–Suceava line, guaranteed by the Austro-Hungarian government,[49] and therefore had no incentive to reconsider the layout of the network. It was Kogălniceanu who had the final say in the matter, eager as he was to expedite a rail link between Iași and Europe after eight years of failed attempts.[50] Kogălniceanu ultimately accepted the Ofenheim plan, and Ofenheim accepted a compromise: a link would also be built across the hills of central Moldavia and another to Botoșani (see fig. 2).[51]

43 MOf, no. 120, May 29/June 10, 1868, 772.

44 MOf, no. 120, May 29/June 10, 1868, p. 773. The same compromise was also made in MOf, no. 122, May 31/July 11, 1868, 789.

45 MOf, no. 122, May 31/July 11, 1868, 789.

46 MOf, no. 120, May 29/June 10, 1868, 772.

47 Amendment by P. Suciu. MOf, no. 120, May 29/June 10, 1868, 772.

48 MOf, no. 117, May 25/June 6, 1868, 746.

49 MOf, no. 120, May 29/June 10, 1868, 773.

50 MOf, no. 120, May 29/June 10, 1868, 773.

51 MOf, no. 120, May 29/June 10, 1868, 772.

Compared to Moldavia, Wallachia ultimately benefited from a more rational rail network layout—particularly in terms of how well-connected major towns were to the main line. This was one of the main reasons why, by the end of the nineteenth century, urban demographic and economic hierarchies had drastically shifted[52]. Moldavia experienced a marked decline in urbanization (23 percent in 1859 versus 18–19 percent in 1899), with an urban population growth of only 55 percent over the same period, compared to 131 percent in Wallachia.[53] Towns located either on a branch line or on no line at all suffered the most;[54] Iași (with only an 18 percent population growth) and Botoșani (20 percent) were the most notable cases, but even the port town of Galați (140 percent) lagged far behind its Wallachian counterpart, Brăila (257 percent). Against the backdrop of a deficient railway infrastructure, the divergence between Moldavia's process of deurbanization and Wallachia's urban expansion was determined by a multidimensional causal complex.[55] This was catalyzed primarily by the decline in economic and commercial attractiveness induced by administrative and political centralization and by internal migratory flows directed toward Bucharest. This phenomenon affected the Jewish community as well, a demographic pillar of local urbanization: the dismantling of the network of Jewish market towns, historically regarded as proto-urban nuclei, was doubled by the community's emigration, either toward Bucharest or abroad. Finally, an underdeveloped healthcare system was reflected in high infant mortality rates.[56]

Path Dependence

Building the core of the railway network through the Ofenheim and Strousberg concessions created a pattern of path dependence that significantly constrained subsequent attempts at correcting or expanding

52 Toader Popescu, "Un aller simple pour la modernité," *New Europe College Yearbook* (2009–2010): 280.

53 Leonida Colescu, *Analiza rezultatelor recensământului general al populației României dela 1899* (Bucharest: Institutul Central de Statistică, 1944)

54 David Turnock, "Railways and Economic Development in Romania before 1918," *Journal of Transport Geography* 9, no. 2 (2001): 137–150.

55 George Țurcănașu, *Evoluția și starea actuală a sistemului de așezări din Moldova* (Iași: Casa Editorială Demiurg, 2006), 265–288.

56 Ecaterina Negruți, "Rolul evreilor în formarea târgurilor din Moldova în prima jumătate a secolului al XIX-lea," *Studia et Acta Historiae Iudaerum Romaniae* 1 (1996): 139–151.

rail infrastructure in Moldavia.[57] The initial decision of bypassing major towns such as Iași, Botoșani, or Galați effectively isolated the breadbasket of northern Moldavia and severed its historical connections with its traditional harbor, Galați. In particular, the Ofenheim concession—designed as an extension of the Lemberg/Lviv–Czernowitz line along a route favorable to the Austro-Hungarian economy—proved highly disadvantageous to Moldavia. Its original objective, from the Austrian standpoint, was to reach a port on the Black Sea; however, failed negotiations with the Russians meant that Odessa was not an option from the outset. As a result, the Ofenheim route followed the Siret Valley down to Galați and the mouths of the Danube. A planned extension toward Wallachia, and ultimately Bucharest, was abandoned once the Strousberg concession took over the southern Moldavian branch line.[58] As the Romanian government was not in a position to dictate the terms of the negotiations, it had to consent to the routes drawn up by the concessionaires—along with their consequences for regional urban development. The rapid decline of Botoșani, situated at the endpoint of a branch line, is a case in point.[59] Still, the railway also facilitated the emergence of some relatively important urban settlements, such as the town of Pașcani, where the main line linked up with the Iași branch.

Moldavia would remain a focus of railway development projects for some time. Until 1879, when the Romanian network joined up with that in the Kingdom of Hungary at Predeal, Romania's only link to the European network remained at Suceava, via the Czernowitz–Lemberg/Lviv–Krakau/Cracow line. Further south, the Danube was a more convenient transport link. Yet neighboring empires conditioned development by following their own strategic interests: Russia was disinclined to build a junction at Reni, near Galați, while Austria-Hungary conditioned the opening of a junction in the west at Orșova–Vârciorova (Wallachia's shortest connection to Vienna) on the simultaneous opening of a link further east. After regaining three counties in southern Bessarabia at the end of the Crimean War, Moldavia

57 Douglas J. Puffert, "Path Dependence in Spatial Networks: The Standardization of Railway Track Gauge," in *Exploration in Economic History* 39, no. 3 (2002): 282–314; Douglas J. Puffert, "The Standardization of Track Gauge on North American Railways, 1830–1890," *Journal of Economic History* 60, no. 4 (2000): 933–960; S. J. Liebowitz and Stephen E. Margolis, "Path Dependence, Lock-In, and History," *Journal of Law, Economics, and Organization* 11, no. 1 (1995): 205–226; see also James Mahoney and Daniel Schensul, "Historical Context and Path Dependence," in *The Oxford Handbook of Contextual Political Analysis*, ed. Robert E. Goodin and Charles Tilly (Oxford: Oxford University Press, 2006), 454–471.

58 Popescu, *Proiectul feroviar românesc*, 43–44.

59 Popescu, *Proiectul feroviar românesc*, 44.

now was no longer landlocked and had the trump card of a potential Black Sea port. Bucharest had hoped, as of 1868, to build the so-called Carol I seaport at Jibrieni, though this never materialized.[60] The seaport would have also benefited from a rail link built by Marshall & Sons, connecting it to the rest of Moldavia.[61] These concessionaires had envisioned the creation of a network node of European importance, as it would have provided for a shorter route between Europe and the Near East.

However, geopolitics ultimately proved unfavorable to Moldavia. When the Ofenheim and Strousberg concessions were first agreed upon, the reasoning was that the region would benefit from international trade. Yet the last of the Russo-Turkish wars, in 1877–1878 (Romania's "War of Independence"), ended up reconfiguring the priorities of the Romanian state. The loss of the three southern counties in Bessarabia and the awarding of Dobrudja as compensation meant that the Austria–Black Sea transport corridor and the proposed seaport at Jibrieni were abandoned in favor of redeveloping the newly acquired harbor at Constanța in Dobrudja. Even so, for northern Moldavia, the eastward connection to the Russian railway network toward Odessa remained more convenient than southbound transit via the harbors of Galați or Brăila—something that displeased policymakers in Bucharest. To redirect grain exports to Brăila and bolster the national economy, the Ion C. Brătianu government doubled tariffs on the route between Iași and the Russian Empire in 1885. This had a catastrophic effect on northern Moldavian producers by undermining the profitability of grain transport.[62] Moreover, a tariff war with Austria-Hungary drove a shift from cattle husbandry to agricultural output,[63] which, in the absence of cost-effective transit, forced agriculturalists in northern Moldavia to enter a cycle of debt and bankruptcy. The main losers were the great landowners whose latifundia exceeded 100 hectares and were exploited either under their own management or through

60 See Constantin Ardeleanu, "'Commercial and Political Needs Demand the Establishment of a Black Sea Port': Infrastructure Development, Opportunities and Anxieties in an Eastern European Periphery (1860s–1870s)," *The Journal of Transport History* 44, no. 3, (2023): 436–456; Vitalie Văratic, "Jibrieni: proiectul primului port românesc la Marea Neagră în a doua jumătate a secolului al XIX-lea," in *160 de ani de la Unirea Principatelor: Oameni, fapte și idei din domnia lui Alexandru Ioan Cuza,* ed. Petronel Zahariuc and Adrian-Bogdan Ceobanu (Iași: Editura Universității "Alexandru Ioan Cuza" din Iași), 539–563.

61 Popescu, *Proiectul feroviar românesc,* 49–50.

62 For instance, the price rose from 32 to 54 francs per wagon on the Iași–Ungheni route, which made one agriculturalist lose 281 francs if they chose Galați over Odessa. See "Jocul tarifelor de căi ferate," *Patria,* July 14, 1885, 1.

63 Alexandru G. Djuvara, *Tariful general al drepturilor de vamă: Raport și discursuri parlamentare* (Bucharest: Litotipografia Carol Göbl, 1892), 74–75.

the leasehold system. The negative consequences of this decline directly impacted the peasantry, which had long been considerably more precarious here than in Wallachia. It was precisely in this agrarian area of northern Moldavia that the 1907 peasant revolt broke out—an event scholars have dubbed the last *jacquerie* in Europe.[64]

Starting with 1882–1886 and up until World War I, attempts at "correcting" the route along the Siret Valley heavily influenced the regional economic agenda through the opening of new lines.[65] To this was added the attempt at rectifying the problems caused by the so-called "Barboşi curve" that had skirted around Galaţi harbor—the traditional outlet for Moldavian exports—as part of the Strousberg concession (see fig. 4).[66] Due to lax contractual stipulations, Strousberg was allowed to cut costs by redirecting the main line along the Siret River toward Brăila. Galaţi's harbor was linked only by a secondary line branching off at Barboşi, a fourteen-kilometer detour that, according to contemporary accounts, meant even a horse-drawn carriage could arrive in Galaţi station faster than the train. In fact, the line did not reach the harbor or even its vicinity, and goods had to be unloaded and transported by carriage, which impeded commercial flows. The consequences of the curve and the isolation of the harbor from the main line were regarded as key factors behind Galaţi's urban and commercial decline.[67] Partisan accusations painted the curve as part and parcel of the "famous Strousberg *gheşeft*,[68] through which a gaggle of German princelings enriched themselves at the expense of this country's sweat."[69] A straight line to Galaţi, denounced former mayor G. Ventura, "would have required expensive works, shouldered by the German entrepreneurs, which would have diminished the profits of shareholders under Carol I's protection. Galaţi had to pay for the millions swallowed up by swindling German princelings."[70]

64 Philip Gabriel Eidelberg, *The Great Rumanian Peasant Revolt of 1907: Origins of a Modern Jacquerie* (Leiden: Brill Archive, 1974); Irina Marin, *Peasant Violence and Antisemitism in Early Twentieth-Century Eastern Europe* (Cham: Palgrave Macmillan, 2018).

65 Bârlad–Vaslui (1886); Vaslui–Iaşi (1892); Dorohoi–Iaşi (1896); Bârlad–Galaţi (1912).

66 Paul Păltănea, *Istoria oraşului Galaţi: De la origini până la 1918: Partea a II-a* (Galaţi: Editura "Porto-Franco," 1995), 121–122.

67 Păltănea, *Istoria oraşului Galaţi*, 121–122.

68 In Romanian, *gheşeft* comes from the Yiddish/German *Geschäft*, originally meaning "business" or "commercial transaction." However, it took on a pejorative meaning (sometimes with anti-Semitic undertones) and eventually referred to a dealing that is dishonest, dubious, and in one's own interest.

69 "Adevărul la Galaţi. Întrunirea de la Alcazar," *Adevărul*, March 28, 1891, 2.

70 "Adevărul la Galaţi. Întrunirea de la Alcazar."

The configuration of the Strousberg and Ofenheim concessions routes left many of the MPs' demands from Moldavia and Oltenia unfulfilled. Though the overall plan had been to traverse the two halves of the country in a way that could link all its major urban settlements, the "backbone"[71] of the Romanian rail network ultimately failed to reach the most important and populous Moldavian urban centers. As already noted, Iași, Galați, and Botoșani—the three largest towns in Romania after Bucharest—became isolated ends of branch railways.

Deepening Lag in Moldavia, "Patriotic *Gheșeft*" in Wallachia

The mirage of technology and the hope that development would eventually reach initially overlooked or disadvantaged areas only intensified after the completion of the railway network's core. As of 1882, a secondary network of lines "meant to meet the internal interests of the country"[72] was on the table. Yet this phase of expansion soon proved even more arbitrary—both in terms of how routes were planned and how they were built. As in 1868, except for a technical council tasked with calculating construction costs, no expert opinion from economists nor governmental studies guided the expansion of the network. In fact, the shift from foreign concessions to state funding meant that political interference became even more intense—whether by bending routes toward certain estates or by prioritizing some routes over others. As examined by Toader Popescu in his chapter, lobbying and discontent among local citizens and notables regarding railway routes and priorities remained important nationwide.[73] Party groupings or local and regional alliances thus pushed for a given route by pleading for their particular visions of development, and investment became "political." Representatives from various counties submitted proposals for local branch lines. Moldavian MPs, for instance, introduced amendments for building a second line that would cross the region from north to south (Iași–Vaslui–Bârlad), a proposal that had been initially mooted in 1868. Conversely, MPs from Wallachia and Oltenia were interested in local lines that would connect to the existing main network. Once again, a group of major "dealers" formed

71 Popescu, *Proiectul feroviar românesc*, 47.

72 Dezbaterile Corpurilor Legiuitoare (DCL), No. 97, April 18/30, 1882, 1662. See the final law in MOf, no. 47, May 28/June 9, 1882, 1116–1117.

73 Toader Popescu, "Hoping and Coping: Local Anxieties and Railway Development in Romania" in this volume.

and attempted to deny the influence of any private interest or "any estates" along the routes they advocated.

The government submitted to the Chamber a proposal for 370 km's worth of lines, to which 245 km were later added, scheduled to be built over a three-year period (1883–1886).[74] Delays emerged, and once again Wallachian railways were prioritized. In 1885, Parliament voted another law on technical studies for lines of local and county importance.[75] These two laws did foster the growth of the railways, but they also irked the Moldavians: adding to delays, some lines remained at the planning stage, others lacked funds, while still others were simply blocked for various reasons (including military ones). Wallachia appeared to be unduly favored once again by rerouting the new lines toward various estates; corruption had now become a visible and integral feature of unequal regional development. Perhaps the most controversial case was that of Prime Minister Ion C. Brătianu's estate at Florica, in 1887.[76] What was initially meant to be a narrow-gauge line was ultimately built at standard gauge and at remarkable speed,[77] while technical studies and money for urgent, major lines toward Iași remained pending. Even at the planning stage, the press already warned of a "great patriotic *gheșeft*" planned by the prime minister, who refused to offer any strategic or economic justification for the route to a delegation from the nearby town of Pitești, which was negatively affected by the rerouting.[78] Such public demands were, in the opinion of the early social-democratic newspaper *Drepturile omului* [The Human Rights], warranted, as "the line from Golești, instead of going straight to Câmpulung, is given an enormous bend [...] only so as to reach Florica, Mr. Brătianu's vineyard."

The fact that the city of Pitești was disadvantaged by the redrawing of the line was also at the heart of the most important political intervention against it, in a question raised by Conservative MP I. Rădulescu to P. S. Aurelian (1833–1909), economist and minister of public works in Brătianu's Liberal government. Based on a study by engineer G. Cantacuzino, Rădulescu noted that the line through Pitești would have been shorter, cheaper, and more "natural," unlike the approved route, which featured stations every

74 Dumitru P. Ionescu, "Primele căi ferate înguste în România," *Revista istorică* 1, no. 7–8 (1990): 744.

75 DCL, no. 89, April 2, 1885, 1706–1707.

76 *Patria*, October 6, 1885, 2; Sc. Pastia, *Un răspuns unor târâtoare veninoase* (Iași: Tipolitografia H. Göldner, 1890), 8.

77 For M. Kogălniceanu's plea for railways in northern Moldavia, see *Dezbaterile Adunării Deputaților* (DAD), no. 12, November 30, 1888, 183, sitting of November 29, 1888.

78 *Drepturile omului*, October 5, 1885, 3.

four kilometers—"stopping at Florica almost as soon as you get on the train at Golești."[79] Aurelian replied to this "insinuation" by asking whether "Mr. Brătianu is the only current minister with a station or waystation built near their estate," yet refused to name others, and turned this into a question of trans-partisan precedent, as "never before have previous ministers been accused of a crime" under similar circumstances.[80] The opening of the line became both a spectacle of Liberal propaganda and an occasion for the opposition press to unveil new details of the *gheșeft*. Thus, one article in the radical (and formerly Liberal) *Lupta* [The Struggle] alleged that local stakeholders had been pacified by being offered a station each and that of the seven stops along the line, only two served public interest: "it is only logical that the poor MPs for Muscel County get their fair share, given that the *honest* PM, Mr. Brătianu, landed a huge slice of the pie."[81]

Meddling with routes and the placement of stations was perceived as an abusive form of serving private interest and a scandalous infringement on public transportation needs. Underfinanced and marginalized in no small part due to Wallachian political groups' abuse of power, projects in Moldavia faced mounting delays. The Committee for the Defense of the Economic and Cultural Interests of Moldavia[82] rallied behind a platform of protest, decrying the persistent stagnation of the region's secondary railway network between 1886–1890. Wallachian lines were correspondingly decried as "the lines of favoritism, eating from the trough of useful lines" (Nicolae Ceaur Aslan), built without preliminary studies, while northern Moldavia's agrarian basin remained isolated and deprived of rail links with harbors.[83] In 1891, statistical evidence of preferential railway investment as a source for economic development in Wallachia was brought before Parliament. MP Dimitrie A. Sturdza (1833–1914), former Liberal minister of public works, showed that of 684 km still under construction, 269 (39 percent) were located in Wallachia and 415 (61 percent) in Moldavia, reflecting massive delays in the north.[84] To stem the tide of Moldavian protestation, the Conservative government's minister of public works, Titu Maiorescu (1840–1917), countered

79 DAD, no. 19, December 12, 1887, 207–208.

80 DAD, no. 19, December 12, 1887, 209–210.

81 "Bacon, Politica la Câmpulung (corespondența particulară a *Luptei*)," *Lupta*, August 5, 1888, 1.

82 For a fuller account, see Mihai Chiper, "Când Unirea devine o decepție: Comitetul pentru apărarea intereselor economice și culturale ale Moldovei (1891)," in *Frământările modernității: discursuri, oameni, instituții*, ed. Marius Chelcu and Dorin Dobrincu (Iași: Editura Universității "Alexandru Ioan Cuza" din Iași, 2019), 215–274.

83 DAD, no. 31, January 16, 1891, 311–313.

84 *Dezbaterile Senatului* (DS), no. 39, February 15, 1891, 328.

with false information, claiming that only 990 km were already in place in Moldavia, 1757 in Wallachia, and 64 in the Dobrudja.[85] In actual fact, Wallachia counted 1884 km of railways versus 774 in Moldavia, which, based on official documentation, proves that Moldavia was indeed at a disadvantage and the discontent of its MPs was justified. Between 1886 and the fall of Brătianu's Liberal government in 1888, a total of 710 km of railways were built in Wallachia, compared to only 127 in Moldavia—an almost 6:1 ratio.[86] The lines that had been promised to Moldavia remained at a planning stage.

Fortifications and the Vetoing of the Galați–Bârlad Line: National Gains through Regional Sacrifices

To complete the second north-south railway corridor, connecting Iași–Vaslui–Bârlad–Galați, the final stretch between Bârlad and Galați was included in the law of March 29, 1885 (see fig. 2 and fig. 4). It was not until 1912, however, that the line opened, due to various delays, among which military matters prominently featured.[87] The Galați–Bârlad line was, among the commercial elites of Galați, considered a "semi-reparation" meant to offset the consequences of the Barboși curve and to reconnect Galați harbor to the Moldavian economy after the town had lost its status as a freeport and access to the agricultural hinterland of southern Bessarabia. The line was seen as important not only for local grain producers but also for those further north, whose economic ties to Odessa had been disrupted by the Brătianu government's tariff policies. Now, the greatest hindrances to the line were neither politics nor private interests but military strategy. Between 1886 and 1891, Romania's system of territorial defense was reorganized. Building fortifications became central to the new doctrine, with forts established around Bucharest and along the former border between the principalities on the Focșani–Nămoloasa–Galați (FNG) line (see fig. 2). The FNG fortification line was conceived to defend Bucharest and Wallachia from a potential Russian invasion from the northeast, effectively leaving Moldavia exposed and implicitly designating it as a future battlefield in the event of war with the Russian Empire. Fortifications thus undermined "equality" between Moldavia and Wallachia, turning the former into a buffer zone for the protection of the latter. All infrastructure

85 DS, no. 39, February 15, 1891, 330–331.

86 A total based on data in Popescu, *Proiectul feroviar românesc*, 250–251.

87 *Traseul direct al liniei ferate Galați–Bârlad propus de George Matei Cantacuzino* (Galați: Tipolitografia G. Bălășescu, 1891); Ion Plesnilă, *Istoricul liniei Galați–Bârlad* (1898).

located beyond the fortification line, including the Galați–Bârlad railway, was subordinated to the military priorities of defending Wallachia—thus reaffirming its centrality within the Romanian state.

Debates on its construction started with a deadlock. The bill had been submitted to Parliament without first obtaining the approval of the Mixed Commission on Communications.[88] This was the sole institutional body authorized to decide on the opening of infrastructure relevant to "the needs of mobilization and troop concentrations, and especially defense." In 1889, the Commission concluded that, regardless of the commercial importance of the Galați–Bârlad line for Moldavia, the line was "altogether dangerous" in terms of defense and was vulnerable to Russian attacks. As such, the Commission recommended an alternative route to be built *behind* the FNG line, in northern Wallachia—a decision that directly contradicted the priorities and the demands of Moldavian MPs.[89]

News of the deadlock caused a wave of discontent not only in Galați but throughout Moldavia. In early 1891, the Committee for the Defense of the Economic and Cultural Interests of Moldavia was established in Iași and represented several other towns, including Galați. Drafting a set of demands that was hoped to gain trans-partisan support among MPs, the committee counted as point number 10 on its program "the opening of the Bârlad–Galați line, which will make Galați into the main outlet for upper Moldavia." Even after the bill for the railway passed the Senate, it was blocked once again by Alexandru Marghiloman (1854–1925), minister of public works and a known defender of Brăila's interests, who demanded more supporting studies.[90] Yet the main adversary of the line, as parliamentary debates revealed, was Iacob Lahovary (1846–1907), minister of war.[91] He vehemently opposed the project on grounds of "national defense and security," given that it lay outside the FNG fortification line.[92] To the disregard of Moldavian interests, Lahovary declared that such questions were above "petty interests and local influence," no matter how "respectable such local interests—commercial or industrial—may be." This perspective, whereby the political center further cemented the peripheral status of Moldavia, was firmly rejected by M. Kogălniceanu. He argued that military officials had already acknowledged

88 Decree no. 2170 established the Commission on July 7, 1886, MOf, no. 79, July 11/23, 1886, 1754–1755.

89 D. I. Vasiliu, *Studii și documente în legătură cu fortificațiile române din secolul al 19-lea*, vol. 1 (Bucharest: Tipografia "Revistei Geniului," 1934), 235–236.

90 DAD, no. 39, January 27, 1891, 395.

91 "Moldova," *Adevărul*, December 3–4, 1890, 1.

92 DAD, no. 44, February 6, 1891, 486.

the strategic value of the Bârlad–Galați railway, as well as that of other lines such as Czernowitz–Lemberg/Lviv or Ițcani–Mihăileni.[93] These debates show how, besides commerce and economy, Moldavia and Wallachia also diverged in terms of military interests, insofar as, in theory, the center was willing to sacrifice the north.

The pressure exerted by the Committee for the Defense of the Economic and Cultural Interests of Moldavia ultimately managed to exempt the line from the restrictions imposed by the "state security law," and a vote was passed on the preliminary technical studies. Still, two decades of further delays followed. Though Decree no. 1577 of April 28, 1892 acknowledged its public utility, the 108-km railway—including the longest tunnel built in Romania before World War I—faced significant technical difficulties and was opened only gradually, reaching completion by 1912 (see fig. 4).

Conclusion

Compared to Wallachia, Moldavia was a relative loser in the railway boom of the nineteenth century. The Ofenheim concession in its north had been designed to serve Austrian interests and reach the Black Sea or the mouths of the Danube on the shortest route; yet it was a route that bypassed the major cities of Iași or Botoșani, which would have profited the regional economy. By comparison, the Strousberg network connected Wallachia's major towns with the Brăila harbor, which, due to the Barboși curve, also became the main export outlet for Moldavia, thereby displacing Galați from its traditional role. The government's weak bargaining position during the negotiations for the Ofenheim concession set the scene for path dependence: it took almost four decades to complete a new north-south line enabling the transport of grain from northern Moldavia to the Danubian ports.

The expansion of the railway network, initially envisioned as a catalyst for equitable modernization, ultimately revealed the systematic vulnerabilities of state-led infrastructure development. Rather than adhering to economic rationality or expert planning, the secondary phase of construction (1882–1885) became deeply mired in political patronage and regional favoritism. The shift from foreign concessions to state funding paradoxically intensified arbitrary decision-making as local elites and parliamentary factions manipulated route planning to serve the private interests of major landowners rather than national cohesion. This politicization of infrastructure not only

93 DS, no. 40, March 7, 1891, 340.

exacerbated regional disparities but also institutionalized corruption as a central mechanism of governance. Consequently, the railway project serves as a historical testament to how technological progress can be co-opted by political power, reinforcing existing inequalities instead of bridging them. Thus, the era's development narrative was less about technical advancement and more about the consolidation of oligarchic influence through state resources.

Finally, the prolonged delay in the construction of the Galați–Bârlad line (1885–1912) exemplifies the subordination of regional economic infrastructure to national military imperatives. While the project was initially conceived as a compensatory measure that would reintegrate Galați into the Moldavian economy following the loss of its free-port status, its execution was obstructed by the strategic reorganization of territorial defense. The prioritization of the Focșani–Nămoloasa–Galați (FNG) fortification line effectively designated Moldavia as a sacrificial buffer zone in order to protect the political center in Wallachia. Ultimately, this case underscores the hegemonic role of military logic in shaping the spatial and economic hierarchy within Romania, reinforcing the centralization of power in Bucharest at the expense of the peripheries' development.

Translated from Romanian by Andrei-Dan Sorescu

Bibliography

Adevărul. "Adevărul la Galați: Întrunirea de la Alcazar." March 28, 1891.

Adevărul. "Moldova." December 3–4, 1890.

Ardeleanu, Constantin. "'Commercial and Political Needs Demand the Establishment of a Black Sea Port': Infrastructure Development, Opportunities and Anxieties in an Eastern European Periphery (1860s–1870s)." *The Journal of Transport History* 44, no. 3 (2023): 436–456. https://doi.org/10.1177/00225266231180284.

Arhivele Naționale Istorie Centrale (ANIC), fond Parlament, dosar 351.

Barițiu, G. *Memorialu pentru cel dintâi drum de feru în România unită*. Brașov: Cu Tipariulu lui Ioanu Gött, 1862.

Botez, Constantin, Dem. Urmă, and Ion Saizu. *Epopeea feroviară românească*. Bucharest: Editura Sport-Turism, 1977.

Buletin statistic general al României 1 (1892), noiembrie-decembrie. Bucharest: Imprimeria Statului, 1893.

Bușe, Constantin. *Comerțul exterior prin Galați sub regimul de port franc (1837–1883)*. Bucharest: Editura Academiei Republicii Socialiste România, 1976.

Chiper, M. "Când Unirea devine o decepţie: Comitetul pentru apărarea intereselor economice şi culturale ale Moldovei (1891)." In *Frământările modernităţii: discursuri, oameni, instituţii*, edited by Marius Chelcu and Dorin Dobrincu, 215–274. Iaşi: Editura Universităţii "Alexandru Ioan Cuza" din Iaşi, 2019.

Chiper, M. "'Valahii ne-au înşelat': Chestiunea compensaţiilor Iaşului în acţiunea politică moldovenească după unire (1862–1874) (1)." *Archiva Moldaviae* 10 (2018): 111–174.

Christodorescu, G. *Economia naţională şi căile noastre ferate*. Constanţa: Tipografia "Ovidiu" H. Vurlis, 1906.

Colescu, Leonida. *Analiza rezultatelor recensământului general al populaţiei României de la 1899*. Bucharest: Institutul Central de Statistică, 1944.

Dezbaterile Adunării Deputaţilor (DAD). No. 19 (1887); No. 12 (1888); No. 11 (1890); No. 31, 39, 44 (1891).

Dezbaterile Corpurilor Legiuitoare (DCL). No. 97 (1882); No. 89 (1885).

Dezbaterile Senatului (DS). No. 39, 40 (1891).

Djuvara, Alexandru G. *Tariful general al drepturilor de vamă: Raport şi discursuri parlamentare*. Bucharest: Litotipografia Carol Göbl, 1892.

Drepturile omului. "—." October 5, 1885.

Edwards, Paul N. "Infrastructure and Modernity: Force, Time, and Social Organization in the History of Sociotechnical Systems." In *Modernity and Technology*, edited by Thomas J. Misa, Philip Brey, and Andrew Feenberg, 185–225. Cambridge: MIT Press, 2003.

Eidelberg, Philip Gabriel. *The Great Rumanian Peasant Revolt of 1907: Origins of a Modern Jacquerie*. Leiden: Brill Archive, 1974.

Ghica, Ion. *Scrieri economice*. Vol. 1. Bucharest: Editura Asociaţiei Generale a Economiştilor din România, 1937.

Ghica, Ion. *Scrieri economice*. Vol. 3. Bucharest: Editura Asociaţiei Generale a Economiştilor din România, 1937.

Ionescu, Dumitru P. "Construirea liniei ferate Iaşi–Ungheni." *AIIXI* 23, no. 1 (1986): 283–291.

Ionescu, Dumitru P. "Primele căi ferate înguste în România." *Revista istorică* 1, no. 7–8 (1990): 743–754.

Ionescu, Dumitru P. "Proiecte şi concesiuni feroviare în vremea lui Cuza Vodă." *AIIX* 31 (1994): 239–272.

Liebowitz, S. J., and Stephen E. Margolis. "Path Dependence, Lock-In, and History." *Journal of Law, Economics, and Organization* 11, no. 1 (1995): 205–226.

Lupta. "Politica la Câmpulung (corespondenţa particulară a *Luptei*)." August 5, 1888.

Mahoney, James, and Daniel Schensul. "Historical Context and Path Dependence." In *The Oxford Handbook of Contextual Political Analysis*, edited by Robert E. Goodin and Charles Tilly, 454–471. Oxford: Oxford University Press, 2006.

Marton, Silvia. "Shades of Dependency and the Discourse on 'Corruption.' Railway Concessions in Romania in the Nineteenth Century." *East European Politics and Societies* 39, no. 3 (2025): 691–716. https://doi.org/10.1177/08883254251352113.

Marton, Silvia, and Sorescu, Andrei. "Introduction: Colonial Anxieties, Corruption Scandals and Xenophobia in Nineteenth-Century Infrastructure Development in Romania." *East European Politics and Societies* 39, no. 3 (2025): 680–690. https://doi.org/10.1177/08883254251352114.

Marţian, D. P. *Opere economice*. Edited by N. Marcu and Z. Ornea. Bucharest: Editura Ştiinţifică, 1961.

Mavrogheni, D. P. *Răspuns adressat de D. P. Mavrogheni la articolul publicat în "Românul" sub titlul "Căile ferate: Partea a V-a."* Bucharest: Typografia Stephan Rassidescu, 1864.

Mănescu, Const. C. *Istoricul căilor fierate din România*. Vol. 1. Bucharest: Ed. Direcţiunii Generale a Căilor Fierate Române, 1906.

Marin, Irina. *Peasant Violence and Antisemitism in Early Twentieth-Century Eastern Europe*. Cham: Palgrave Macmillan, 2018.

Monitorul Oficial al României (MOf). No. 113, 115, 116, 117, 118, 120, 122 (1868); No. 47 (1882).

Negruţi-Munteanu, Ecaterina. "Dezvoltarea agriculturii în Moldova între anii 1848–1864." In *Dezvoltarea economiei Moldovei 1848–1864*, edited by V. Popovici, C. C. Anghelescu, L. Boicu, N. Corivan, E. Negruţi-Munteanu, and C. Turcu. Bucharest: Editura Academiei Republicii Socialiste România, 1963, 9–167.

Negruţi-Munteanu, Ecaterina, "Rolul evreilor în formarea târgurilor din Moldova în prima jumătate a secolului al XIX-lea." *Studia et Acta Historiae Iudaerum Romaniae* 1 (1996): 139–151.

Pastia, Sc. *Un răspuns unor târâtoare veninoase*. Iași: Tipolitografia H. Göldner, 1890.

Patria. "Jocul tarifelor de căi ferate." July 14, 1885.

Patria. "—." October 6, 1885.

Păltănea, Paul. *Istoria orașului Galaţi: De la origini până la 1918: Partea a II-a*. Galaţi: Editura "Porto-Franco," 1995.

Plesnilă, Ion. *Istoricul liniei Galaţi–Bârlad*. 1898.

Popescu, Toader. *Proiectul feroviar românesc (1842–1916)*. Bucharest: Simetria, 2014.

Popescu, Toader. "Un aller simple pour la modernité." *New Europe College Yearbook* (2009–2010): 273–305.

Porter, Theodore M. *Trust in Numbers: The Pursuit of Objectivity in Science and Public Life*. Princeton: Princeton University Press, 1995.

Puffert, Douglas J. "Path Dependence in Spatial Networks: The Standardization of Railway Track Gauge." *Exploration in Economic History* 39, no. 3 (2002): 282–314. https://doi.org/10.1006/exeh.2002.0786.

Puffert, Douglas J. "The Standardization of Track Gauge on North American Railways, 1830–1890." *Journal of Economic History* 60, no. 4 (2000): 933–960. https://doi.org/10.1017/S0022050700026322.

Românul. "Drumurile de fer in România." March 20–21, 1867.

Sava, Eduard. *Istoricul căilor ferate român: Perioada dintre anii 1885–1902.* Bucharest: Tipografia G. A. Lăzăreanu, 1904.

Schram, Albert. *Railways and the Formation of the Italian State in the Nineteenth Century.* New York: Cambridge University Press, 1997.

Slăvescu, Victor. *Vechi proiecte de căi ferate în Moldova 1846–1851.* Bucharest, 1942.

Stan, Apostol. *Agricultura românească în faza finală a clăcășiei 1831–1864.* Bucharest: Editura Academiei Române, 1994.

Țincu, Bucur. "Contribuții la istoria căilor ferate din România: Probleme în perioada 1859–1869." *Studii. Revistă de istorie* 24, no. 5 (1975): 951–962.

Țurcănașu, George. *Evoluția și starea actuală a sistemului de așezări din Moldova.* Iași: Casa Editorială Demiurg, 2006.

Traseul direct al liniei ferate Galați–Bârlad propus de George Matei Cantacuzino. Galați: Tipolitografia G. Bălășescu, 1891.

Turnock, David. "Railways and Economic Development in Romania before 1918." *Journal of Transport Geography* 9, no. 2 (2001): 137–150. https://doi.org/10.1016/S0966-6923(00)00046-6.

Vasiliu, D. I. *Studii și documente în legătură cu fortificațiile române din secolul al 19-lea.* Vol. 1. Bucharest: Tipografia "Revistei Geniului," 1934.

Văratic, Vitalie. "Jibrieni: proiectul primului port românesc la Marea Neagră în a doua jumătate a secolului al XIX-lea." In *160 de ani de la Unirea Principatelor,* edited by Petronel Zahariuc and Adrian-Bogdan Ceobanu, 539–563. Iași: Editura Universității "Alexandru Ioan Cuza" din Iași, 2020.

About the Author

Mihai Chiper is a researcher of the social and economic history of Moldavia at the "A. D. Xenopol" Institute of History of the Romanian Academy. His interests include internal Orientalism, regional disparities, and centralization. His most recent publication is the chapter "Moldavia and Oltenia after the Union: Symbolic Geographies, Historical Imagination, and Regional Prestige" in the collection *Națiuni, naționalism și perspective interetnice în Transilvania* [Nations, Nationalism, and Interethnic Perspectives in Transylvania], edited by Constantin Bărbulescu et al., 301–316. (Cluj-Napoca: Mega, 2025).

10. Conduits of Violence: Railway Infrastructure and Student Anti-Semitism in Interwar Romania

Raul Cârstocea

Abstract: Railway infrastructure played a vital role in connecting the different provinces of Greater Romania while also revealing a potentially surprising function as a prominent conduit of violence. As anti-Semitism mutated after World War I from parliamentary rhetoric to grassroots violence against Jews, student organizations were often at the forefront of the latter. Consequently, as students primarily traveled by train (an aspect facilitated by state-sponsored discounted fares), trains and railway stations increasingly became sites of anti-Semitic violence. Drawing on police reports from the 1920s attesting to both individual attacks against Jews on trains and to organized violence in and around railway stations, this chapter argues that railway infrastructure played a prominent role in both the imagination and actions of Romania's interwar (ultra-)nationalists.

Keywords: Railways, Anti-Semitic violence, Student activism, Fascism, Ultra-nationalism, Interwar Greater Romania

Introduction

The building of the Romanian railways was one of the most significant infrastructure projects the state undertook in the late nineteenth century. Borne out of a broader modernization process and initially enjoying quasi-unanimous support in Parliament, the Romanian railway project was pivotal to building what Toader Popescu has termed "'nation-space,' a

Marton, Silvia, and Andrei-Dan Sorescu, eds. *Infrastructural Development, Corruption, Xenophobia, and Colonization in Central and Southeastern Europe*. Amsterdam: Amsterdam University Press/Central European University Press, 2026.

https://doi.org/10.5117/9789048577279-11

territorial reflection of the nation-state."[1] As shown in earlier scholarship as well as by previous research undertaken within the present project, the debates surrounding it were markedly entangled with the emergence of a "native" version of modern anti-Semitism in Romania, synchronous with similar developments abroad while also characterized by local specificities.[2] Among these, the scandal of the so-called Strousberg affair witnessed a further imbrication of anti-Semitic discourses not just with infrastructure development but also with allegations of "corruption" and colonial anxieties related to both a feared German encroachment (where Jews were seen to act as a potential "vanguard") and Jewish "colonization" proper.[3]

Once established, this nexus continued into the very different context of the significantly expanded Romanian state after World War I, while also witnessing mutations that saw anti-Semitism transformed from parliamentary rhetoric to street violence against Jews. At the forefront of the latter were student organizations, rising to prominence against the background of yet another major state project, that of expanding higher education for the purposes of nationalizing a suddenly very diverse country that had doubled its territory and population, while the number of national minorities within its borders quadrupled.[4] In this context, railway infrastructure at once

1 Toader Popescu, *Proiectul feroviar românesc, 1842–1916* (Bucharest: Simetria, 2014), 19.

2 See, for example, Fritz Stern, *Gold and Iron: Bismarck, Bleichröder, and the Building of the German Empire* (London: Allen & Unwin, 1977), 351–393; Silvia Marton, "'La Belgique de l'Orient' et les chemins de fer: les raisons d'une comparaison," *Studia Politica: Romanian Political Science Review* 8, no. 1 (2008): 27–44; Silvia Marton, "De la construction de l'Etat au racisme: judéophobie et antisémitisme en Roumanie avant la Grande Guerre," in *Antisemitismus im 19. Jahrhundert aus internationaler Perspektive / Nineteenth Century Anti-Semitism in International Perspective*, ed. Mareike König and Oliver Schulz (Göttingen: Vandenhoeck & Ruprecht, 2019), 203–225; Constantin Iordachi, *Liberalism, Constitutional Nationalism, and Minorities: The Making of Romanian Citizenship, c. 1750–1918* (Leiden: Brill, 2019), 306–351. More generally, on how Romanian anti-Semitism first became an issue of international concern in the nineteenth century, see Carole Fink, *Defending the Rights of Others: The Great Powers, the Jews, and International Minority Protection, 1878–1938* (Cambridge: Cambridge University Press, 2004), and Lisa Moses Leff, "Liberalism and Antisemitism: A Reassessment from the Peripheries," in *Jews, Liberalism, Antisemitism: A Global History*, ed. Abigail Green and Simon Levis Sullam (Cham: Palgrave Macmillan, 2020), 23–46.

3 Silvia Marton and Andrei Sorescu, eds., "Colonial Anxieties, Corruption Scandals and Xenophobia in Nineteenth-Century Infrastructure Development in Romania," *East European Politics and Societies* 39, no. 3 (2025): 680–690.

4 With the annexation of Transylvania and Bukovina, two universities (in Cluj and Cernăuţi) were added to the existing ones in Bucharest and Iaşi, and the number of students increased from 8,632 in 1913–1914 to 22,379 in 1924–1925 and 37,314 in 1929–1930. Irina Livezeanu, *Cultural Politics in Greater Romania: Regionalism, Nation-Building, and Ethnic Struggle, 1918–1930* (Ithaca: Cornell University Press, 1995); Raul Cârstocea, "Students Don the Green Shirt: The Roots

proved its vital role in connecting the different provinces of a still loosely integrated Greater Romania and revealed a potentially surprising function as a prominent conduit of violence. As students primarily traveled by train (an aspect facilitated by state-sponsored discounted fares or even free travel), trains and railway stations increasingly became sites of anti-Semitic violence, especially when student delegations would attend congresses at other universities or other nationalist manifestations. Numerous police reports from the 1920s attest to both individual attacks against Jews on trains and to organized violence in and around railway stations, frequently used as meeting points by traveling students. As my chapter argues, railway infrastructure was the main conduit of mobility for anti-Semitic student violence, facilitating both its spread and escalation into ever-more-serious incidents that came to be seen as a threat to public order. The chapter will first address some of the early incidents of anti-Semitic violence occurring on trains and at or around railway stations to then discuss its escalation in 1927, especially in connection with the students' travel to and from the student congress at Oradea Mare. These are followed by some general conclusions regarding the importance of railway infrastructure for an emerging interwar Romanian far right.

Railway Infrastructure and Student Anti-Semitic Violence

If the nineteenth century had witnessed the development of a native version of modern anti-Semitism in Romania, whereby anti-Jewish rhetoric was doubled by legal exclusion from citizenship and accompanied by occasional pogroms, the end of the World War I brought about the long-delayed Jewish emancipation. This was formalized in the Constitution of 1923, making Romania the last country in Europe to emancipate its Jewish minority.[5] Reactions to Jewish emancipation antedated the voting of the new constitution, with student organizations coming to the fore of a new, grassroots anti-Semitic movement. At the first student congress in interwar

of Romanian Fascism in the Anti-Semitic Student Movements of the 1920s," in *Alma Mater Antisemitica: Akademisches Milieu, Juden und Antisemitismus an den Universitäten Europas zwischen 1918 und 1939*, ed. Regina Fritz, Grzegorz Rossoliński-Liebe, and Jana Starek (Vienna: New Academic Press, 2016), 39–66 (here, 49).

5 Andrei-Dan Sorescu and Raul Cârstocea, "The (Great) Numbers Game: Demographic Anxieties and Quotas in Nineteenth- and Twentieth-Century Romania and the Global Antisemitic Imaginary," in *Quotas: The "Jewish Question" and Higher Education in Central Europe, 1880–1945*, ed. Michael Miller and Judith Szapor (New York: Berghahn, 2024), 67–109 (here, 80).

Romania, held in Cluj, on September 4–6, 1920, Jews were excluded from student societies following the proposal of a relatively small but radical group of nationalist students from the universities of Iaşi and Cernăuţi led by Corneliu Zelea Codreanu (1899–1938), the future founder and leader of Romania's fascist movement.[6] Escalating calls for a *numerus clausus* at Romanian universities, where Jews were the most visibly over-represented national minority, especially in certain faculties such as law or medicine,[7] eventually led to a general strike of university students and the establishment on December 10, 1922 of the National Union of Christian Students from Romania (*Uniunea Naţională a Studenţilor Creştini din România*, UNSCR), a nation-wide anti-Semitic student organization.[8] This was followed by the establishment on March 4, 1923, of the League of National-Christian Defense (*Liga Apărării Naţional-Creştine*, LANC), led by A. C. Cuza (1857–1947)—a veteran of the first Romanian political party with an explicitly anti-Semitic platform, the Democratic Nationalist Party, which he had co-led before the war with historian Nicolae Iorga (1871–1940)—and having Codreanu as the national leader of its youth section. In time, the leaderships of UNSCR and LANC Youth came to overlap to a significant degree.[9] With the establishment of LANC, the press in Romania wrote of "the birth of Romanian fascism," noting also that "it would seem that fascism and anti-Semitism in our country have the same origin."[10] However, it should be noted that this pattern was not specific to Romania—as Ezra Mendelsohn notes in his sweeping comparative study, "universities all over East Central Europe were centers of anti-Semitism" after World War I.[11]

Anti-Semitic street violence was one of the markers of student politics, drawing the attention of the state authorities, which struggled to contain it. It was eventually this context, of student violence being met with police brutality, that led to Codreanu's murder of the Iaşi police prefect, Constantin Manciu, on October 25, 1924, leading to a highly publicized trial that

6 Corneliu Zelea Codreanu, *Pentru legionari* (Sibiu: Totul pentru Ţară, 1936), 40; Cârstocea, "Students Don the Green Shirt," 51–52.

7 Livezeanu, *Cultural Politics*, 238, 259; Lucian Nastasă, ed., *Antisemitismul universitar în România (1919–1939)* (Cluj-Napoca: Kriterion, 2011), 68–69.

8 Cârstocea, "Students Don the Green Shirt," 53.

9 Cârstocea, "Students Don the Green Shirt," 54–55.

10 Constantin Mille, "Fascismul român," *Pressa*, June 14, 1923.

11 Ezra Mendelsohn, *The Jews of Central Europe* (Bloomington: Indiana University Press, 1983), 188. On the politics of anti-Semitism in an academic context, see, for instance, Linda Erker, *Die Universität Wien im Austrofaschismus: Österreichische Hochschulpolitik 1933 bis 1938, ihre Vorbedingungen und langfristigen Nachwirkungen* (Göttingen: V&R Unipress, 2021).

propelled him into the position of national leader of the student movement.[12] His acquittal on May 28, 1925, by a jury whose members "wore swastikas in [Romanian] tricolor bands on their lapels" is also one of the earliest instances when student anti-Semitism came to travel along the routes of the Romanian railways.[13] As a special celebratory train had been arranged by A. C. Cuza for Codreanu's triumphal return from the court in Turnu-Severin, in southwestern Romania, to Iaşi, in northeastern Romania, it stopped in most stations it passed, including the capital, Bucharest, being greeted everywhere with enthusiastic student demonstrations. Newspapers reported that at the small station of Slatina, over 5,000 people waited for the train—whose "locomotive was decorated with tricolor flags and greenery," with Codreanu and Ion I. Moţa (1902–1937), his future lieutenant in the Legion of the Archangel Michael movement, standing on its front platform—at the railway station.[14] In Bucharest, the jubilant students clashed with the army battalions mobilized for the occasion, who were trying to cordon off the area around the station; their attempt to break through the army lines resulted in a fight in the course of which several students were wounded.[15] En route to Iaşi, the students' enthusiasm degenerated into anti-Semitic violence: at the railway station in Buzău, where the train arrived at 1:15 a.m. on the night of May 28–29, students carried Codreanu on their shoulders while shouting "Down with the Jews!," destroyed the advertising panels of left-leaning newspapers long identified as "Jewish" (*Dimineaţa* [The Morning], *Adevărul* [The Truth], and *Lupta* [The Struggle]) on station platforms, and vandalized several shops belonging to Jews, breaking their windows.[16] The time of this incident, in the middle of the night, probably explains why there was no meaningful police presence to prevent the violence at the station. At the same time, it is also proof of the commitment and enthusiasm of the nationalist students, who were present in large numbers despite the ungodly hour. At the arrival of the train at its final destination, in Iaşi, Codreanu and the students were escorted by units of the police, gendarmerie, and army, and the demonstrations remained peaceful; additionally, a railway car with gendarmes had been attached to the train at Chitila way station.[17]

12 Cârstocea, "Students Don the Green Shirt," 55–58.

13 Leontin Iliescu and N. Ciocârdia, "Procesul Corneliu Zelea Codreanu: Achitarea," *Universul*, May 28, 1925.

14 "Manifestaţii studenţeşti la Slatina, Piteşti şi în Capitală," *Universul*, May 31, 1925.

15 "Manifestaţiile de eri," *Adeverul*, May 30, 1925.

16 Arhiva Consiliului Naţional pentru Studierea Arhivelor Securităţii (ACNSAS), Fond Documentar, File D 10767, vol. 2, 58.

17 "Manifestaţii studenţeşti"; "Manifestaţiile de eri."

A spate of violent incidents taking place on trains or at railway stations followed, as Codreanu, Moţa, and other students continued their propaganda tour following the acquittal. During their return to Iaşi from a visit to Ungheni, the students threw two 17-year-old Jewish girls, Marla Wenger and Malca Riner, off the train, which was subsequently stopped and turned back to recover the victims, who were hospitalized.[18] The following day, during the return of A. C. Cuza, Corneliu Şumuleanu, Ion Zelea Codreanu, Corneliu Zelea Codreanu, "and other anti-Semitic leaders" from another visit, "anti-Semitic excesses occurred at [Tecuci] station, provoked by the pupils of the high school in that town, joined by numerous other persons. Four Jews who were in the station—Segal, Marcu Lupu, and Scrutman from Tecuci, and Herşcovici from Focşani—were beaten up and wounded, while the majority [of others] escaped unharmed, managing to flee."[19] Once again, the attack occurred at night, at 3:30 a.m. The same report further noted that "anti-Semitic disturbances will commence with more acute intensity than up to now, due to the propaganda that was and is being made in schools and on trains," citing a previous report to this effect.[20] It concluded that the disinterest of the personnel of the Romanian Railways (*Căile Ferate Române*, CFR) in preventing such incidents was partly responsible for them, as was that of the school teachers.

It should be noted at this point that the Romanian railway project was no longer the uncontested, prized marker of Romania's modernity that had been supported by the majority of the Romanian parliament in the mid-nineteenth century. Already by the turn of the century, A. C. Cuza—a one-time socialist[21] turned would-be leader of an emerging radical right—was critical of a project that he thought "cost enormous amounts of money, extorted mostly from the work of our peasants, [and] profited mostly the ruling classes, landowners, and leaseholders for transporting bread, the rich for more confort in the long journeys they undertake, while causing the peasants a double prejudice, both material and moral."[22] This was not only a critique of the railway project—an instance of "infrastructure shaming" on the broadest scale, questioning its very usefulness. It was equally articulated through the consistent populist dichotomy between a putatively corrupt, alienated ruling

18 ACNSAS, Fond Documentar, File D 10767, vol. 2, 43–44; Cor., "Sălbăteciile dela Iaşi," *Dimineaţa*, June 3, 1925; "Cum s'a produs 'accidentul' de la Iaşi," *Dimineaţa*, June 4, 1925.

19 ACNSAS, Fond Documentar, File D 10767, vol. 2, 45.

20 ACNSAS, Fond Documentar, File D 10767, vol. 2, 45.

21 Horia Bozdoghină, *Antisemitismul lui A. C. Cuza în politica românească* (Bucharest: Curtea Veche, 2012), 21–23.

22 A. C. Cuza, *Ţăranii şi clasele dirigente* (Iaşi: Tipografia Naţională, 1895), XXXIII.

elite and an idealized peasant seen as the "authentic" representative of the nation. This was consonant with a more general contestation of Romania's modernization, undertaken initially through emulation of the "Western," liberal model.[23] In turn, as I have written elsewhere, this contention was by no means specific to Romania or to "the Balkans" as a region but was rather synchronous with similar contestations of the liberal model in "the West," just as Cuza's politics were synchronous with the emergence of a European radical right.[24] Now, universal suffrage, increased literacy, and interconnectivity made for a broader and ever more politicized public that was ready to hold the political class accountable and ascribe blame for the perceived disregard of the public interest.[25] Specifically, by the 1920s, with train becoming accidents a frequent occurrence and blamed on both faulty infrastructure and insufficient personnel, criticism of the CFR was widespread, while the rhetoric of alienated, corrupt elites being harmful to the nation and "in the service of Jews" had become a staple feature of anti-Semitic discourse as a distinct means of "Othering" corruption.[26]

Following his acquittal, Codreanu capitalized on his popularity by organizing his wedding with Elena Ilinoiu on June 14, 1925, a grandiose event held in the town of Focşani and attended by a crowd of 80,000 to 100,000 people.[27] This occasion saw the mobilization not just of regular police and gendarmerie units but of the *Siguranţa*, the Romanian state's secret police, given they had intercepted LANC communiques demanding that "every anti-Semitic organization from every part of the country should send as many delegates as possible to this ceremony, as the festivity will be accompanied by an anti-Semitic demonstration in the town."[28] Trains and railway stations were of particular concern, as a ciphered telegram sent by

23 Diana Mishkova and Roumen Daskalov, "'Forms without Substance': Debates on the Transfer of Western Models to the Balkans," in *Entangled Histories of the Balkans*, vol. 2, *Transfer of Political Ideologies and Institutions*, ed. Roumen Daskalov and Tchavdar Marinov (Leiden: Brill, 2014), 1–97.

24 Raul Cârstocea, "Synchronous Nationalisms: Reading the History of Nationalism in South-Eastern Europe between and beyond the Binaries," *National Identities* 24, no. 5 (2022): 481–503.

25 See also Silvia Marton, "Transportation Infrastructure and Corruption Scandals" in this volume.

26 Raul Cârstocea, "Anti-Semitism at the Intersection of Corruption and Colonialism: Continuities of Political Rhetoric in Romania from the Nineteenth Century to the Interwar Period," *East European Politics and Societies* 39, no. 3 (2025): 762–83.

27 It is worth noting that according to the 1930 census, Focşani had a population of 32,481, of which 26,407 persons declared their nationality as Romanian and 4243 as Jewish. Sabin Manuilă, ed., *Recensământul general al populaţiei României din 29 decemvrie 1930*, vol. 2, *Neam, Limbă Maternă, Religie* (Bucharest: Imprimeria Naţională, 1940), 351.

28 ACNSAS, Fond Urmărire Penală, File P 11784 "Corneliu Zelea Codreanu," vol. 2, 18.

the central directorate of the *Siguranţa* to its local branches and the local police of all the towns on the routes to Focşani indicates. Fearing that delegates coming to the wedding "will engage in acts of violence against Jews during their journey on trains," it ordered these local chapters to investigate by questioning both train conductors and passengers if such incidents occurred while summoning special mobile gendarme units active on trains and station platforms to prevent them.[29] Even as the local Jewish population of Focşani was extremely worried, with many of them planning to leave town for the occasion, and as the left-leaning press feared a "bloodbath," the massive presence of the authorities eventually prevented any violence.[30] The pro-government daily *Universul* denounced such fears as unfounded, citing even the former president of the local chapter of the Union of Native Jews and the leader of the local Jewish community to this effect, while describing the wedding, where the procession was led "by a giant swastika made of greenery and flowers," as "a wedding according to ancestral customs."[31] In turn, the left-leaning press retorted by accusing the government of shielding the nationalists and of preparing "for Mr. Zelea Codreanu, for anti-Semitism, for fascism, its past and future ally, a demonstration that will take place under the protection of bayonets and gendarmes."[32]

The position of the authorities toward the student movement was indeed ambiguous. On the one hand, there is considerable evidence of consistent efforts to prevent violent anti-Semitic incidents associated with the railway infrastructure. Following up on the allegations of anti-Semitic propaganda being distributed by LANC on trains, on the Iaşi–Câmpulung and Vatra Dornei lines, a *Siguranţa* report ordered its immediate confiscation.[33] This was by no means an isolated occurrence: another police report noted in January 1927 the distribution and sale by LANC members of anti-Semitic postcards at Ploieşti station, enclosing the confiscated materials. These are also interesting due to the fact that the postcard in question depicted an alleged scene from the "horrible Jewish crime at Tisza-Eslar [sic!] in Hungary in 1882," a notorious nineteenth-century blood libel trial from neighboring Austria-Hungary that had concluded with the acquittal of the

29 ACNSAS, Fond Urmărire Penală, File P 11784 "Corneliu Zelea Codreanu," vol. 2, 22; U., "Nunta dela Focşani," *Adeverul*, June 12, 1925.

30 ACNSAS, Fond Urmărire Penală, File P 11784 "Corneliu Zelea Codreanu," vol. 2, 19. "Ce se petrece la Focşani," *Dimineaţa*, June 13, 1925.

31 "Nunta lui Corneliu Zelea Codreanu," *Universul*, June 17, 1925.

32 "Nunta dela Focşani."

33 Arhivele Naţionale Istorice Centrale (ANIC), Fond Ministerul de Interne, Diverse, File 1/1925, 4.

accused, but not before prompting a wave of anti-Semitism that apparently still reverberated decades after being debunked, in interwar Romania.[34]

The gravity of such incidents and the critical nature of the railway infrastructure are also proven by the extensive preparations made by the authorities ahead of student congresses. Regarding the "criticality" of certain infrastructures, as Jens Ivo Engels argues in his chapter in this volume, "infrastructure is also 'critical' to the state itself: if it fails or is threatened, part of the state's *raison d'être* is called into question."[35] This was certainly true for a relatively "young" state that had also recently expanded its territory and was still striving to prove its "standard of civilization" internationally. Preventing anti-Semitic violence was itself a critical component of such attempts.

Anticipating the congress that was to be held in Iaşi on November 28–30, 1926, a ciphered telegram signed by the minister of the interior and addressed to all police prefects, all *Siguranţa* brigades, and all city and railway station police units in the country warned about the very high number of students traveling in special cars and ordered them to take the appropriate measures to prevent "excesses and aggressions against Jewish passengers that will be either on trains or at railway stations on the occasion of the passing of the participants to the congress, both on their way to Iaşi and during their return."[36] Despite these preparations, numerous violent incidents occurred, as the students traveling to the congress "mistreated all the Jewish passengers on trains, forcing some to get off the trains and preventing others from getting on them." On their return, a violent attack took place at Bârlad station, where the students "invaded the station platform and restaurant, chasing away all the Jews."[37]

However, the same report blamed the inability of the local police to prevent the violence not only on the very large numbers of students and

34 ANIC, Fond Ministerul de Interne, Diverse, File 9/1927, 54–55. For some background on the Tiszaeszlár blood libel trial, see Miloslav Szabó, "Zentrum oder Peripherie? Der Antisemitismus in Ungarn um 1900," in *Modern Antisemitisms in the Peripheries: Europe and its Colonies, 1880–1945*, ed. Raul Cârstocea and Éva Kovács (Vienna: New Academic Press, 2019), 371–389; Daniel Véri, "The Tiszaeszlár Blood Libel: Image and Propaganda," in *Antisemitismus im 19. Jahrhundert aus internationaler Perspektive: Nineteenth Century Anti-Semitism in International Perspective*, ed. Mareike König and Oliver Schulz (Göttingen: Vandenhoeck & Ruprecht, 2019), 263–292; Hillel J. Kieval, *Blood Inscriptions: Science, Modernity, and Ritual Murder at Europe's Fin de Siècle* (Philadelphia: University of Pennsylvania Press, 2022).

35 Jens Ivo Engels, "Writing the History of Critical Infrastructures: Some Conceptual Remarks" in this volume.

36 ACNSAS, Fond Documentar, File D 10767, vol. 1, 62.

37 ACNSAS, Fond Documentar, File D 10767, vol. 2, 185.

the insufficient numbers of police agents but also on "the imprudence of some of the foreign travelers."[38] This formulation presents us not only with a *topos* familiar from nineteenth-century Romania, where Jews were legally "foreigners"[39] (although this was no longer the case in 1926), but also with a pattern of blaming the victims, which is not altogether uncommon in police reports on anti-Semitic student violence. An extensive synthesis report on the violence that occurred in trains and at railway stations in the aftermath of the student congress of November 28–30, 1926, when the students traveled to Bessarabia on a propaganda tour, is replete with statements indicative of the anti-Semitism of the authorities. While the students were "singing patriotic songs and in their elation were far from any aggressive intervention," "*numerous Jews though—the majority of them young—members of the Maccaby* [sic!] *society and of the Jewish Cultural League, knowing from the newspapers that the students are coming, came out in groups at intermediary stations, having a provocative attitude.*"[40] The author of the report went on to describe these two Jewish organizations as follows: "The *Maccabeyi* [sic!], as a sporting association, is preparing Jewish youth for resistance, agitation, and even combat against the Christian population, [and] truly represents the armed power of the Jews, while the *Jewish Cultural League*, under the guise of an institution of culture, hides subversive ideas, plans, and actions."[41] Following their exclusion from student societies, Jewish students had indeed formed their own, and these were engaged in resistance against UNSCR and LANC violence, frequently clashing with the latter. Nevertheless, the direct association of these Jewish associations with "terrorist and communist" groups—including some that had allegedly crossed the border from the Soviet Union over the Dniester River bringing "revolvers and cartridges, destined for the subversive Jewish elements in Chişinău who were preparing

38 ACNSAS, Fond Documentar, File D 10767, vol. 2, 185.

39 The designation of Jews as "foreigners" applied in nineteenth-century Romania both to native Jews who were barred from citizenship, and thus effectively stateless, and to "subjects of a foreign country," which held another citizenship. Raul Cârstocea, "Uneasy Twins? The Entangled Histories of Jewish Emancipation and Anti-Semitism in Romania and Hungary, 1866–1913," *Slovo* 21, no. 2 (2009): 64–85; Iordachi, *Liberalism, Constitutional Nationalism, and Minorities*, 265–399.

40 ACNSAS, Fond Documentar, File D 10767, vol. 1, 120 (All emphases are in the original unless otherwise indicated). The Maccabi were the main Jewish sporting association in Romania. See Diego-Maricel Ciobotaru, "Jewish Cultural and Sporting Associations between "Romanianization" and "Unionization": Episodes of Antisemitic Violence in the Stadiums of the Capital and the Provinces before and after World War II (1922–1948)," *Holocaust. Studii şi cercetări* 14, no. 1 (2022): 87–124.

41 ACNSAS, Fond Documentar, File D 10767, vol. 1, 120.

for eventual clashes with the students"—leaves no doubt as to the authorities' political views.[42] When siding with the nationalist students, invoking the familiar *topos* of Judeo-Bolshevism and the Romanian state's geopolitical and ideological archenemy, the Soviet Union, represented a convenient way for anti-Semitic police staff to frame reports to their superiors.

The report in question does mention several instances of violence carried out by Jewish students against their Christian counterparts, and it would be problematic to dismiss these outright. A quick search through the digitized collection of the foremost and most long-lived Jewish periodical at the time, *Egalitatea* [Equality], reveals that out of 131 occurrences of the keyword *trenuri* [trains] for the period between 1919 and 1936, 123 explicitly refer to trains as loci of violence and anti-Semitic propaganda. Even without a closer reading, this suggests how the Jewish experience of railway travel in interwar Romania was perceived by the community as a likely dangerous one, and how the community mouthpiece was in no doubt that trains were indeed a privileged location for anti-Semitic violence, both physical and symbolic.[43] It is therefore likely that, in this context, Jewish student organizations did occasionally retaliate, although this was quite clearly an uneven fight, with anti-Semitic students having not only the advantage of numbers but also the occasional support of the authorities. At the passage of the train through Călăraşi station, "students were booed by members of the Jewish Cultural League, who were on the platform; *the student Ştefănescu Gheorghe, the president of his theologian colleagues from Chişinău, was hit with a cane, at the departure of the train from the station, by the Jew Nuhim Schwartz, while other young Jews threw rocks at the cars occupied by the students*."[44] The report mentions numerous other instances of nationalist students being attacked by groups of Jewish youth at railway stations, one such incident resulting in the wounding of their leader, Ilie Gârneaţă, who suffered a head injury. "At the passing of the train through Visterniceni station, shots were fired by a group of young Jews at the congress students' train, and rocks were thrown, breaking the window of the restaurant car."[45] The nationalist students retaliated, breaking the windows of shops near the railway station

42 ACNSAS, Fond Documentar, File D 10767, vol. 1, 120–121.

43 Compiled on the basis of: https://www.nli.org.il/en/newspapers/egalit

44 ACNSAS, Fond Documentar, File D 10767, vol. 1, 121.

45 ACNSAS, Fond Documentar, File D 10767, vol. 1, 124–125. It is worth noting that, when confronted with widespread anti-Semitic violence, many young Jews in interwar Romania—and Europe more generally—turned to sporting activities as a means of cultivating their capacity for physical self-defense. See Michael Brenner and Gideon Reuveni, eds., *Emancipation through Muscles: Jews and Sports in Europe* (Lincoln: University of Nebraska Press, 2006).

and attacking young Jews. The synthesis report concludes by squarely placing the blame on Jewish provocations, naming the persons who had been caught and indicted—all Jewish—and underlining that, outside of the violence associated with the train, "there has been no incident in any town in Bessarabia or Moldavia, outside of Călăraşi, an eminently Jewish city, where the students had been molested and provoked in the railway station by the Jewish league members and the Maccabeyi [sic!]."[46] Another police report from the same town, Călăraşi, in January 1927 provides an isolated case of an attack undertaken by four members of a Zionist organization against Christian passengers on a train, as well as against the station master, following the latter's intervention. In this case as well, and unlike in many of the cases of anti-Semitic student violence detailed above, the report mentions the arrest and indictment of the culprits, yet another indication that police authorities were much more likely to pursue legal action against Jewish students than against their Romanian nationalist counterparts.[47]

The Student Congress and Pogrom at Oradea Mare (1927)

The year 1927 witnessed a split within LANC, followed by Corneliu Zelea Codreanu's decision to establish his own political organization, the Legion of the Archangel Michael. This would develop into Romania's native fascist movement, which eventually grew to surpass LANC in popularity, especially among student circles. There had been disagreements between Cuza and Codreanu ever since the beginning of their joint activity within LANC, with the disputes being over organization and strategy rather than ideology. Cuza, a member of Parliament since the late nineteenth century, wanted LANC to function as a political party and aimed at a reversal of Jewish emancipation through legal means. Codreanu, born in 1899, was inspired instead by the new politics of Mussolini's fascism, despised parliamentary politics, and wanted to organize the student movement as a disciplined "army" privileging action over ideas and based on unconditional obedience to its leader.[48] Writing from Grenoble on January 21, 1926, while pursuing his doctoral studies, Codreanu sent Cuza an unpublished letter acknowledging the latter's merits as a "theoretician" of the "national idea." However, he also accused Cuza of being incapable of organizing a "national movement" for

46 ACNSAS, Fond Documentar, File D 10767, v ol. 1, 125–128.

47 ACNSAS, Fond Documentar, File D 10767, vol. 2, 334.

48 Codreanu, *Pentru legionari*, 124.

concrete action, even claiming the original initiative behind LANC was his—Codreanu's.[49] His declared preference for a "combat organization" grounded in "integral discipline" and engaged in "the world of action, not of ideas, but of deeds" was to serve as the basis of the Legion, distinguishing it from LANC.[50] The difference was not lost on its members, with one Constantin Dănilă, a member of both LANC and the Legion, noting in a letter that "A. [C.] Cuza, the supreme president of LANC, wants the national defense [organized] on party lines, to be imposed through projected laws, which he hopes that, once in power, will get voted"; the Legion has a different purpose—"attack, 'Ala fascismus marcia su Roma' [*sic!*], making sure the country is saved, *depoliticianized*."[51]

Politicianism was the legionary movement's main conceptual antagonist. Having started as a critique of the endemic corruption in Romanian politics, the concept eventually developed into a wholesale rejection of parliamentarism and democracy, seen as a system that inevitably bred corruption through the opportunities and incentives it provided to rent-seeking elites holding temporary, potentially nonrenewable office.[52] As an "-ism" that was able to subsume and accommodate disparate semantic content, *politicianism* was also inextricably linked to the movement's anti-Semitism. In Codreanu's words, "money, the press, and votes decide over life and death in democracy. Jews have all these, and, through them, the Romanian political parties become simple tools in the hands of Judaic power."[53] Party politics was thus subject to wholesale condemnation—despite the paradox that legionaries eventually participated in elections with a political party of their own. Consistent with the legionary redefinition of political "questions" as ontological, *politicianism* was eventually associated with an entire human typology, seen as responsible for "creating this filth, this moral scum: the politician who no longer has anything in common with the nobility of our race, who dishonors us and kills us [...]. Of all the evils that the Jewish invasion has brought us, this is the most terrible!"[54] In accordance with

49 ACNSAS, Fond Urmărire Penală, File P 11784 "Corneliu Zelea Codreanu," vol. 18, 173–175. "Before LANC, you had other organizations that crumbled; it is true, for many reasons, but also for this one: the lack of any sense of organization on your part," 174.

50 ACNSAS, Fond Urmărire Penală, File P 11784 "Corneliu Zelea Codreanu," vol. 18, 174–175.

51 ANIC, Fond Ministerul Justiţiei, Direcţia Judiciară, File 111/1930–1932, Vol. 2, 684.

52 For corruption in pre–1914 electoral politics, see Silvia Marton, "La 'corrupción' electoral en Rumania: Los comienzos titubeantes de la democracia," *Ayer: Revista de Historia Contemporánea* 115, no. 3 (2019): 77–104.

53 Codreanu, *Pentru legionari*, 156.

54 Codreanu, *Pentru legionari*, 306. For an excellent conceptual history of "questions" as a form of nineteenth- and early twentieth-century politics, see Holly Case, *The Age of Questions:*

this ontological status, "corruption" itself became much more fluid and pluri-semantic, being understood not just as an electoral phenomenon but as moral, simultaneously individual and systemic degeneration. This, in turn, rendered the concept into a veritable nexus for all of the movement's antagonists and into the prime motivation for engaging in what the legionaries alternatively described as apolitical or anti-political politics.[55]

The establishment of a fascist movement explicitly committed to "action" and "combat" was bad news in the context of escalating anti-Semitic violence in 1927 Romania. The competition between LANC and the Legion over student sympathies led to further radicalization, and the violence associated with rail infrastructure continued over the course of the summer. Spa resorts in Bukovina appeared to be frequent targets of such incidents, and authorities took measures to prevent this, ordering the deployment of additional gendarmes at the stations that did not have their own police units.[56] Despite such preventive actions, student violence against Jewish passengers on trains continued throughout the summer; on the Vatra Dornei–Dărmăneşti line, there were repeated reports of LANC students attacking Jews, who were beaten up or tortured, as well as thrown off trains. They also attacked railway stations, chasing Jews away from station platforms or waiting rooms.[57] The student violence eventually culminated in the largest, most violent pogrom to that date in postwar Romania, on the occasion of the student congress at Oradea Mare. Originally scheduled for November 27–29, the congress was postponed due to the death of Prime Minister Ionel I. C. Brătianu (1864–1927) and eventually took place on December 4–7, 1927.[58] While the violence in the city itself is extensively documented,[59] relatively less attention has been paid to the importance of railway infrastructure for the logistics of

Or, A First Attempt at an Aggregate History of the Eastern, Social, Woman, American, Jewish, Polish, Bullion, Tuberculosis, and Many Other Questions over the Nineteenth Century, and Beyond (Princeton: Princeton University Press, 2018).

55 Cârstocea, "Anti-Semitism at the Intersection of Corruption and Colonialism."

56 ANIC, Fond Ministerul de Interne, Diverse, File 9/1927, 3. For a comparable case of antisemitism in tourist resorts in Weimar Germany, see Jacob Borut, "Antisemitism in Tourist Facilities in Weimar Germany," *Yad Vashem Studies* 28 (2000): 7–50, https://www.yadvashem.org/articles/academic/antisemitism-in-tourist-facilities.html.

57 ANIC, Fond Ministerul de Interne, Diverse, File 9/1927, 26–27.

58 Lucian Nastasă-Kovács, ed., *Pogromul itinerant sau decembrie antisemit: Oradea, 1927* (Bucharest: Curtea Veche, 2014), 45. On defining pogroms, see David Engel, "What's in a Pogrom? European Jews in the Age of Violence," in *Anti-Jewish Violence: Rethinking the Pogrom in East European History*, ed. Jonathan Dekel-Chen, David Gaunt, Natan M. Meir, and Israel Bartal (Bloomington: Indiana University Press, 2011), 19–40.

59 Nastasă-Kovács, *Pogromul itinerant*; Roland Clark, *Holy Legionary Youth: Fascist Activism in Interwar Romania* (Ithaca: Cornell University Press, 2015), 39–42.

the congress, as well as to the systematic violence that occurred along the routes, both to and, especially, on the students' return from Oradea, in what Lucian Nastasă-Kovács aptly called an "itinerant pogrom."[60]

Unlike the usual travel of students on the Romanian railways, where they benefited from a 50 percent discount, or the special cars for students that were occasionally attached to regularly scheduled trains for other events, this time the state authorities provided the students attending the congress with special trains free of charge.[61] This was, on the one hand, in anticipation of the very large numbers of students traveling and, on the other, in an attempt to prevent the recurring incidents of violence against Jewish passengers that had occurred in the past. There were good reasons for the authorities' concern ahead of the congress, in addition to the anticipated high turnout, far exceeding that at previous such gatherings. Two articles in the most recent issue of *Cuvântul Studenţesc*—the newspaper of UNSCR—published before the congress ominously articulated the union's views. The first, signed by Lorin Popescu, UNSCR president, stated that "the antinationalism of the Jews is the cause of our movement. We do not attack. We defend. The offensive war is carried out by the Jews. And then the students are not responsible. The state should eliminate the cause, and the effect will extinguish itself directly." Referring to the controversial location of the congress, it added, "At Oradea, in the heart of chauvinist Magyarism and at the border of the Bethlens' homeland, the student body must prove that it is united: that it loves its country, even at the price of anyone's sacrifice, and that it won't allow any changes [to it], no matter how much blood flows from the heart of Romanians."[62] Another article in the same issue, this time—understandably unsigned—was even more explicit in its instigation to violence and in threatening that the anticipated bloodletting would not just be of the self-sacrificial nationalist students. Titled "The Tally" and listing 30 alleged attacks, individual and collective, by Jews against Romanian nationalists—including the one on the student train in Bessarabia presented above—it concluded that "the student body has a tally, and on it, it etches the facts. This tally is as long as Romanian patience. When it comes to an end, though... Fire!"[63] Railway infrastructure was once

60 Nastasă-Kovács, *Pogromul itinerant.*

61 Announcement of UNSCR, November 1927, cited in Nastasă-Kovács, *Pogromul itinerant*, 125–126.

62 Lorin Popescu, "La Oradea," *Cuvântul Studenţesc*, November 24, 1927. The Bethlens were a Hungarian noble family who had been important in Transylvanian politics ever since the Middle Ages. In 1927, Count István Bethlen de Bethlen was prime minister of Hungary.

63 "La Răboj," *Cuvântul Studenţesc*, November 24, 1927.

again central to the authorities' concerns. A circular signed by the head of the *Siguranţa*, Romulus P. Voinescu, and distributed to all police inspectorates, *Siguranţa* brigades, and station police units in the entire country noted that "individuals are traveling on trains selling various brochures with an anti-Semitic character, while also instigating the Christian population against Jews," ordering confiscation of the materials and prosecution of those selling them.[64]

Travel to and from the congress was meant to be strictly regulated on special trains not meant to stop at way stations before Oradea, especially in "sensitive" locations that had witnessed violent incidents in the past. However, not everything went according to plan, and occasional violence already ensued on the way to the congress—at Adjud station, the students accompanying Codreanu assaulted two Jewish traveling salesmen as well as another passenger on a train. Given their declared intentions to return via the same route, the local police chief asked for army reinforcements, noting that his seven sergeants and three policemen were no match for the 500 students and would not be able to hold them on the station grounds, preventing devastations.[65] At the passage of another train of theology students through the Salina station in Târgu-Ocna during the night of December 1–2, 1927, the moving train was "bombarded" with stones by a group of Jews who managed to break the cars' windows. Pulling the emergency brake to stop the train, students disembarked and chased their assailants but were unable to catch them; as such, they vowed revenge on their return.[66]

An estimated 6,000 students took part in the congress, where the introduction of *numerus clausus* at universities, the "cadaver question" (a demand that Jewish students only be allowed to perform autopsies on Jewish cadavers rather than Christian ones), and anti-Semitism more generally featured prominently on the agenda.[67] The street violence started from the first day of the congress, instigated also by the speakers, with the program featuring all the prominent anti-Semitic leaders.[68] It would eventually degenerate into the largest pogrom in interwar Romania, with the students "destroying all shop windows and businesses, breaking shutters and destroying shops," as

64 ANIC, Fond Ministerul de Interne, Diverse, File 9/1927, 34.

65 Note of the Adjud police, December 4, 1927, cited in Ioan Scurtu, Cristian Troncotă, Natalia Tampa, and Dana Beldiman, eds., *Ideologie şi formaţiuni de dreapta în România*, vol. 2, *25 Iunie 1927—2 Ianuarie 1931* (Bucharest: Institutul Naţional pentru Studiul Totalitarismului, 2000), 89–90.

66 ANIC, Fond Ministerul Justiţiei, Direcţia Judiciară, File 71/1927, 189–190.

67 "Congresul studenţesc dela Oradea," *Adeverul*, December 8, 1927.

68 "Turburări studenţeşti la Oradea," *Adeverul*, December 8, 1927.

well as devastating four synagogues.[69] Torah scrolls and other cult objects were desecrated, with the damages the Romanian state paid to individual Jews amounting to 15 million lei, while those paid to the Jewish community of Oradea Mare for the destruction in the synagogues amounted to 8.8 million lei.[70] The city was placed under a state of siege enforced by army units as they attempted to ensure that the students were "kept isolated from the workers"; beyond the pogrom itself, authorities feared that the joining of forces by rioting students and workers could represent "the beginning of a rebellion and civil war."[71]

With students locked in army barracks during the night of December 6–7 and taken to the railway station under army escort in the morning, the pogrom continued along the routes of the special trains returning students to their homes. Although the trains were not meant to stop in Cluj—a major university town and location of anti-Semitic student riots in 1922—"students climbed on the locomotives, removed the driving staff, and stopped the trains."[72] After being held back by the army for three hours, and with the arrival of reinforcements from other student trains that students had forced the CFR personnel to stop at the station, the rioters eventually broke through the army lines and devastated Jewish shops, cafes, and businesses in the city, as well as several synagogues; "the great synagogue was completely destroyed." The violent clashes saw 890 police, gendarmes, and army soldiers fighting more than 3,000 students; swords were used by the army, resulting in numerous wounded on both sides.[73] Violent incidents were widespread along the routes of the returning trains, and attacks on individual Jews or incidents of pogromist mass violence occurred at Ciucea, Huedin, Aiud, Ghimeş, Teleagd, Şoricani, Timişul de Jos, and Predeal.[74] The promised retaliation at the Salina station in Târgu-Ocna took place on the return trip. Even though the authorities had canceled the respective scheduled stop as a result of the incident that had occurred on the way to the congress, returning students once again pulled the emergency brakes when in town,

69 Scurtu et al., *Ideologie şi formaţiuni de dreapta*, vol. 2, 99; ANIC, Fond Ministerul Justiţiei, Direcţia Judiciară, File 71/1927, 168–176. See also Clark, *Holy Legionary Youth*, 39–42.

70 CNSAS, Fond Documentar, File D 10767, vol. 1, 108; Cârstocea, "Students Don the Green Shirt," 62.

71 ANIC, Fond Ministerul Justiţiei, Direcţia Judiciară, File 71/1927, 174.

72 "Devastările dela Cluj," *Dimineaţa*, December 10, 1927.

73 "Devastările dela Cluj," *Dimineaţa*, December 10, 1927; ANIC, Fond Ministerul Justiţiei, Direcţia Judiciară, File 71/1927, 178–180.

74 ANIC, Fond Ministerul Justiţiei, Direcţia Judiciară, File 71/1927, 45–46, 152–153, 178–180, 197, 205; Nastasă-Kovács, *Pogromul itinerant*, 134–135, 139–142.

stopped the train, and embarked on a large-scale pogrom that saw the devastation of some 18–20 Jewish shops, two taverns owned by Jews, and the two local synagogues in Târgu-Ocna.[75] As the report from the small station of Ghimeş indicates, even a 10- to 15-minute stop to change the train engine was sufficient for mass violence; the same report notes that students were well armed "with revolvers, axes, lances, and clubs" and that they were in control of the situation—as the train attempted to leave, a student positioned himself in front of the locomotive, club in hand, preventing its departure until all students had returned from the pogrom at and around the station.[76] In fact, attacks against CFR personnel that tried to stop or prevent the student violence were not infrequent.[77]

At the same time, there is also considerable evidence of authorities intervening successfully to prevent the violence, whether through significant police and army presence at stations, diverting trains from potentially dangerous localities, or escorting students back to their dormitories once they reached their destinations. These can be encountered mostly in the railway stations that had already experienced such student violence in the past, predominantly in Moldavia, Bessarabia, and Bukovina: at Bacău, Paşcani, Muncelul, Cernăuţi, Chişinău, and Iaşi.[78] Upon the arrival of the students at the main Bucharest railway station, Gara de Nord, 196 students were arrested in the three special student trains that arrived from the congress, together with 72 students traveling on regular trains and 25 students arrested on trams at Calea Griviţei, near the station, having presumably escaped the police presence on the platform.[79]

A question that is not easy to answer based on the available evidence is how "the Jews" who suffered attacks on trains were identified as such by their assailants. That might have been easy in the case of Orthodox Jews wearing traditional clothing but much more difficult for secular ones. If anti-Semitic violence in a given locality can rely on preexisting knowledge of the Jewish community there, the same cannot be said for passengers on a train, who are for the most part strangers to one another. When it comes to violence at or around train stations, the targeting of certain businesses could be based on the names displayed on the windows of shops or taverns as well as on the knowledge of the local population, who often joined in the

75 ANIC, Fond Ministerul Justiţiei, Direcţia Judiciară, File 71/1927, 189–190.

76 ANIC, Fond Ministerul Justiţiei, Direcţia Judiciară, File 71/1927, 197.

77 ANIC, Fond Ministerul Justiţiei, Direcţia Judiciară, File 71/1927, 45–46, 152, 178.

78 Nastasă-Kovács, *Pogromul itinerant*, 154–155.

79 Nastasă-Kovács, *Pogromul itinerant*, 185–194.

destruction, occasionally also providing students with axes and hatchets.[80] At times, authorities tried to thwart students' plans by "disguising" Jewish businesses as Christian ones. One such "stratagem ahead of the arrival of the congress participants" was employed by the *pretor* of the *plasă* to which Ciucea station belonged when he proceeded "to the removal of merchants' signboards, placing peasants in the shops to give the impression that they are Christian stores, thus avoiding the devastation of some of them."[81] In turn, the occasional vandalism and looting of a business owned by a Christian, mistaken for a Jewish one, also occurred. During the pogrom at Huedin, anti-Semites broke five windows of the shop of one Mr. Kubaszek—"upon learning that he is a Christian, the students, through their committee, apologized to the owner, paying the damages through a collection made among the students, after which they left the station, boarding [the trains] bound for Cluj."[82] The considerate, apologetic attitude of the students toward the wronged shop owner—whose name indicates his belonging to a national minority, albeit a "Christian" one—and the immediate payment of reparations are a clear indication of the ideological nature of their violence. Far from being the actions of hot-blooded riotous youth looking for cheap thrills or attracted by the prospect of looting—which is sporadically reported to have occurred, although on rare occasions—the student attacks appear to have been methodical, planned, and motivated by ideological anti-Semitism, which was rendered compatible with a nationalist "moral code" to which its perpetrators claimed to adhere.

Conclusion

In the aftermath of the congress at Oradea Mare, an address to the Senate by Dr. Emil Iacobi—senator of Bihor County and an inhabitant of Cluj who had witnessed the pogrom there firsthand—accused the Romanian authorities of failing to do their duty to protect public order, especially in view of evidence that the students had come to Oradea armed and thus that the pogrom was premeditated.[83] In the Romanian Parliament, MP Wilhelm Filderman (1882–1963), the leader of the Jewish community in Romania,

80 Cârstocea, "Students Don the Green Shirt," 62.

81 ANIC, Fond Ministerul Justiţiei, Direcţia Judiciară, File 71/1927, 154. A *plasă* was a territorial division unit in interwar Romania, above a commune and below a county. It was led by a *pretor*, appointed by the county prefect.

82 ANIC, Fond Ministerul Justiţiei, Direcţia Judiciară, File 71/1927, 152.

83 ANIC, Fond Ministerul Justiţiei, Direcţia Judiciară, File 71/1927, 203–207.

asked the government in a session dedicated to the student congress and pogroms "what measures had been taken to prevent the disturbances on trains, at railway stations, and in Oradea Mare."[84] An opposition MP, Virgil Madgearu (1887–1940), from the National Peasant Party, blamed the pogrom on the dismal situation of Romanian students and accused the National Liberal Party government of "exploiting the miserable state of the student body, having an interest to provoke a diversion in public opinion by reawakening the anti-Semitic movement."[85] The answer of I. G. Duca (1879–1933), minister of the interior, restated the government's commitment to taking "the most severe sanctions" against the students responsible, reminding Filderman of its long-standing opposition to *numerus clausus* proposals, implemented in Hungary and temporarily also in Poland but categorically rejected in Romania by the Liberal government.[86] He also dismissed Madgearu's claims that this was a diversion meant to distract from the country's real problems or to increase a police presence that could be used against all political opponents.

In turn, Madgearu reminded Duca of the international implications of the pogrom. In a session of the Hungarian Parliament on December 7, 1927, MPs were outraged by the violence, seeing it as proof that Romania could not guarantee minority rights and asking that the issue be raised by Prime Minister István Bethlen in Geneva at the League of Nations.[87] That this was indeed the case is confirmed by a telegram sent from Geneva by the Romanian foreign minister, Nicolae Titulescu (1882–1941), who claimed that the case was one of "a simple street incident on the occasion of a student congress, provoked by irredentist and communist *agents provocateurs*, leading to the breaking of a few windows and the destruction of some shops."[88] This, however, was not a view shared by Jewish organizations abroad, nor indeed by foreign governments, as numerous dispatches of Romanian diplomats indicate.[89] As on many other occasions in the past,

84 Adunarea Deputaţilor, Sesiunea Ordinară 1927–1928, Şedinţa de joi, [1]8 decemvrie 1927, *Monitorul Oficial, Desbaterile Adunării Deputaților* (hereafter MOf DAD), *octombrie-decembrie 1927*, December 22, 18, 1927, 303.

85 Adunarea Deputaţilor, Sesiunea Ordinară 1927–1928, Şedinţa de joi, [1]8 decemvrie 1927, MOf DAD), *octombrie-decembrie 1927*, December 22, 18, 305.

86 Adunarea Deputaţilor, Sesiunea Ordinară 1927–1928, Şedinţa de joi, [1]8 decemvrie 1927, MOf DAD), *octombrie-decembrie 1927*, December 22, 18, 307–308.

87 "Ungurii exploatează în străinătate chestia turburărilor studenţeşti," *Dimineaţa*, December 10, 1927.

88 Ciphered telegram no. 71009, December 8, 1927, cited in Nastasă-Kovács, *Pogromul itinerant*, 152.

89 Nastasă-Kovács, *Pogromul itinerant*, 153–154, 169–171, 202–209, 218–219, 226–227, 236–242.

anti-Semitic student violence had become an international scandal damaging Romania's reputation abroad.

With the pogrom at Oradea Mare, the anti-Semitic student violence that had been largely confined to Moldavia and Bessarabia—LANC strongholds—became a national crisis. While previous pogroms were invariably led by students from the University of Iaşi, the protagonists this time were largely students from the universities of Cluj and Bucharest,[90] both of which had earlier opposed Codreanu's proposal to exclude Jews from student societies at the 1920 congress. In the intervening period, student anti-Semitism had migrated, primarily along the routes of the Romanian railways, which witnessed an escalation of violence despite the authorities' significant efforts to prevent it. The reasons why railway infrastructure occupied such a central part in the proliferation of student violence, therefore, deserve some concluding reflections.

First, it is testament to the success of the Romanian railway project in constituting the main conduit of mobility within the country, even when such mobility turned violent, subverting or hijacking the state's expectations.[91] The Legion's second "battle" after its establishment (following the first, issuing the newspaper *Pământul Strămoşesc* [The Ancestral Land] and securing the subscriptions that allowed its—irregular—publication) was the purchase of a Chevrolet pickup truck, affectionately nicknamed "The Deer"—yet the road system was underdeveloped, treacherous, or impracticable in winter, and automobiles remained rare outside the main cities.[92] As such, electoral propaganda in the countryside was often undertaken on foot or on horseback, and trains remained the main transport infrastructure allowing mass mobility over long distances and nationwide.[93]

Second, the very fact the railway was the only means of transportation having such a "mass" character helps explain why anti-Semitic propaganda was often distributed or sold on trains or at railway stations, as we have seen above. In addition to the potential for reaching a large audience, the

90 Scurtu et al., *Ideologie şi formaţiuni de dreapta*, vol. 2, 99.

91 Though post-1920 Romania had rather limited resources, it developed several railway projects to connect the newly acquired provinces across the former imperial frontiers and across the Carpathians. See David Turnock, "Railway Network Development in Inter-war Romania: Economic and Strategic Motives," *Geographica Pannonica*, no. 8 (2004): 16–24. The student movement fully used and understood the national integrative capacity of the expanded railway network.

92 "Rezultatul muncii noastre. A doua bătălie," *Pământul Strămoşesc*, February 1, 1928; "Bătălia a 2-a s-a câştigat," *Pământul Strămoşesc*, March 1, 1928.

93 Codreanu, *Pentru legionari*, 361–367; ANIC, Fond Ministerul de Interne, Diverse, File 9/1927, 319–320.

nature of train travel—involving relatively long journeys and, paradoxically, immobility of the passengers inside the railway cars—offered the advantage of a "captive public," whether it consisted of the Jews students frequently attacked or the Christian population they tried to sway to their cause. The time that rail journeys helped shorten in reaching one's destination was in ample supply and on the propagandists' side when inside the trains.

Third, the discounted train travel students benefited from, or even the free special trains provided by the state on the occasion of student congresses, certainly facilitated the anti-Semitic violence, as representatives of the Jewish community duly noted in their complaints.[94] This raises the question of the complicity of the Romanian authorities, one that is not easy to answer, given the conflicting reports showing on the one hand extensive police, *Siguranţa*, and army mobilization to prevent violence and, on the other, clear evidence of anti-Semitism among both police authorities and (albeit more rarely) CFR personnel.[95]

Fourth, the relatively predictable train schedule—although significant delays were a common occurrence—allowed planning, whether for the anti-Semites waiting in the middle of the night for Codreanu's triumphal return from Turnu-Severin at Buzău station or for the Jewish youth attacking the nationalists' trains at Călăraşi, Visterniceni, or Târgu-Ocna. Finally, all of these aspects appear to have been evident to the protagonists involved, from the anti-Semitic students and the authorities preparing for the arrival of their trains or deploying "mobile" agents on board to the young members of Maccabi or the Jewish Cultural League engaging in organized resistance to the attacks.

If Cuza's leadership of the student movement had been reaffirmed once more at the Oradea congress, student organizations eventually came to gravitate toward the legionary movement and established themselves as its backbone in the 1930s. By 1935, the Legion controlled 37 local chapters of UNSCR out of a total of 42, against LANC's five.[96] As the legionary movement grew beyond its initial student following and attracted ever more followers from different social groups into its ranks, the importance of the railway infrastructure for Romania's fascists also became visible in their (increasingly

94 ANIC, Fond Ministerul Justiţiei, Direcţia Judiciară, File 71/1927, 205.

95 Such an incident occurred at Adjud station, where several CFR mechanics insulted "the Jew Iosef Şmilovici," who then retaliated by shooting one of them in the chest; the victim was then visited at the hospital by Codreanu and Ilie Gârneaţă. ANIC, Fond Ministerul de Interne, Diverse, File 9/1927, 228.

96 CNSAS, Fond Documentar, File D 12761, vol. 1, 158–166; Cârstocea, "Students Don the Green Shirt," 62–63.

successful) attempts to attract, organize, and control CFR workers and personnel.[97] For a movement that intermittently contemplated the possibility of coming to power via a violent takeover, railway infrastructure held the dual promise of facilitating connections between different locations and of blocking the transport of potentially hostile forces.

Beyond such plans, which eventually never materialized, the railway system remained the most prominent conduit of mobility for the Romanian fascists. This was visible in such high-profile acts of violence as the assassination of Prime Minister I. G. Duca by a team of three legionaries (all of them students) at Sinaia railway station on December 29, 1933, or in the reenactment of Codreanu's triumphal return from Turnu Severin with a much more grandiose, international funeral train arranged for Ion I. Moţa and Vasile Marin (1904–1937) in January and February 1937. The bodies of the two legionary leaders, killed fighting in the Spanish Civil War, were returned from Spain on a special train that was honored in both Fascist Italy and Nazi Germany before entering and symbolically crisscrossing Romania as part of the movement's bid to mobilize public support for its anti-establishment politics.[98] By this time, with the movement a serious contender in the Romanian political arena, the railways were reconceptualized within a broader, alternative project of nation-space-making based on grassroots mobilization and voluntarism.[99] This was but the prelude to the final and most tragic chapter in the entanglement of railway infrastructure, anti-Semitism, and violence against Jews: the prominence of trains as a means of transportation in the Holocaust, in Romania and elsewhere.[100] That story is the subject of a future study; its roots, however, lay in the potential shown by railway infrastructure as a conduit of anti-Semitic student violence in the 1920s

97 CNSAS, Fond Documentar, File D 8913, vol. 3, 239.

98 Valentin Săndulescu, "Sacralized Politics in Action: The February 1937 Burial of the Romanian Legionary Leaders Ion Moţa and Vasile Marin," *Totalitarian Movements and Political Religions* 8, no. 2 (2007): 259–269; Raul Cârstocea, "Bringing Out the Dead: Mass Funerals, Cult of Death and the Emotional Dimension of Nationhood in Romanian Interwar Fascism," in *Emotions and Everyday Nationalism in Modern European History*, ed. Andreas Stynen, Maarten Van Ginderachter, and Xosé Manoel Núñez Seixas (Abingdon: Routledge, 2020), 134–163.

99 Raul Cârstocea, "Building a Fascist Romania: Voluntary Work Camps as Mobilisation Strategies of the Legionary Movement in Interwar Romania," *Fascism* 6, no. 2 (2017): 163–195.

100 For an excellent study tracing the role of the railways in the encounter between Germans and Jews and their respective visions of modernity, see Todd Samuel Presner, *Mobile Modernity: Germans, Jews, Trains* (New York: Columbia University Press, 2007). For the Holocaust more specifically, see Simone Gigliotti, *The Train Journey: Transit, Captivity, and Witnessing in the Holocaust* (New York: Berghahn, 2009).

and in the national integration of students and anti-Semitism it allowed for.[101]

Bibliography

Adeverul. "Manifestaţiile de eri." May 30, 1925.

Adeverul. "Nunta dela Focşani." June 12, 1925.

Adeverul. "Congresul studenţesc dela Oradea." December 8, 1927.

Adeverul. "Turburări studenţeşti la Oradea." December 8, 1927.

Arhiva Consiliului Naţional pentru Studierea Arhivelor Securităţii (ACNSAS). Fond Documentar.

Arhiva Consiliului Naţional pentru Studierea Arhivelor Securităţii (ACNSAS). Fond Urmărire Penală.

Arhivele Naţionale Istorice Centrale (ANIC). Fond Ministerul de Interne, Diverse.

Arhivele Naţionale Istorice Centrale (ANIC). Fond Ministerul Justiţiei, Direcţia Judiciară.

Borut, Jacob. "Antisemitism in Tourist Facilities in Weimar Germany." *Yad Vashem Studies* 28 (2000): 7–50. https://www.yadvashem.org/articles/academic/antisemitism-in-tourist-facilities.html.

Bozdoghină, Horia. *Antisemitismul lui A. C. Cuza în politica românească*. Bucharest: Curtea Veche, 2012.

Brenner, Michael, and Reuveni, Gideon, eds. *Emancipation through Muscles: Jews and Sports in Europe*. Lincoln: University of Nebraska Press, 2006.

Case, Holly. *The Age of Questions: Or, A First Attempt at an Aggregate History of the Eastern, Social, Woman, American, Jewish, Polish, Bullion, Tuberculosis, and Many Other Questions over the Nineteenth Century, and Beyond*. Princeton: Princeton University Press, 2018.

Cârstocea, Raul. "Anti-Semitism at the Intersection of Corruption and Colonialism: Continuities of Political Rhetoric from the Nineteenth Century to the Interwar Period." *East European Politics and Societies* 39, no. 3 (2025): 762–783 https://doi.org/10.1177/08883254251352116.

Cârstocea, Raul. "Bringing Out the Dead: Mass Funerals, Cult of Death and the Emotional Dimension of Nationhood in Romanian Interwar Fascism." In *Emotions and Everyday Nationalism in Modern European History*, edited by

101 On the continued relevance of rail infrastructure for anti-Semitic mobilization and official reactions to it in the late 1930s, see Andreea Kaltenbrunner, "Anti-Semitic Violence in Eastern Romania: The National Christian Party's Congress, 8 November 1936," *Contemporary European History* 33, no. 1 (2024): 117–136.

Andreas Stynen, Maarten Van Ginderachter, and Xosé Manoel Núñez Seixas, 134–163. Abingdon: Routledge, 2020.

Cârstocea, Raul. "Building a Fascist Romania: Voluntary Work Camps as Mobilisation Strategies of the Legionary Movement in Interwar Romania." *Fascism* 6, no. 2 (2017): 163–195. https://doi.org/10.1163/22116257-00602002.

Cârstocea, Raul. "Students Don the Green Shirt: The Roots of Romanian Fascism in the Anti-Semitic Student Movements of the 1920s." In *Alma Mater Antisemitica: Akademisches Milieu, Juden und Antisemitismus an den Universitäten Europas zwischen 1918 und 1939*, edited by Regina Fritz, Grzegorz Rossoliński-Liebe, and Jana Starek, 39–66. Vienna: New Academic Press, 2016.

Cârstocea, Raul. "Synchronous Nationalisms: Reading the History of Nationalism in South–Eastern Europe between and beyond the Binaries." *National Identities* 24, no. 5 (2022): 481–503. https://doi.org/10.1080/14608944.2021.2007367.

Cârstocea, Raul. "Uneasy Twins? The Entangled Histories of Jewish Emancipation and Anti-Semitism in Romania and Hungary, 1866–1913." *Slovo* 21, no. 2 (2009): 64–85.

Ciobotaru, Diego-Maricel. "Jewish Cultural and Sporting Associations between "Romanianization" and "Unionization": Episodes of Antisemitic Violence in the Stadiums of the Capital and the Provinces before and after World War II (1922–1948)." *Holocaust. Studii şi cercetări* 14, no. 1 (2022): 87–124.

Clark, Roland. *Holy Legionary Youth: Fascist Activism in Interwar Romania.* Ithaca: Cornell University Press, 2015.

Cor., "Sălbăteciile dela Iaşi." *Dimineaţa*, June 3, 1925.

Cuvântul Studenţesc. "La Răboj." November 24, 1927.

Cuza, A. C. *Ţăranii şi clasele dirigente.* Iaşi: Tipografia Naţională, 1895.

Dimineaţa. "Cum s'a produs 'accidentul' de la Iaşi." June 4, 1925.

Dimineaţa. "Ce se petrece la Focşani." June 13, 1925.

Dimineaţa. "Devastările dela Cluj." December 10, 1927.

Dimineaţa. "Ungurii exploatează în străinătate chestia turburărilor studenţeşti." December 10, 1927.

Engel, David. "What's in a Pogrom? European Jews in the Age of Violence." In *Anti-Jewish Violence: Rethinking the Pogrom in East European History*, edited by Jonathan Dekel-Chen, David Gaunt, Natan M. Meir, and Israel Bartal, 19–40. Bloomington: Indiana University Press, 2011.

Erker, Linda. *Die Universität Wien im Austrofaschismus: Österreichische Hochschulpolitik 1933 bis 1938, ihre Vorbedingungen und langfristigen Nachwirkungen.* Göttingen: V&R Unipress, 2021.

Fink, Carole. *Defending the Rights of Others: The Great Powers, the Jews, and International Minority Protection, 1878–1938.* Cambridge: Cambridge University Press, 2004.

Gigliotti, Simone. *The Train Journey: Transit, Captivity, and Witnessing in the Holocaust*. New York: Berghahn, 2009.

Iliescu, Leontin and N. Ciocârdia. "Procesul Corneliu Zelea Codreanu: Achitarea." *Universul*, May 28, 1925.

Iordachi, Constantin. *Liberalism, Constitutional Nationalism, and Minorities: The Making of Romanian Citizenship, c. 1750–1918*. Leiden: Brill, 2019.

Kaltenbrunner, Andreea. "Anti-Semitic Violence in Eastern Romania: The National Christian Party's Congress, 8 November 1936." *Contemporary European History* 33, no. 1 (2024): 117–136. https://doi.org/10.1017/S0960777321000825.

Kieval, Hillel J. *Blood Inscriptions: Science, Modernity, and Ritual Murder at Europe's Fin de Siècle*. Philadelphia: University of Pennsylvania Press, 2022.

Leff, Lisa Moses. "Liberalism and Antisemitism: A Reassessment from the Peripheries." In *Jews, Liberalism, Antisemitism: A Global History*, edited by Abigail Green and Simon Levis Sullam, 23–46. Cham: Palgrave Macmillan, 2020.

Livezeanu, Irina. *Cultural Politics in Greater Romania: Regionalism, Nation-Building, and Ethnic Struggle, 1918–1930*. Ithaca: Cornell University Press, 1995.

Manuilă, Sabin, ed. *Recensământul general al populaţiei României din 29 decemvrie 1930*. Vol. 2, *Neam, Limbă Maternă, Religie*. Bucharest: Imprimeria Naţională, 1940.

Marton, Silvia. "'La Belgique de l'Orient' et les chemins de fer: les raisons d'une comparaison." *Studia Politica: Romanian Political Science Review* 8, no. 1 (2008): 27–44.

Marton, Silvia. "La 'corrupción' electoral en Rumania: Los comienzos titubeantes de la democracia." *Ayer: Revista de Historia Contemporánea* 115, no. 3 (2019): 77–104. https://doi.org/10.55509/ayer/115-2019-04.

Marton, Silvia. "De la construction de l'Etat au racisme: judéophobie et antisémitisme en Roumanie avant la Grande Guerre." In *Antisemitismus im 19. Jahrhundert aus internationaler Perspektive / Nineteenth Century Anti-Semitism in International Perspective*, edited by Mareike König and Oliver Schulz, 203–225. Göttingen: Vandenhoeck & Ruprecht, 2019.

Marton, Silvia and Andrei Sorescu, eds. "Introduction: Colonial Anxieties, Corruption Scandals and Xenophobia in Nineteenth-Century Infrastructure Development in Romania." *East European Politics and Societies* 39, no. 3 (2025): 680–690. https://doi.org/10.1177/08883254251352114.

Mendelsohn, Ezra. *The Jews of Central Europe*. Bloomington: Indiana University Press, 1983.

Mille, Constantin. "Fascismul român." *Pressa*, June 14, 1923.

Mishkova, Diana, and Roumen Daskalov. "'Forms without Substance': Debates on the Transfer of Western Models to the Balkans." In *Entangled Histories of the Balkans*. Vol. 2, *Transfer of Political Ideologies and Institutions*, edited by Roumen Daskalov and Tchavdar Marinov, 1–97. Leiden: Brill, 2014.

Monitorul Oficial, Desbaterile Adunării Deputaților, Octombrie-Decembrie 1927. Adunarea Deputaţilor, Sesiunea Ordinară 1927–1928, Şedinţa de joi, December 18, 22, 1927.
Nastasă-Kovács, Lucian, ed. *Pogromul itinerant sau decembrie antisemit: Oradea, 1927*. Bucharest: Curtea Veche, 2014.
Nastasă, Lucian, ed. *Antisemitismul universitar în România (1919–1939)*. Cluj-Napoca: Kriterion, 2011.
Pământul Strămoşesc. "Rezultatul muncii noastre. A doua bătălie." February 1, 1928.
Pământul Strămoşesc. "Bătălia a 2-a s-a câştigat." March 1, 1928.
Popescu, Lorin. "La Oradea." *Cuvântul Studenţesc*, November 24, 1927.
Popescu, Toader. *Proiectul feroviar românesc, 1842–1916*. Bucharest: Simetria, 2014.
Presner, Todd Samuel. *Mobile Modernity: Germans, Jews, Trains*. New York: Columbia University Press, 2007.
Săndulescu, Valentin. "Sacralised Politics in Action: The February 1937 Burial of the Romanian Legionary Leaders Ion Moţa and Vasile Marin." *Totalitarian Movements and Political Religions* 8, no. 2 (2007): 259–269.
Scurtu, Ioan, Cristian Troncotă, Natalia Tampa, and Dana Honciuc Beldiman, eds. *Ideologie şi formaţiuni de dreapta în România*. Vol. 2, *25 Iunie 1927—2 Ianuarie 1931*. Bucharest: Institutul Naţional pentru Studiul Totalitarismului, 2000.
Sorescu, Andrei-Dan, and Raul Cârstocea. "The (Great) Numbers Game: Demographic Anxieties and Quotas in Nineteenth- and Twentieth-Century Romania and the Global Antisemitic Imaginary." In *Quotas: The "Jewish Question" and Higher Education in Central Europe, 1880–1945*, edited by Michael Miller and Judith Szapor, 67–109. New York: Berghahn, 2024.
Stern, Fritz. *Gold and Iron: Bismarck, Bleichröder, and the Building of the German Empire*. London: Allen & Unwin, 1977.
Szabó, Miloslav. "Zentrum oder Peripherie? Der Antisemitismus in Ungarn um 1900." In *Modern Antisemitisms in the Peripheries: Europe and its Colonies, 1880–1945*, edited by Raul Cârstocea and Éva Kovács, 371–389. Vienna: New Academic Press, 2019.
Universul. "Manifestaţii studenţeşti la Slatina, Piteşti şi în Capitală." May 31, 1925.
Universul. "Nunta lui Corneliu Zelea Codreanu." June 17, 1925.
Véri, Daniel. "The Tiszaeszlár Blood Libel: Image and Propaganda." In *Antisemitismus im 19. Jahrhundert aus internationaler Perspektive: Nineteenth Century Anti-Semitism in International Perspective*, edited by Mareike König and Oliver Schulz, 263–292. Göttingen: Vandenhoeck & Ruprecht, 2019.
Zelea Codreanu, Corneliu. *Pentru legionari*. Sibiu: Totul pentru Ţară, 1936.

About the Author

Raul Cârstocea is Professor of History at Maynooth University and Principal Investigator of the ERC Consolidator Grant "Insurgent Temporalities: Fascism as a Global Anti-Universalist Project (INTEMPO)." His research interests focus on anti-Semitism, fascism, and, more broadly, state formation and nation-building processes in nineteenth- and twentieth-century Central and Eastern Europe.

Index

Aali Pasha, Mehmed Emin, 167, 168
Abdülaziz (Sultan), 56, 171
Abdülhamid II, 53, 54
Adjud, 231, 282, 288
Aegina, 151
Ahmed Essad Pasha, 178
Aiud, 283
AKP (Justice and Development Party, Turkey), 32
Algeria, 75
Alléon, Jacques, 208, 209
Alliance Israélite Universelle, 114
Alvinc (Vințu de Jos), 196-201, 203-205
Alvinc-Vöröstoronyi HÉV RT, 201
America, 109
Anatolia, 193, 194, 210, 211, 216
Anatolian Railways, 53, 54
Anderson, James, 38, 117, 120, 126, 127
Ankara, 31, 41, 42, 53, 57
Antalya, 36
Antonova-Goleva, Boriana, 41, 59, 79, 165, 182, 194, 207, 208, 209, 215, 216
Arab provinces, 194
Arad, 190, 220
Ardeleanu, Constantin, 12, 13, 19, 22, 41, 42, 65, 68, 72, 159, 160, 162, 164, 180, 206, 207, 221, 242, 254
Asia (*also* provinces), 47, 98, 165, 194
Aslan, Nicolae Ceaur, 258
Assan brothers, 229
Atatürk, Mustafa Kemal, 39, 57
Athens, 30, 31, 143, 145, 146, 154-156, 178
Aurelian, P. S., 108, 257
Austin, Charles Edward, 177, 178
Austria-Hungary (*also* Austria), 39, 64, 116, 188, 191, 192, 196, 207, 212, 221, 222, 244, 253, 254, 274
Austrian Danube Navigation Company (Donau-Dampfschiffahrts-Gesellschaft), 161
Austrian Lloyd, 161
Avni Pasha, 177
Ayasoluk, 45, 46
Aydın, 45, 46, 56, 172, 193

Bacău, 231, 284
Bagdadbahn, 194
Balkan(s), 12, 122, 163, 186, 194
Bánffy, Dezső, 200
Barboși, 255, 259, 261
Barițiu, Gheorghe, 190, 240, 262
Barkley, Henry Charles, 166
Barkley, John Trevor, 63, 165, 169-171, 174, 175, 208, 209, 233
Barkley, T. – Staniforth, J. (concession), 245, 248
Bârlad, 231, 249, 250, 255, 256, 259, 260, 261, 262, 264, 265, 275
Baross, Gábor, 191, 197, 199, 200
Basra, 193
Bedeus, Gustav, 204-206
Bedeus, Josef von, 204
Belgium, 65, 80, 121, 127, 144, 194, 209, 215
Belgrade, 30, 31
Berlin (*also* Airport), 34, 36, 46, 49, 53, 62, 89, 90
Berlin, Congress of, 62, 75, 194
Berlin, Treaty of, 179, 181
Bessarabia, 115, 164, 179, 253, 254, 259, 276, 278, 281, 284, 287
Beszterce (Bistritz / Bistrița), 195
Beszterce, county, 200
Bethlen, András, 197
Bethlen, István Bethlen de, 281, 286
Bibescu Basarab Brâncoveanu, Grigore, 228
Bihor, 285
Bismarck (chancellor), 75, 85, 268, 293
Bjala, 41, 42
Black Sea, 19, 20, 51, 65, 77, 79, 84, 159, 160-161, 164-167, 169, 171-172, 174, 177, 178, 182-184, 194, 208-209, 212, 215, 220, 222, 244, 248, 253-254, 261-262
Bodencreditanstalt, 204, 210
Boğazköy (*see also* Tchernavoda, Cernavodă), 160
Böhme, Hartmut, 91, 101
Borcia, Lucian, 202-203
Botoșani, 248-249, 251-253, 256, 261
Bouchon, Sara, 96, 101
Brăila, 160-161, 182, 244, 246, 248-249, 252, 254-255, 260-261
Brașov (Kronstadt/Brassó), 190
Bratislava (Pozsony), 190
Brătianu, Dimitrie, 250
Brătianu, Ion C., 120, 247, 250, 254, 258-259
Brătianu, Ionel I. C., 280
Britain, 8, 38, 57, 69, 70, 73, 74, 76, 80, 81, 161, 165, 180, 184, 222
Brote, Aurel, 197
Brussels, 47
Bucharest City Hall, 229, 233, 234, 235
Bucharest, 20-22, 68, 105, 180, 227-229, 232-234, 239, 242, 244-245, 248-254, 256, 259, 262, 271, 284, 287
Budapest, 30, 191, 196, 202, 210-211
Bukovina, 192, 221, 248, 268, 280, 284
Bulgaria, 29, 34, 41, 42, 50, 51, 52, 57, 78, 81, 166, 182
Bulwer, Henry, 168, 169
Bursa, 44, 210, 211, 212, 213
Buzău, 231, 245, 271, 288

Căile Ferate Române (Romanian Railways Company) (CFR), 221, 222, 230, 272, 273, 283, 284, 288, 289
Călăraşi, 277, 278, 288
Calea Griviţei, 284
Câmpulung, 231, 257, 258, 263, 274
Cantacuzino, G., 257, 259, 265
Carp, Petre P., 123-127, 250
Carpathians, 193, 222, 244, 287
Cârstocea, Raul, 21, 22, 68, 75, 76, 114, 121, 267-271, 273, 275, 276, 280, 283, 285, 288, 289
Cavid Bey, 54
CDU (Christian Democratic Union, Germany), 36
Ceauşescu, Nicolae, 40, 58
Central America, 117
Central Europe, 7, 97, 105, 126, 131, 160, 219, 244, 248, 267, 270
Central-South-East Europe, 26, 86, 162
Cernăuţi, 268, 270, 284
Cernavodă (*see also* Tchernavoda, Boğazköy), 51, 160, 220, 221
Charles VI, emperor, 38, 55
China, 112, 137, 155
Chiper, Mihai, 20, 68, 71, 77, 206, 220, 222, 225, 239, 245, 258
Chişinău, 276, 277, 284
Chitila, 271
Ciucea, 283, 285
Clarendon, George Villiers, Earl, 172, 173, 183
Clémentine of Orléans, 52
Cluj (*see also* Kolozsvár), 270, 283, 285, 287
Codrescu, Ion, 249
Committee of Union and Progress (CUP, Ottoman Empire), 53
Compagnie des Eaux de Constantinople, 56
Conservative Party (Romania), 64, 107, 123, 257-258
Constant, Benjamin, 139, 140, 154
Constanţa (*see also* Kustendjie), 51, 65, 159, 208, 220-221, 254
Constantinople (*see also* Istanbul), 44, 49, 53, 56, 59, 140, 160, 167, 168, 169, 170, 171, 178, 179, 183, 194, 208, 209, 210, 214, 222
Çorlu, 31
Cracow, 253
Craiova, 231, 249, 250, 251
Crawley, George B., 64
Crimean War (1853-1856), 12, 63, 78, 122, 160, 164, 193, 253
Cunard, Samuel, 165
Čuprov, Aleksandr I., 50
Cuza, Alexandru C., 270-272, 278, 279
Cuza, Alexandru Ioan, 228, 244
Czernowitz, 220, 248, 251, 253, 261

Dănilă, Constantin, 279
Danube, 12, 19, 20, 51, 124, 160, 164, 176, 180, 208, 220, 222, 227, 244, 246, 250, 251, 253, 261
Danube Delta, 19, 160, 164, 180
Danube Mouths, 160, 164
Danube Province, 41, 42
Danubian Principalities, 139
Dărmăneşti, 280
Davoudian, Garabet Artin (Davud Pasha), 172
DBSR (Danube and Black Sea Railway and Kustendjie Harbour Company), 19, 160, 209
De Pradt, Dominique Dufour, 139-140
Delavrancea, Barbu Ştefănescu, 106, 108, 109, 128
Democratic Nationalist Party (Partidul Naţionalist-Democrat – PND), 270
Dersaâdet Su Şirketi, 49
Deutsche Bank, 51, 53, 54
Djavid, Mehmet bey, 210
Dniester River, 276
Dobrudja, 19, 68, 112, 124, 159, 160, 161, 162, 163, 164, 165, 168, 169, 174, 175, 179, 180, 181, 208, 221, 223, 254, 259
Duca, Ion G., 286, 289
Dussaud Frères, 48
Dutrône, Henri, 143, 149, 152, 153, 155

Eastern Europe, 22, 72, 76, 95, 254
Eastern Germany, 94, 102
Eastern Hungary, 73, 190
Eastern Rumelia, 51
Edhem Pasha, 159, 175
Edirne, 41, 47, 59
Egry, Gábor, 20, 71, 73, 77, 174, 185, 189, 196, 197, 199, 200, 211, 221, 244
Egypt, 35, 58, 112
Elliot, Henry George, 171, 173, 177, 178
Engels, Jens-Ivo, 17, 18, 35, 64-66, 69, 70, 71, 73, 74, 76, 87, 90, 94, 95, 97-99, 206, 241, 242, 275
England, 46, 136
Enns, Austria below the, 39
Ephesos (Efes), 45, 46
Epidaurus, 141
Erdély-Délvidéki HÉV Rt, 202
Erdoğan, Recep Tayyip, 31-33, 36, 37, 56
Erzurum, 47, 59
Essad Pasha, 178
Europe, 8, 14, 16, 22, 36, 37, 38, 40, 47, 88, 92, 98, 99, 110, 141, 142, 144, 148, 150, 151, 185, 194, 224, 225, 240, 251, 254, 255, 269, 277
European Commission of the Danube, 164, 167
European Union, 88
Eynard, Jean-Gabriel, 150, 151

Făgăraş, county, 203
Fawcett, John Henry, 179
Fehmi Effendi, 179
Fekete, Alexander, 98, 101
Felek (Feleac), 197-200, 202-203
Ferdinand of Bulgaria, 52
Fičeto, Kolju (also Nikola Ivanov Fiçoğlu or Nikola Ivanov Fičev or Kolju Usta) 41-43

Filderman, Wilhelm, 285, 286
Filiași, 231
Fiume, 39, 58, 190
Florica, 257, 258
Focșani, 229, 231, 259, 262, 272-274
Focșani–Nămoloasa–Galați (FNG) line, 259, 262
Fogaras (Fogarasch, Făgăraș), 197, 198, 202-203
Foreign Office, 166, 167, 171, 178, 184
Forsthoff, Ernst, 94
France, 8, 74, 88, 93, 107, 120, 121, 135, 136, 138, 139, 143, 149, 150, 153, 222, 249
Frankfurt (also Airport), 36, 58, 90, 95, 103, 186, 216
Fraport, 36
Fuad Pasha, Mehmed, 167, 171
Fuhrmann, Malte, 18, 29, 41, 43, 52, 77, 93, 95, 160, 173, 186, 207, 221, 227

Galați, 161, 228, 243, 244, 246, 248, 249, 252-256, 259, 260-262, 264, 265
Galicia, 221
Gara de Nord, 284
Gârneață, Ilie, 277, 288
Gemlik, 210, 212
Geneva, 143, 286
Gentz, Friedrich von, 141
Germany (*also* German Empire), 62, 63, 74, 75, 88, 122, 196, 222, 280, 289
Geyikdagi, M. Yasar, 163, 183
Geyikdagi, V. Necla, 183
Gezi Uprising, 36
Ghica, 240, 244, 247, 263
Ghica, Ion, 263
Ghimeș, 193, 283, 284
Ginsburg, Poljakov, Utin, and Černi Consortium, 50
Giurgiu, 227, 228, 232, 233, 245, 248
Gladstone, William, 208
Golești, 231, 257, 258
Gordon, Lewis D. B., 165
Gosse, Louis-André, 143, 146, 147, 155
Gratz, Gusztáv, 205
Great Britain, 70, 81, 161, 165, 180, 184
Greater Romania, 21, 115, 267, 268, 269, 292
Greece, 19, 29, 30, 33, 116, 135-154
Grenoble, 278
Groedel, baron, 212, 213
Grosse, Minna, 48, 49
Gülhane Edict (1839), 40
Gyimes (*see also* Ghimeș), 193

Habsburg Empire, 62, 75
Halil Rifat Pasha, 42
Hamburg, 36, 37, 43, 90, 91, 103
Harris, Edward, 112, 129, 175, 176
Hasan Fehmi Effendi, 179
Hașdeu, B. P., 7-9, 23
Hejaz Railway, 187

Helfferich, Karl, 53-55
Hermannstädter Allgemeine Sparkasse, 197
Hermannstädter Bodencreditanstalt, 204
Hirsch, Maurice de, 47, 50, 51, 194
Hobbes, Thomas, 138
Hohenzollern-Sigmaringen, Carol (prince) 62-64, 69, 71, 74, 75, 247, 254, 255
Homestead Act, 108
Honecker, Erich, 56
Horváth, Mihály, 192, 197, 214, 215, 217
Huedin, 283, 285
Huguenin, Edouard, 54, 55
Hungarian General Bank (Magyar Általános Hitelbank – MÁH), 200
Hungary (*also* Kingdom of), 20, 73, 77, 187, 188, 190, 192, 193, 196, 198, 200, 202, 206, 207, 209, 212, 213, 253, 281, 286
Huseyin Avni Pasha, 177
Hyde Clarke, Henry, 45

Iacobi, Emil, 285
Iași, 9, 231, 242, 244, 245, 248, 249, 250-254, 256, 257, 259-261, 268, 270-272, 274, 275, 284, 287
İbrahim Edhem Pasha, 159
Ilinoiu, Elena, 273
İmamoğlu, Ekrem, 31, 32, 33, 57
India, 9, 70, 79, 186, 217
Iorga, Nicolae, 203, 217, 270
Iraq, 194
Istanbul (*see also* Constantinople), 31, 33, 36, 40, 41, 43, 46-49, 56, 68, 121, 140, 161-165, 168, 169, 171, 173, 177-179, 181, 182, 244
Italy, 121, 180, 184, 289
Ițcani, 261
Izmir (*see also* Smyrna), 46-49, 172, 193, 210

Jantra, 41
Japan, 112
Jensen, John H., 78, 82, 165, 166, 183, 208, 217
Jibrieni, 254, 265
Jiu Valley (Zsil-völgy), 193, 197, 228
Judea, 114

Kâmil Bey, 49
Kapodistrias, Ioannis, 19, 137, 142-153
Kapodistrias, Augustinos, 144
Kapodistrias, Viaros, 144
Karadeniz Ereğli, 165
Karavelov, Petko, 50
Kassa (Košice), 190,
Kassa-Oderberg Railway, 192
Kastamonu, 212, 213
Katrina (hurricane), 88
Keleti Vasút (Eastern Railways), 190
Kogălniceanu, Mihail, 113, 244, 250, 251, 257, 260
Kokkonis, I.P., 151
Kolokotronis, Theodoros, 152
Kolozsvár (*see also* Cluj), 190
Kornfeld, Zsigmond, 200

Kossuth, Lajos, 190, 216
Kustendjie (*see also* Constanța), 19, 42, 68, 72, 159-182, 207-208, 242

Laak, Dirk van, 12, 14, 88, 89, 90, 94, 98
Laconia, 148
Lagarde, Auguste, 147, 150, 151, 152
Lahovary, Alexandru, 197
Lahovary, Iacob, 260
Lánczy, Leó, 199, 200, 201
Layard, Henry, 169, 178, 179, 180
Legiunea "Arhanghelul Mihail" – LAM (Legion of the Archangel Michael), 271, 278-280, 287-289
Lemberg, 73, 220, 242, 253, 261
Lemberg-Cernowitz-Jassy Railway Company, 73
Leopold, King of Belgium, 144
Liberal Party (Romania), 61, 62, 64, 67, 74, 75, 77, 106, 108, 123-125, 257-259, 286
Liddell, Charles, 165
Liga Apărării Naţional-Creştine – LANC (League of National-Christian Defence), 270
Liga Culturală Evreiască – LNC (Jewish Cultural League), 276-278, 288
London, 45, 68, 143, 162, 163, 165, 168, 171-173, 176-178, 181
Lower Danube, 9, 75, 161, 166, 174, 180
Lupaş, Ioan, 202
Lviv, 253, 261
Lyons, Richard, 169, 170, 171

Maccabi, 276, 288
Machiavelli, 138
Madagascar, 107, 109
Madgearu, Virgil, 286
Magyar Államvasutak (MÁV), 191-193, 195, 200, 201, 205
Magyar Helyi Érdekű Vasutak Rt, 202, 205
Maiorescu, Titu, 258
Manciu, Constantin, 270
Mărăşeşti, 231, 245
Marghiloman, Alexandru, 260
Marin, Vasile, 255, 264, 289, 293
Mariupol (Zhdanov), 40
Marmara Sea, 44, 194, 210
Maros, 191
Marseille, 48
Marshall & Sons, 254
Marțian, D. P., 107, 114, 129, 240, 264
Marton, Silvia, 7-9, 17, 18, 22, 33, 50, 61, 62, 64, 69, 72, 75, 76, 97, 110, 114, 121, 122, 123, 162, 179, 219, 222, 230, 241, 245, 247, 268, 273, 279
Mavrokordatos, Alexandros, 141, 142
Mavromichalis family, 149
Mavromichalis, Petros, 144, 148, 149, 153
Mecca, 195
Medgidia (*see also* Medjidieh), 175
Medgyes (*see also* Mediasch / Mediaş), 195
Mediaş (*see also* Medgyes / Mediasch), 195, 196
Medina, 195
Mediterranean (Sea), 40, 41, 60, 248
Medjidieh (*see also* Medgidia), 175
Mehmed Emin Aali Pasha, 167
Mehmed Fuad Pasha, 167
Mehmed Reşid, 46
Meltzl, Oscar, 199, 200, 205
Merkel, Angela, 36, 49
Metternich, Klemens von, 40, 141
Micu, Andrei, 202
Midhat Pasha, 41-43
Mihăileni, 261
Minard, Charles-Joseph, 249, 251
Mitchell, Timothy, 34, 35, 58
Moldavia, 9, 20, 63, 68, 77, 110, 113, 114, 164, 191, 193, 220, 221, 223, 228, 239, 240, 242-262, 278, 284, 287
Montesquieu, 138
Mordtmann, Andreas D., 43, 44, 45, 47, 49, 50, 56, 58
Moţa, Ion I., 271, 272, 289, 293
Moustoxydis, Andreas, 151
Muncelul, 284
Munich (University of), 67
Murzinos family, 144
Mussolini, Benito, 278

Nagy-Küküllő, 191, 203, 205
Nagyszeben (Hermannstadt / Sibiu), 195-200, 203, 205
Nagyszombat (Trnava), 190
Nagyvárad (Oradea), 190, 220
Napflio, 151
Napoleon III, 120
Napoleon, 139, 142
National Bank of Greece, 150
National Peasant Party (Partidul Naţional Ţărănesc – PNŢ), 286
National Union of Christian Students from Romania (Uniunea Naţională a Studenţilor Creştini din România – UNSCR), 270
Nationsuniversität, Saxon, 199
New York, 94
Noradunkyan, Gabriel, 53
North America, 185
Northern Europe, 40
Novi Sad, 29, 30, 33

Odessa, 221, 253, 254, 259
Ofenheim concession, 63, 221, 239, 244, 245, 247-254, 256, 261
Ofenheim, Viktor von, 73, 251
Olkhin, A. A., 174
Olt, 193, 196, 197, 203, 205, 228
Oltenia, 250, 256, 265
Oradea (Oradea Mare), 269, 278, 280-283, 285-288
Orient, 75, 120, 140, 244, 248

Orient Express, 194
Orient Railways, 50-52
Orléans, 52
Orsova (Orschowa, Orşova), 190, 192, 244, 253
Ottoman Empire, 20, 29, 34, 40-42, 45, 48, 50, 52-54, 63, 73, 77, 111, 112, 116, 121, 136, 137, 141, 159, 162-164, 168, 170, 172, 177, 179, 181, 185-189, 191, 193, 195, 206, 211, 213, 214, 215, 220

Palestine, 99, 114
Pallavicini, János, 210, 211, 212, 213
Pamukova, 29, 31
Panama (scandal), 66
Panu, George, 107, 108, 109
Paris, Treaty, 164
Paşcani, 284
Pellion, Jean Pierre, 143, 144, 145, 149, 156
Perdonnet, Albert A., 249
Persian Gulf, 193
Peru, 113
Pest, 190
Pesti Hungarian Commercial Bank (Pesti Magyar Kereskedelmi Bank – PMKB), 200
Phanariots, 139, 156
Piteşti, 231, 257
Ploieşti, 230, 236, 274
Poland, 2, 286
Popescu, Lorin, 281
Popescu, Toader, 20, 22, 51, 63, 64, 66, 73, 191, 219, 221-223, 226, 230, 232, 234, 244, 246, 248, 252-254, 256, 259, 267, 268, 281
Popovici, Andrei, 113, 130, 229, 243, 248, 264
Posno, Joseph, 208, 209
Pozsony (Bratislava), 190
Prague, 71
Predeal, 193, 244, 253, 283
Price, William Philip, 165, 167, 169, 171, 172, 177
Prussia, 62, 75

Quays Company, Izmir, 48

Rădulescu, I., 257
Rasova, 164, 180, 183
Reform Edict (1856), 41
Republican People's Party (CHP, Turkey), 32
Reşid, Mustafa Pasha, 208
Richter, Melvin, 67, 138, 139, 153
Rifat Pasha, 42
Riner, Malca, 272
Rize, 47
Rola, Ljudmil, 42
Roman, 93, 250
Romania, 8, 9, 18, 20-22, 51, 61-66, 68, 72- 74, 76, 77, 88, 105-126, 159, 160, 179, 180, 191-193, 196, 197, 198, 219-235, 239, 241-262, 267-290
Romanian Principalities, 9, 72, 83, 111, 139, 156
Rosegger, Gerhard, 78, 82, 165, 166, 183, 208, 217
Roşiori, 231
Ruschuk, 195, 207, 208, 212, 214
Ruse–Varna Railway, 50
Russell, John, 167, 168, 169, 170
Russia, 63, 111, 115, 120, 131, 142, 160, 163, 179, 180, 184, 222, 253
Russo-Turkish war (1877–1878), 110, 111, 221, 254

Saligny, Anghel, 124
Salina, 282, 283
San Stefano, Treaty of, 179, 181
Sankey, Frank Fremoult, 166
Schenk, Frithjof Benjamin, 12, 77, 90, 95
Schenk, Gerrit J., 90, 100
Segesvár (Schäßburg / Sighişoara), 196, 203, 205
Selçuk, 45
Sencea, Ioan, 202
Senegal (French Colony), 95
Şerban, Nicolae, 202, 203
Serbia, 29, 30, 33, 72, 80
Sfınţescu, Cincinat, 234
Siguranţa, 273, 274, 275, 282, 288
Silistra, 194, 195, 207, 208, 223
Sinaia, 289
Siret Valley, 251, 253, 255
Sivas, 42, 43
Slatina, 228, 271, 293
Smyrna (*see also* Izmir), 48, 193, 194, 210
Smyrna–Aydın Railway Company, 45, 46, 56
SNS (Serbian Progressive Party), 30
Sorescu, Andrei-Dan, 7-9, 13, 16, 17, 19, 62, 69, 74, 75, 105, 107, 110, 112-116, 162, 219, 227, 241, 243, 262, 268, 269
Şoricani, 283
South America (*also* Latin), 117, 121
Southeast Europe, 40, 50, 79, 137, 180
Southeastern Europe, 7, 10, 11, 12, 14, 15, 17, 19, 105, 137, 186
Southern Bessarabia, 179
Soviet Communist Party, 39
Soviet Union, 2, 276, 277
Spain, 11, 25, 74, 136, 289
SPD (Social Democratic Party of Germany), 34
Stalin, Joseph, 39
Stratford Canning, Viscount Stratford de Redcliffe, 45
Strousberg (affair, concession, scandal), 7-9, 21, 50, 62, 64, 66, 67, 72, 74, 75, 122, 221, 222, 233, 245, 247, 248, 249, 250, 252, 253, 254, 255, 256, 261, 268
Strousberg, Bethel Henry, 62, 122
Sturdza, Dimitrie A., 61, 62, 63, 64, 65, 67, 69, 70, 71, 73, 74, 77, 78, 85, 258
Sublime Porte, 45, 47, 161, 162, 163, 165, 166, 167, 168, 169, 170, 173, 175, 177, 178, 180, 181
Subotica, 30
Suceava, 248, 249, 251, 253
Suciu, P., 251
Suez Canal, 193, 248
Sulina, 19, 160, 164, 167, 180, 182
Şumuleanu, Corneliu, 272

Șuțu–Salamanca (concession), 244
Switzerland, 2, 108, 136, 143
Syria, 194
Szászrégen (Sächsisch-Reen/Reghin), 195
Szászváros (Broos/Orăștie), 195
Szeben, county, 197, 201
Székelyföld, 186, 192
Szentágota (Agnetheln/Agniţa), 203, 205

Talât Bey, 53
Târgu Jiu, 231
Târgu-Frumos, 250
Târgu-Ocna, 282, 283, 284, 288
Tchernavoda (*see also* Cernavodă, Boğazköy), 160, 161, 164, 174, 176
Tecuci, 250, 272
Teleagd, 283
Tempi, 29, 30, 35
Terkos, Lake, 49
Ternau, 49
Thalmann, Gustav, 197, 198, 200
Tiberius, 141
Timișoara (Temesvár/Temeschwar), 190
Timişul de Jos, 283
Tipei, Alex R., 16, 19, 68, 70, 71, 119, 121, 135, 136, 137, 140, 142, 143, 148, 154
Tipping, William, 172, 178, 182
Tiszaeszlár, 275, 293
Titulescu, Nicolae, 286
Titus, 141
Trabzon, 47, 59
Transfăgărășan, 40
Transylvania, 20, 73, 77, 174, 185, 186, 187, 188, 190, 191, 192, 195, 196, 202, 209, 213, 214, 244, 265, 268
Tripolitsa, 141
Tsarist Empire (Russia), 95
Tuckerman, Charles Keating, 178, 179, 180
Türesay, Özgür, 214
Turkey, 29, 31, 33, 36, 37, 45, 159, 171
Turner, Victor, 55
Turnu-Severin, 248, 271, 288

Ukraine, 40
Ungheni, 244, 254, 263, 272
Union of Native Jews (Uniunea Evreilor Pământeni – UEP), 274
United States of America, 88, 95, 108, 178
Upper Rhine (region), 100
Ursache, C., 250
USSR, 39, 40

Vác, 190
Vârciorova, 244, 253
Varna, 50, 172, 174, 195, 207, 208, 212, 214, 215, 221
Vaslui, 231, 236, 250, 255, 256, 259
Vasu, Octavian, 202, 203
Vatra Dornei, 274, 280
Veliko Tărnovo, 41
Venice, 149
Ventura, G., 255
Vianden, Benedikt Friedhelm, 99
Vienna, 73, 161, 190, 210, 253
Visterniceni, 277, 288
Vladimirescu, Tudor, 113, 115
Voinescu, Romulus P., 282
Volkspartei, Saxon, 199
Vondra, Béla (*also* Vondra Bey), 211, 212
Vöröstorony (Rotenturm/Turnu Roșu), 193
Voulgaris, Stamatis, 150, 151, 157
Vučić, Aleksandar, 30, 33, 56
Vulturescu, Grigore, 107

Wallachia, 20, 63, 110, 113, 124, 196, 220, 223, 228, 239, 240, 243-253, 255-262
Walpole, Frederick, 173
Wekerle, Sándor, 200
Wenger, Marla, 272
Western Europe, 44, 135, 136, 137, 139, 142, 146-149, 151-154, 162, 219
Wiener Bankverein, 210, 212
Wilson, Thomas, 164, 165
Wittfogel, 93, 104
Wittfogel, Karl August, 104
Wolff, Karl, 196-202, 206, 215
Wowereit, Klaus, 34, 35, 46

Youssouf Bey (Yusuf Bey), 175, 176
Ypsilantis, Demetrios, 141

Zelea Codreanu, Corneliu, 270-274, 278, 279
Zelea Codreanu, Ion, 272
Zimnicea, 231, 232